Check-List Of
Birds Of The World

A Continuation Of The Work Of

James L. Peters

(Volume X)

Editors

Ernst Mayr

Raymond A. Paynter

Alpha Editions

This Edition Published in 2020

ISBN: 9789354308420

Design and Setting By
Alpha Editions
www.alphaedis.com
Email – info@alphaedis.com

CHECK-LIST
OF BIRDS OF THE WORLD
VOLUME X

INTRODUCTION

Peters' Check-list has long been recognized as an indispensable reference for reliable information on the classification, correct names, synonymies, and geographic distribution of birds. The fact that the account of each family is now prepared by a specialist has increased the usefulness and authority of this series. There is nothing like it in the world literature for any other kind of organism. Ornithologists will, therefore, rejoice that this volume follows the publication of volumes IX (1960) and XV (1962) so speedily.

This volume differs little in scope and style from the two preceding volumes (IX and XV), which were the first to be published as a cooperative endeavor after the death of Mr. Peters. One minor change is the elimination of vernacular names. Criticisms of Volumes IX and XV centered on the selection of English names, which had been added in the hope of facilitating the use of the Check-list by the growing number of amateur ornithologists. There was no attempt to propose new vernaculars or to pass judgment on those already in use; they were merely taken from a series of widely-used ornithological works written in English. However, the critics of these names seem to feel that an effort should be made to standardize and supply appropriate English names for all species. Because this would require cumbersome polls and other elaborate procedures beyond the scope of a scientific work, the editors requested the advice of the twelve authors of this and forthcoming volumes. Ten of the authors said unequivocally that vernacular names should be abandoned.

It is perhaps well to review briefly the functions of the editors. Our duties, as we perceive them, are to arrange for revisions of taxa by specialists, to coordinate their work in order to avoid duplications and omissions, to achieve a degree of uniformity in written style but not necessarily in taxonomic treatment and thoroughness of coverage, and, to the extent our knowledge permits, to ferret out errors that slip past even the most conscientious author. It is not the task of the editors to check on the completeness and accuracy of synonymies, the validity of subspecies and genera, the accuracy of the range descriptions, and the other strictly taxonomic matters for which the authors alone are responsible.

The editors are grateful for continued cooperation from many collaborators. We appreciate that the authors, and those who have read the manuscript and offered suggestions, have given much of their research time to these tasks. We are glad they agree that completing James L. Peters' monumental project is worth the considerable personal sacrifice.

Manuscript for this volume was completed on 1 July 1963. Only minor modifications were made after this date.

We thank the National Science Foundation for support toward the publication of this volume, under grant number G-3124.

The index was prepared by Mrs. Michael D. McBride, who generously contributed her time and skill in order to speed the completion of the series.

ERNST MAYR

RAYMOND A. PAYNTER, JR.

21 August 1963

CONTENTS

NEW NAMES PROPOSED IN VOLUME X

CHECK-LIST
OF BIRDS OF THE WORLD
VOLUME X

Order PASSERIFORMES
Suborder OSCINES
Family PRUNELLIDAE
S. Dillon Ripley

cf. Hartert, 1910, Vög. pal. Fauna, 1, pp. 761-775.
Baker, 1924, Fauna Brit. India, Birds, ed. 2, 2, pp. 187-198.
————, 1930, Fauna Brit. India, Birds, ed. 2, 7, pp. 126-128; 8, p. 626.
Dementiev, 1935, Systema Avium Ross., Oiseau Rev. Franç. Orn., pp. 266-270.
Hartert and Steinbacher, 1935, op. cit., Ergänzungsb., pp. 333-339.
Marien, 1951, Amer. Mus. Novit., no. 1482, pp. 1-28 (southern Eurasia).
Ripley, 1952, Postilla, Yale Univ., no. 13, pp. 35-36 (list).
Dementiev and Gladkov (editors), 1954, Birds Soviet Union, 6, pp. 624-660.
Johansen, 1955, Journ. f. Orn., 96, pp. 75-80 (western Siberia).
Vaurie, 1955, Amer. Mus. Novit., no. 1751, pp. 1-25 (systematic notes).
————, 1959, Birds Pal. Fauna, Passeriformes, pp. 208-219.
Ripley, 1961, Synopsis Birds India Pakistan, pp. 540-545.

Genus PRUNELLA Vieillot

Prunella Vieillot, 1816, Analyse, p. 43. Type, by monotypy, "Fauvette de haie" Buffon = *Motacilla modularis* Linnaeus.
Aprunella Bianchi, 1904, Ann. Mus. Zool. Acad. Imp. Sci. St. Pétersbourg, 9, p. 118. Type, by monotypy, *Accentor immaculatus* Hodgson.

PRUNELLA COLLARIS

Prunella collaris collaris (Scopoli)
Sturnus collaris Scopoli, 1769, Ann. I Hist.-Nat., p. 131 — Carinthia.

Prunella collaris tschusii Schiebel, 1910, Orn. Jahrb., 21,
p. 102 — Corsica.

Prunella collaris nigricans Heim de Balsac, 1925, Rev.
Franç. Orn., 9, p. 170 — Grand Pic l'Ouarsenis, Algeria.

Breeding in southern Europe above tree line in the Pyrenees, Alps, Apennines, Sicilian and southern Spanish
mountains; wintering in western Europe north to Helgoland, Belgium and England, and south to the western
Mediterranean islands (Sardinia) and north Africa (Algeria and Morocco).

Prunella collaris subalpina (Brehm)

Accentor subalpinus Brehm, 1831, Handb. Naturgesch.
Vög. Deutschl., p. 1009 — Dalmatia.

Breeds in the mountains of southeast Europe, Yugoslavia, Bulgaria, Greece, Samothrace, Crete and in northern
(Bursa) and western Turkey at the tree line; wintering at
lower elevations.

Prunella collaris montana (Hablizl)

Sturnus montanus Hablizl, 1783, Neue Nord. Beyträge, 4,
p. 53 — Gilan, northern Iran.

Accentor collaris caucasicus Tschusi, 1902, Orn. Monatsb.,
10, p. 186 — Wladikawskas, Terek District, northern
Caucasus.

Breeds in the Caucasus, south to the southern coast of
the Caspian Sea (Taurus Mts.?), east to the Kopet Mountains; wintering in southern Iran and northern Iraq.

Prunella collaris rufilata (Severtzov)

Accentor rufilatus Severtzov, 1879, Izvestia Obsht. Liub.
Est. Anthr. Ethnog., Turkest. Otd., 1, fasc. 1, p. 45 —
Turkestan [reference not verified].

Laiscopus collaris kwenlunensis Buturlin, 1910, Mess.
Orn., 1, p. 188 — Keria range, western Kun Lun, Tibet.

Tadzhikistan, Kirgizstan, and western Sinkiang; wintering as far south as northern Afghanistan (breeding ?),
and West Pakistan in North West Frontier Province, Gilgit, and Astor; Baltistan?; Ladakh?

Prunella collaris whymperi (Baker)

Laiscopus collaris whymperi Baker, 1915, Bull. Brit. Orn.
Club, 35, p. 61 — Garhwal.

From Barai Valley, Kishenganga, and the hills of Kashmir about the Vale east to Lahul, Garhwal, and Kumaon, breeding in the alpine zone, descending in winter as low as the Simla Hills and Naini Tal, Uttar Pradesh.

Prunella collaris nipalensis (Blyth)

A.(ccentor) Nipalensis Blyth (ex Hodgson MS), 1843, Journ. Asiat. Soc. Bengal, 12, p. 958 — Kachar region of Nepal.

Accentor talifuensis Rippon, 1906, Bull. Brit. Orn. Club, 19, p. 19 — east of Talifu, western Yunnan.

Prunella collaris ripponi Hartert, 1910, Vög. pal. Fauna, 1, p. 766 — Gyi-dziu-shán, Tibet.

Breeds from Nepal east along the Himalayas to southeast Tibet, northern Yunnan, southern and eastern Sikang, descending as low as 6,500 feet in winter.

Prunella collaris tibetana (Bianchi)

Ac.(centor) collaris tibetanus Bianchi, 1904, Ann. Mus. Zool. Acad. Imp. Sci. St. Pétersbourg, 9, p. 128 — eastern Tibet.

Breeds in eastern Tibet and northwest China in northern Sikang, southern Tsinghai, Jyekundo, and Kansu.

Prunella collaris erythropygia (Swinhoe)

Accentor erythropygius Swinhoe, 1870, Proc. Zool. Soc. London, p. 124, pl. 9 — Kemeih, Prefecture of Seuenhwafoo.

Laiscopus collaris berezowskii Serebrovskij, 1927, Compt. Rend. Acad. Sci. Leningrad, ser. A, p. 325 — near Lunngan-fu, northwestern Szechwan.

Laiscopus collaris changaicus Tugarinov, 1929, Ann. Mus. Zool. Acad. Imp. Sci. St. Pétersbourg, 29 (1928), p. 269 — Uliassutai.

Russian Altai, Mongolia, east in north China to the Sea of Ochotsk, Korea (breeding ?), Japan (Honshu) and Formosa; breeding on rocky slopes, descending to the adjacent plains in winter.

PRUNELLA HIMALAYANA

Prunella himalayana (Blyth)

Accentor Himalayanus? Blyth, 1842, Journ. Asiat. Soc. Bengal, 11, p. 187 — Himalaya range.

Central Asia from Kirghizstan east along the Tian Shan, Sayan and Bolshoi ranges to the eastern and northern sides of Lake Baikal, south to Afghanistan, West Pakistan, and India from Gilgit and Kashmir along the outer Himalayan ranges to Nepal, Darjeeling, and Sikkim, at high altitudes to 10,000 feet.

PRUNELLA RUBECULOIDES

Prunella rubeculoides rubeculoides (Moore)

Accentor rubeculoides Moore (ex Hodgson MS), 1854, Cat. Birds Mus. East India Co., 1, p. 361 — Nepal.

Prunella rubeculoides muraria R. and A. Meinertzhagen, 1926, Bull. Brit. Orn. Club, 46, p. 99 — Astor, northwestern Kashmir.

At high altitudes in the Himalayas of West Pakistan and India from Astor, Baltistan, and Ladakh, southeast to Nepal, Sikkim, and southeast Tibet.

Prunella rubeculoides fusca Mayr

Prunella rubeculoides fusca Mayr, 1927, Orn. Monatsb., 35, p. 148 — Bamutang, two days southwest of Batang, Szechwan [= central Sikang].

Prunella rubeculoides beicki Mayr, 1927, Orn. Monatsb., 35, p. 149 — near Lau-hu-kou, northern Kansu.

Mountains of east Tibet and west China in Sikang, Kansu, Kuku-Nor, and Shensi.

PRUNELLA STROPHIATA

Prunella strophiata jerdoni (Brooks)

Accentor Jerdoni Brooks, 1872, Journ. Asiat. Soc. Bengal, 41, p. 327 — Dharmsala and Kashmir; restricted to Kashmir by Baker, 1921, Journ. Bombay Nat. Hist. Soc., 27, p. 95.

Prunella strophiatus sirotensis Koelz, 1939, Proc. Biol. Soc. Washington, 52, p. 67 — Sirotai [= Saroti, Safed Koh], Afghanistan.

At high altitudes from the Afghanistan-West Pakistan border, the Safed Kohs, south Waziristan, and Gilgit to Kashmir and Baltistan southeast along the Himalayas to Kumaon; wintering as low as 4,300 feet.

Prunella strophiata strophiata (Blyth)

A.(ccentor) strophiatus Blyth (ex Hodgson MS), 1843, Journ. Asiat. Soc. Bengal, 12, p. 959 — Nepal.

Accentor multistriatus David, 1871, Ann. Mag. Nat. Hist., ser. 4, 7, p. 256 — Moupin [= Paohing], western Szechwan.

Nepal east along the Himalayas in Sikkim, Bhutan, southern Tibet, northern Burma, and mountains of western China, in Yunnan, Chwanben, Sikang, Szechwan, eastern Tsinghai, western Kansu, eastern Koko Nor, and Shensi; straggling to lower altitudes in winter.

PRUNELLA MONTANELLA

Prunella montanella (Pallas)

Motacilla montanella Pallas, 1776, Reise versch. Prov. Russ. Reichs, 3, p. 695 — Dauria.

Prunella montanella badia Portenko, 1929, Compt. Rend. Acad. Sci. Leningrad, ser. A, p. 220 — Chukchi, North Siberia.

Siberia from the Urals east to the Chukchi Peninsula and Wrangel Island, south to the Stanovoi Mountains and southwest to the Barbuzinski and Sayan range in Mongolia and the border of Tannu-Tuva; migrates to Mongolia, Alashan, the lower Amur, Korea, and rarely in northern Japan (Hokkaido and Honshu).

PRUNELLA FULVESCENS

Prunella fulvescens fulvescens (Severtzov)

Accentor fulvescens Severtzov, 1873, Vertikal . . . Turkest. Zhivotn., (1872), p. 132 — Turkestan [reference not verified].

Prunella fulvescens juldussica Sushkin, 1925, Proc. Boston Soc. Nat. Hist., 38, p. 50 — defile of Naryn, Tian-shan.

Prunella fulvescens karlykensis Sushkin, 1925, Proc. Boston Soc. Nat. Hist., 38, p. 50 — Ortun-tam, south slope of Karlyktag (near Hami).

Prunella fulvescens hissarica Sushkin, 1925, Proc. Boston Soc. Nat. Hist., 38, p. 53 — Zerafshan, Lake Dschai, Gusun-pass.

Breeds in the mountains of Tadzhikistan and Tien Shan

range, south to northern Afghanistan (?), and east Ladakh; migrates through eastern Afghanistan and Wakhan; winters, above 4,000 feet, in West Pakistan from Chitral to Baltistan.

Prunella fulvescens dahurica (Taczanowski)

Accentor dahuricus Taczanowski, 1874, Journ. f. Orn., 22, p. 320 — Stary Tsuruchaitui (Argun River, Transbaicalia).

Prunella fulvescens mongolica Sushkin, 1925, Proc. Boston Soc. Nat. Hist., 38, p. 52 — near Kobdo town, northwestern Mongolia.

From the Dzhungarski Ala Tau range of eastern Alma Ata along the ranges of the Russian-Chinese border to the Altai of Tannu-Tuva and Mongolia, Irkutsk, Targabatai, and the Sayan mountains; wintering at adjacent lower altitudes.

Prunella fulvescens dresseri Hartert

Prunella fulvescens dresseri Hartert, 1910, Vög. pal. Fauna, 1, p. 770; new name for *A.(ccentor) fulvescens* var. *pallidus* Dresser, 1895, Birds Europe, 9, p. 105 — Karakash Valley; preoccupied by *A. pallidus* Menzbier, 1887.

From the Muztagh Ata range on the border of Tadzhikistan east in the Kunlun mountains of southwest Sinkiang and northern Tibet, east to the Tsinghai-Kansu border.

Prunella fulvescens nanschanica Sushkin

Prunella fulvescens nanschanica Sushkin, 1925, Proc. Boston Soc. Nat. Hist., 38, p. 51 — Kwei-Tê-ting (or Guidui), eastern Nan Shan, northern Kansu.

Prunella fulvescens nadiae Bangs and Peters, 1928, Bull. Mus. Comp. Zool., 68, p. 355 — Tao River valley, near Choni, southwestern Kansu.

From the Nan Shan Mountains of the Tsinghai-Kansu border south as far as the Tasurkai Shan.

Prunella fulvescens khamensis Sushkin

Prunella fulvescens khamensis Sushkin, 1925, Proc. Boston Soc. Nat. Hist., 38, p. 54 — Re-chii River, Kham, northeastern Tibet.

Northeastern Tibet and western China, in northern Sikang and southern Tsinghai, in the drainage area of the upper Mekong and Yangste rivers.

Prunella fulvescens sushkini Collin and Hartert

Prunella fulvescens sushkini Collin and Hartert, 1927, Novit. Zool., 34, p. 52; new name for *Prunella fulvescens tibetana* Sushkin, 1925, Proc. Boston Soc. Nat. Hist., 38, p. 53 — Khamba-jong, Tibet; preoccupied by *Accentor collaris tibetanus* Bianchi.

Southern and southeast Tibet, usually at tree line; wintering in the same hills down to 7,500 feet.

PRUNELLA OCULARIS

Prunella ocularis ocularis (Radde)

Accentor ocularis Radde, 1884, Ornis Caucasica, pp. 33, 244, pl. 14 — Küsjurdi Mt., Talysh, Iranian border.

Breeds in the Armenian and Azerbaijan mountains of the border between southern U.S.S.R., northeastern Turkey, and Iran, east in the Elburz mountains of northern Iran to eastern Khurasan.

Prunella ocularis fagani (Ogilvie-Grant)

Accentor fagani Ogilvie-Grant, 1913, Bull. Brit. Orn. Club, 31, p. 88 — Yemen.

Breeds in the high mountains of Yemen; rare straggler in winter to lower elevations in the Aden Protectorate.

PRUNELLA ATROGULARIS

Prunella atrogularis atrogularis (Brandt)

Accentor atragularis (*sic*) Brandt, 1844, Bull. Phys. Acad. Imp. Sci. St. Pétersbourg, col. 140 — Semipalatinsk [reference not verified].

Breeds in the northern Ural Mountains east as far as the Petchora River; wintering in Turkmenistan, Iran, and Afghanistan; straggler to Gilgit.

Prunella atrogularis huttoni (Moore)

Accentor huttoni Moore, 1854, in Horsfield and Moore, Cat. Birds Mus. East India Co., 1, p. 360; new name for *A. atrogularis* Blyth (ex Hutton MS), 1849, Journ. Asiat. Soc. Bengal, 18, p. 811 — ranges above Simla; preoccupied by *A. atrogularis* Brandt.

Prunella atrogularis menzbieri Portenko, 1929, Compt.
Rend. Acad. Sci. Leningrad, ser. A, p. 216 — Katon-
Karagai, southwestern Altai.
Prunella atrogularis lucens Portenko, 1929, Compt. Rend.
Acad. Sci. Leningrad, ser. A, p. 217 — Tzagma River,
central Tian Shan.

Breeds in Russian Altai, in the Dzhungarski Mountains on
the borders of Alma Ata and Sinkiang, and in Tadzhikistan;
winters in Tadzhikistan, northeast Afghanistan, and in
West Pakistan and the Indian Himalayas east to East Punjab and northern Uttar Pradesh.

PRUNELLA KOSLOWI

Prunella koslowi (Przewalski)

Accentor Koslowi Przewalski, 1887, Zapiski Imp. Akad.
Nauk, 55, p. 83 — Alashan.
Prunella kozlowi tenella Kozlova, 1929, Ann. Mus. Zool.
Acad. Sci. [U.R.S.S.; Leningrad], 29 (1928), p. 275 —
Dzapkhan River, Narvanchy-kure region, Outer Mon-
golia.

Mountains of Mongolia and Ningsia; Khobdo, Gobi, and
Ala Shan.

PRUNELLA MODULARIS

Prunella modularis hebridium Meinertzhagen

Prunella modularis hebridium Meinertzhagen, 1934, Ibis,
p. 57 — South Uist, Outer Hebrides.
Prunella modularis hibernicus Meinertzhagen, 1934, Ibis,
p. 57 — Curragh, Ireland.

Ireland, western Scotland, and the Hebrides.

Prunella modularis occidentalis (Hartert)

Accentor modularis occidentalis Hartert, 1910, Brit. Birds,
3, p. 313 — Tring.
Prunella modularis interposita Clancey, 1943, Bull. Brit.
Orn. Club, 64, p. 14 — Dornoch, Sutherlandshire, northern Scotland.

Scotland (excluding western Scotland), England, and
western France; non-breeding birds reach the Shetlands;
winters in southern half of its range and on the Continent;
introduced in New Zealand.

Prunella modularis modularis (Linnaeus)

Motacilla modularis Linnaeus 1758, Syst. Nat., ed. 10, 1, p. 184 — "Habitat in Europa" [= Sweden, ex Faun. Svec., 1746, p. 89; *vide* Hartert, 1910, Vög. pal. Fauna, p. 772].

Prunella modularis meinertzhageni Harrison and Pateff, 1937, Ibis, p. 612 — Beglik and Rila, Bulgaria.

Prunella modularis arduennus Verheyen, 1941, Bull. Mus. Hist. Nat. Belg., 17 (51), p. 5 — Brumagne, Belgium.

Prunella modularis belousovi Uvarova, 1950, Ornithofauna of the high mountain range Bassez, its ecological and zoogeographical relationships — northern Urals [reference not verified].

Breeds in Europe north to lat. 70° N. in Scandinavia, Finland, and northwest Russia, east to the Urals, south to the central parts of France, Corsica, Sardinia, central Italy, southeastern Europe, and the Black Sea; winters in Europe, Spain, north Africa, northern Turkey, and the Eastern Mediterranean.

Prunella modularis mabbotti Harper

Prunella modularis mabbotti Harper, 1919, Proc. Biol. Soc. Washington, 32, p. 243 — three kilometers south of Saillagouse, Dept. of Pyrénées-Orientales, France.

Prunella modularis lusitanica Stresemann, 1928, Journ. f. Orn., 76, p. 389; new name for (*Tharrhaleus*) *Prunella modularis obscura* Tratz, 1914, Orn. Monatsb., 22, p. 50 — Oporto, Portugal; preoccupied by *Motacilla obscura* Hablizl.

Breeds in Portugal, Spain, the Pyrenees, and southwestern France; intergrading with the nominate form on the borders of the range.

Prunella modularis obscura (Hablizl)

Motacilla obscura Hablizl, 1783, Neue Nord. Beyträge, 4, p. 56 — Gilan.

Accentor orientalis Sharpe, 1883, Cat. Birds Brit. Mus., 7, p. 652 — Batoum, Black Sea.

Accentor modularis blanfordi Zarudny, 1904, Orn. Monatsb., 12, p. 164 — mountain oak-forest southwest of Isfahan, Iran.

Prunella modularis enigmatica Dunajewski, 1948, Bull. Brit. Orn. Club, 68, p. 131 — Yalta, Crimea.

Breeds in the Crimea, Caucasus, and northern and western Iran, east to the Elburz mountains; winters south to the Zagros mountains, Luristan, and west to the mountains of Lebanon.

Prunella modularis euxina Watson

Prunella modularis euxina Watson, 1961, Postilla, Yale Univ., no. 52, p. 9 — Ulu Dag [= Asiatic Mount Olympus], Bursa, northwest Turkey.

Northern Turkey. Eastern range in Asia Minor unknown; may meet preceding form in Transcaucasia or Turkish Armenia.

PRUNELLA RUBIDA

Prunella rubida fervida (Sharpe)

Accentor fervidus Sharpe, 1883, Cat. Birds Brit. Mus., 7, p. 653 — Hakodate, Hokkaido, Japan.

Hokkaido and the southern Kurile Islands, north through Uruppu.

Prunella rubida rubida (Temminck and Schlegel)

Accentor modularis rubidus Temminck and Schlegel, 1848, in Siebold, Fauna Jap., Aves, p. 69. pl. 32 — Japan.

Breeds on Honshu; winters southward as far as northern Kyushu and Shikoku (once, 1937).

PRUNELLA IMMACULATA

Prunella immaculata (Hodgson)

Acc.(entor) immaculatus Hodgson, 1845, Proc. Zool. Soc. London, p. 34 — central and northern regions of the hills, Nepal.

Breeds from the central Himalayas in Nepal, east to Sikkim, Bhutan, southeast Tibet, Yunnan, and Szechuan; winters south to northeast Burma.

FAMILY MUSCICAPIDAE
SUBFAMILY TURDINAE[1]

S. DILLON RIPLEY

cf. Ridgway, 1907, Bull. U. S. Nat. Mus., 50, pt. 4, pp. 1-179 (North and Middle America).

Hartert, 1909-1910, Vög. pal. Fauna, 1, pp. 603-606; 640-761.

Baker, 1924, Fauna Brit. India, Birds, ed. 2, 2, pp. 7-187.

La Touche, 1925, Birds Eastern China, 1, pp. 99-155.

Baker, 1930, Fauna Brit. India, Birds, ed. 2, 7, pp. 96-126; 8, pp. 603-606; 619-626.

Mathews, 1930, Syst. Avium Aust., pt. 2, pp. 577-589.

Hellmayr, 1934, Field Mus. Nat. Hist. Publ., Zool. Ser., 13, pt. 7, pp. 350-484 (North and South America).

Dementiev, 1935, Systema Avium Ross., Oiseau Rev. Franç. Orn., pp. 237-265.

Bannerman, 1936, Birds Trop. West Africa, 4, pp. 309-438.

Jackson, 1938, Birds Kenya Colony Uganda Protect., 2, pp. 946-1013.

Witherby, et al., 1938, Handbook Brit. Birds, 2, pp. 104-204 (biology and distribution).

Ripley, 1952, Postilla, Yale Univ., no. 13, pp. 1-48 [+ 2 pp. addenda, 1953] (review).

Chapin, 1953, Bull. Amer. Mus. Nat. Hist., 75A, pp. 480-593 (Congo region).

Dementiev and Gladkov (editors), 1954, Birds Soviet Union, 6, pp. 398-621.

Mackworth-Praed and Grant, 1955, Birds Eastern and North Eastern Africa, 2, pp. 227-333.

Amer. Orn. Union, 1957, Check-list North Amer. Birds, ed. 5, pp. 430-448.

Vaurie, 1959, Birds Pal. Fauna, Passeriformes, pp. 333-419.

Ripley, 1961, Synopsis Birds India Pakistan, pp. 492-538.

[1] MS read, in whole or part, by E. R. Blake, J. Bond, R. M. de Schauensee, A. L. Rand, M. A. Traylor, Jr., and A. Wetmore.

White, 1961, Bull. Brit. Orn. Club, 81, pp. 117-119; 150-152 (African genera).

Ripley, 1962, Postilla, Yale Univ., no. 63, pp. 1-5 (notes on genera).

Genus BRACHYPTERYX Horsfield

Brachypteryx Horsfield, 1822, Trans. Linn. Soc. London, 13, p. 157. Type, by subsequent designation (G. R. Gray, 1855, Cat. Gen. Subgen. Birds, p. 41), *Brachypteryx montana* Horsfield.

Heteroxenicus Sharpe, 1902, Bull. Brit. Orn. Club, 12, p. 55. New name for *Drymochares* Gould, preoccupied by *Drymochares* Mulsant, 1847, Coleoptera.

Heinrichia Stresemann, 1931, Orn. Monatsb., 39, p. 9. Type, by original designation, *Heinrichia calligyna* Stresemann.

cf. Rothschild, 1926, Novit. Zool., 33, pp. 270-271 (*nipalensis; montana*).

Stresemann, 1940, Journ. f. Orn., 88, pp. 112-114 (*calligyna*).

BRACHYPTERYX STELLATA

Brachypteryx stellata stellata Gould

Brachypteryx (*Drymochares*) *stellatus* Gould, 1868, Proc. Zool. Soc. London, p. 218 — Nepal.

Breeds in eastern Nepal, Sikkim, Bhutan, southeast Tibet, and northeast Burma, from 7,000 to 14,000 feet.

Brachypteryx stellata fusca Delacour and Jabouille

Brachypteryx stellatus fuscus Delacour and Jabouille, 1930, Oiseau Rev. Franç. Orn., 11, p. 397 — Chapa, Tonkin.

Mountains of northern Tonkin.

BRACHYPTERYX HYPERYTHRA

Brachypteryx hyperythra Jerdon and Blyth

Brachypteryx hyperythra Jerdon and Blyth, 1861, Proc. Zool. Soc. London, p. 201 — Darjeeling.

Eastern Nepal, Darjeeling, Sikkim, and Assam, in North Lakhimpur, Daphla Hills, and (?) Naga Hills.

BRACHYPTERYX MAJOR

Brachypteryx major major (Jerdon)

Phoenicura major Jerdon, 1844, Madras Journ. Lit. Sci.,
13, p. 170 — Nilgiris.

Peninsular India, breeding in the Bababudan, Brahmagiri,
and Nilgiri Hills of Mysore and western Madras.

Brachypteryx major albiventris (Blanford)

Callene albiventris Blanford, 1867, Proc. Zool. Soc. London, p. 833, pl. 39 — Palni Hills.

Peninsular India in the hills of Kerala and southwestern
Madras, from the Palnis south to Tirunelveli and Mynall.

BRACHYPTERYX CALLIGYNA

Brachypteryx calligyna simplex (Stresemann)

Heinrichia calligyna simplex Stresemann, 1931, Orn.
Monatsb., 39, p. 81 — Matinan Mountains.

Matinan Mountains, northern Celebes.

Brachypteryx calligyna calligyna (Stresemann)

Heinrichia calligyna Stresemann, 1931, Orn. Monatsb., 39,
p. 9 — Latimodjong Mountains.

Mount Latimodjong, south-central Celebes.

Brachypteryx calligyna picta (Stresemann)

Heinrichia calligyna picta Stresemann, 1932, Orn. Monasb., 40, p. 108 — Mengkoka Mountains.

Tanke Salokko, Mengkoka Mountains, above 4,900 feet,
southeastern Celebes.

BRACHYPTERYX LEUCOPHRYS

Brachypteryx leucophrys nipalensis Hodgson

Brachypteryx nipalensis Hodgson, 1854, in Horsfield and
Moore, Cat Birds Mus. East India Co., 1, p. 397 —
Nepal.

Brachypteryx leucophrys geokichla Koelz, 1952, Journ.
Zool. Soc. India, 4, p. 41 — Karong, Manipur.

Breeds in the mountains of Garhwal, Nepal, Sikkim, Bhutan, Burma (south to Tenasserim), and western Yunnan,
from the edge of the plains to 5,000 feet.

Brachypteryx leucophrys carolinae La Touche

> *Brachypteryx carolinae* La Touche, 1898, Bull. Brit. Orn.
> Club, 8, p. 9 — Kuatun, northwest Fukien.
> *Brachypteryx nipalensis harterti* Weigold, 1922, Orn.
> Monatsb. 30, p. 63 — Omei Shan, Szechwan.
> *Heteroxenicus nangka* Riley, 1932, Proc. Biol. Soc. Wash-
> ington, 45, p. 59 — Pang Meton, northern Siam.

Breeds in northern Thailand, upper Laos, Tonkin, and in
China in Yunnan, Szechwan, and northwestern Fukien.

Brachypteryx leucophrys langbianensis Delacour and Greenway

> *Brachypteryx leucophrys langbianensis* Delacour and
> Greenway, 1939, Bull. Brit. Orn. Club, 59, p. 131 — Pic
> de Langbian, near Dalat, Annam.

The mountains of southern Laos and southern Annam
(Viet Nam).

Brachypteryx leucophrys wrayi Ogilvie-Grant

> *Brachypteryx wrayi* Ogilvie-Grant, 1906, Bull. Brit. Orn.
> Club, 19, p. 10 — Gunong Batu Putch and Gunong
> Tahan, 5,300-7,000 ft.

Mountains of Malaya.

Brachypteryx leucophrys leucophrys (Temminck)

> *Myiothera leucophrys* Temminck, 1827, Pl. Col., livr. 76,
> pl. 448, fig. 1 — Java.

Mountains of Sumatra, Java, Bali, Lombok, Sumbawa,
and Timor Islands.

BRACHYPTERYX MONTANA

Brachypteryx montana cruralis (Blyth)

> *Calliope* ? *cruralis* Blyth, 1843, Journ. Asiat. Soc. Bengal,
> 12, p. 929 — Darjeeling.
> *Heteroxenicus cruralis formaster* Thayer and Bangs, 1912,
> Mem. Mus. Comp. Zool., 40, p. 169 — Mount Washan,
> 9,000-10,000 ft., western Szechwan.
> *Heteroxenicus cruralis laurentei* La Touche, 1921, Bull.
> Brit. Orn. Club, 42, p. 29 — Mengtz, southeast Yunnan.

Breeds in the mountains of Nepal, Sikkim, Bhutan, Assam, and Burma, south as far as the Chin Hills in the west
and Karenni in the east, western China in Sikang, Yunnan

and Szechwan, migrating in winter as far as southwestern Yunnan, extreme northern Thailand (Doi Anka), Viet Nam in Tonkin and Annam (Dalat), and Laos.

Brachypteryx montana sinensis Rickett and La Touche

> *Brachypteryx sinensis* Rickett and La Touche, 1897, Bull. Brit. Orn. Club, 6, p. 50 — Kuatun, northwest Fukien.

Resident in northwest Fukien, China.

Brachypteryx montana goodfellowi Ogilvie-Grant

> *Brachypteryx goodfellowi* Ogilvie-Grant, 1912, Bull. Brit. Orn. Club, 29, p. 108 — Mount Arizan, 8,000 ft., Formosa.

Formosa (Taiwan).

Brachypteryx montana sillimani Ripley and Rabor

> *Brachypteryx montana sillimani* Ripley and Rabor, 1962, Postilla, Yale Univ., no. 73, p. 6 — Magtaguimbong, Mount Mantalingajan, 3,600-4,350 feet, Palawan Island, Philippines.

Mountains of southern Palawan Island, Philippines.

Brachypteryx montana poliogyna Ogilvie-Grant

> *Brachypteryx poliogyna* Ogilvie-Grant, 1895, Bull. Brit. Orn. Club, 4, p. 40 — mountains of Lepanto, northern Luzon.

Highlands of Luzon and Mindoro Islands, Philippines.

Brachypteryx montana brunneiceps Ogilvie-Grant

> *Brachypteryx brunneiceps* Ogilvie-Grant, 1896, Ibis, p. 526 — Negros.

Mountains of Negros Island, Philippines, above 3,500 feet.

Brachypteryx montana malindangensis Mearns

> *Brachypteryx malindangensis* Mearns, 1909, Proc. U. S. Nat. Mus., 36, p. 441 — summit of Grand Malindang Mountain, 9,000 ft., Misamis Province, northwestern Mindanao, Philippine Islands.

Mount Malindang, Mindanao Island, Philippines.

Brachypteryx montana mindanensis Mearns

> *Brachypteryx mindanensis* Mearns, 1905, Proc. Biol. Soc. Washington, 18, p. 3 — Mount Apo, 6,000 ft.

Mount Apo, Mindanao Island, Philippines.

Brachypteryx montana erythrogyna Sharpe

Brachypteryx erythrogyna Sharpe, 1888, Ibis, p. 389, pl. 10 — Kina Balu.

Mountains of northern Borneo, above 4,000 feet.

Brachypteryx montana saturata Salvadori

Brachypteryx saturata Salvadori, 1879, Ann. Mus. Civ. Genova, 14, p. 225 — Padang Highlands, Sumatra.

Mountains of Sumatra.

Brachypteryx montana montana Horsfield

Brachypteryx montana Horsfield, 1822, Trans. Linn. Soc. London, 13, p. 157 —Java.

Mountains of Java.

Brachypteryx montana floris Hartert

Brachypteryx floris Hartert, 1897, Novit. Zool., 4, p. 170 — South Flores.

Mountains of Flores Island, Lesser Sunda Islands.

GENUS **ZELEDONIA** RIDGWAY

Zeledonia Ridgway, 1889, Proc. U. S. Nat. Mus., 11 (1888), p. 537. Type, by monotypy, *Zeledonia coronata* Ridgway.

ZELEDONIA CORONATA

Zeledonia coronata Ridgway

Zeledonia coronata Ridgway, 1889, Proc. U. S. Nat. Mus., 11 (1888), p. 538 — Laguna del Volcán de Póas, Costa Rica.

Zeledonia insperata Ridgway (ex Cherrie MS), 1907, Bull. U. S. Nat. Mus., 50, pt. 4, p. 72 — Volcán de Irazú.

Mountains of Costa Rica and western Panama.

GENUS **ERYTHROPYGIA** SMITH

Erythropygia A. Smith, 1836, Rep. Exped. Centr. Africa, p. 46. Type, by subsequent designation (Sharpe, 1883, Cat. Birds Brit. Mus., 7, p. 72), *Erythropygia pectoralis* Smith = *Sylvia leucophrys* Vieillot.

Agrobates Swainson, 1837, Class. Birds, 2, p. 241. Type, by monotypy, *Sylvia galactotes* Temminck.

Tychaëdon Richmond, 1917, Proc. U. S. Nat. Mus., 53,
p. 575. Type, by original designation, *Cossypha signata*
Sundevall.

cf. Reichenow, 1905, Vög. Afr., 3, pp. 767-776.
Benson, 1946, Bull. Brit. Orn. Club, 67, pp. 32-33 (*barbata* and *quadrivirgata*).
South African Orn. Soc. List. Comm., Second Rept.,
1958, Ostrich, 29, p. 40 (*signata* and *leucophrys*).

ERYTHROPYGIA CORYPHAEUS

√ **Erythropygia coryphaeus coryphaeus** (Lesson)

Sylvia coryphaeus Lesson, 1831, Traité Orn., p. 419, ex
Levaillant, 1824, Oiseaux Afrique, 3, pl. 120, "Le Cor-
iphée" — Sondag and Zwarte-kop Rivers; restricted to
Uitenhage, Sunday River, Cape Province by Macdonald,
1952, Bull. Brit. Orn. Club, 72, p. 91.

Erythropygia coryphaeus abbotti Friedmann, 1932, Proc.
Biol. Soc. Washington, 45, p. 65 — Fish River, 6 miles
from Berseba, South West African Protectorate.

Southern South West Africa in Nama Land, Great Namaqualand east to southern Bechuanaland, and South Africa
in Orange Free State, and Cape Province west to the Touws
River.

Erythropygia coryphaeus cinerea Macdonald

Erythropygia coryphaeus cinereus Macdonald, 1952, Bull.
Brit. Orn. Club, 72, p. 91 — 16 miles north of Port Nolloth, Little Namaqualand.

Lower Orange River, South West Africa south to the Cape
Flats, Cape Province east to Stilbaai and Zoutendalsvallei,
Bredasdorp District.

ERYTHROPYGIA LEUCOPHRYS

√ **Erythropygia leucophrys leucoptera** (Rüppell)

Salicaria leucoptera Rüppell, 1845, Syst. Uebers. Vög.
Nord-ost.-Afr., pl. 15 — Shoa district.

Northern Somalia, southern Ethiopia south to northern
Kenya from Lake Rudolf southeast to the lower Tana River
and northeastern Uganda, southeastern and southern Sudan.

Erythropygia leucophrys eluta Bowen

> *Erythropygia leucophrys eluta* Bowen, 1934, Proc. Biol.
> Soc. Washington, 47, p. 159 — Kismayu, Jubaland.
> *Erythropygia leucoptera pallida* Benson, 1942, Bull. Brit.
> Orn. Club, 63, p. 14 — Serenli, Juba River.

Southern Somalia.

V **Erythropygia leucophrys brunneiceps** Reichenow

> *Erythropygia brunneiceps* Reichenow, 1891, Journ. f.
> Orn., 39, p. 63 — Nguruman, near Lake Natron.
> *Erythropygia ukambensis* Sharpe, 1900, Bull. Brit. Orn.
> Club, 11, p. 28 — Ukambani.

Central Kenya and Tanganyika, from Kidong Valley to
Ukamba, south to Manyara, Uaso Nyiro and Mount Kilimanjaro, intergrading with *zambesiana* in the Simba area.

V **Erythropygia leucophrys vulpina** Reichenow

> *Erythropygia vulpina* Reichenow, 1891, Journ. f. Orn.,39,
> p. 62 — Ndi in Teita.

Kenya in the Teita and south Ukamba districts south to
northeastern Tanganyika.

V **Erythropygia leucophrys zambesiana** Sharpe

> *Erythropygia zambesiana* Sharpe, 1882, Proc. Zool. Soc.
> London, p. 588, pl. 45 — Tete, Lower Zambesi.
> *Erythropygia leucophrys soror* Reichenow, 1905, Vög.
> Afr., 3, p. 774 — East Africa; restricted to Klein Aru-
> scha by Sclater, 1930, Syst. Av. Aethiop., 2, p. 483.
> *Erythropygia ruficauda iubilaea* Grote, 1927, Orn. Mon-
> atsb., 35, p. 103 — Mikindani, south coast German East
> Africa.
> *Erythropygia ruficauda iodomera* Grote, 1927, Orn. Mon-
> atsb., 35, p. 104 — Usegua, Tanganyika.

Coastal Kenya from the Tana River south, and Tanganyika in the coastal area inland to Arusha and Shire Valley,
south to northern Mozambique and Nyasaland, west in
northeastern Southern Rhodesia and Northern Rhodesia,
except the west and south, to southeastern Congo.

Erythropygia leucophrys vansomereni Sclater

> *Erythropygia leucophrys vansomereni* Sclater, 1929, Bull.
> Brit. Orn. Club, 49, p. 62 — Mokia, Ruwenzori.

Erythropygia leucoptera sclateri Grote, 1930, Bateleur, 2, p. 14 — Iringa, Uhehe, Tanganyika.
Erythropygia leucophrys jungens Bowen, 1934, Proc. Biol. Soc. Washington, 47, p. 162 — Kabelolot Hill, Sotik District, Kenya.

Eastern Congo border, Uganda except the northeast, western Kenya, and Tanganyika, west of the range of the preceding form, south to Iringa and Morogoro.

Erythropygia leucophrys munda (Cabanis)
Thamnobia munda Cabanis, 1880, Orn. Centralbl., p. 143 — Malandje, northern Angola [reference not verified].
Erythropygia ruficauda Sharpe, 1882, Proc. Zool. Soc. London, p. 589, pl. 54 — Malimbe, Portuguese Congo.
Erythropygia ansorgii Ogilvie-Grant, 1914, Bull. Brit. Orn. Club, 33, p. 134 — Malange, North Angola.
Erythropygia leucophrys saturata Neumann, 1920, Journ. f. Orn., 68, p. 83 — Yambuja, lower Aruwimi.
Erythropygia munda ovamboensis Neumann, 1920, Journ. f. Orn., 68, p. 83 — Ombongo and Ovankenyama.
Erythropygia leucoptera permutata Grote, 1930, Orn. Monatsb., 38, p. 187 — Huxe, (Benguella, Angola).
Erythropygia leucophrys kabalii White, 1944, Bull. Brit. Orn. Club, 64, p. 49 — Chikonkwelo stream, Balovale.

Congo except the extreme east and southeast, Angola, western Northern Rhodesia, and northern South West Africa in Damaraland and Ovamboland.

Erythropygia leucophrys makalaka Neumann
Erythropygia makalaka Neumann, 1920, Journ. f. Orn. 68, p. 83 — Makalaka Land, north of the Limpopo.

Western Southern Rhodesia, in Matabeleland, south into northern Bechuanaland and Ngamiland.

Erythropygia leucophrys limpopoensis Roberts
Erythropygia leucophrys limpopoensis Roberts, 1932, Ann. Transvaal Mus., 15, p. 30 — Bubye River, Southern Rhodesia.

Southeastern Southern Rhodesia, southern Mozambique to northeastern Transvaal, and Mashonaland.

∨ **Erythropygia leucophrys pectoralis** Smith

Erythropygia pectoralis A. Smith, 1836, Rep. Exped. Centr. Africa, p. 46 — between Orange River and Kurrichaine [=Kurrichane], western Transvaal.

Southwestern Southern Rhodesia, eastern and southern Bechuanaland, Transvaal (except northeast), and Swaziland.

Erythropygia leucophrys leucophrys (Vieillot)

Sylvia leucophrys Vieillot, 1817, Nouv. Dict. Hist. Nat., nouv. éd., 11, p. 191 — Gamtoos River, eastern Cape Province, ex Levaillant, 1802, Oiseaux Afrique, 3, pl. 118, "Le Grivetin."

Southern and eastern Cape Province and Natal, including Zululand.

ERYTHROPYGIA HARTLAUBI

Erythropygia hartlaubi Reichenow

Erythropygia hartlaubi Reichenow, 1891, Journ. f. Orn., 39, p. 63 — Mutjara [= Mutsora, near Ruwenzori, Congo].

Erythropygia hartlaubi kenia van Someren, 1930, Journ. East Africa Uganda Nat. Hist. Soc., 9 (1931), p. 196 — Mount Kenya.

Western Cameroons, Congo, Uganda, Kenya east to Mount Kenya, and northern Angola.

ERYTHROPYGIA GALACTOTES

Erythropygia galactotes galactotes (Temminck)

Sylvia galactotes Temminck, 1820, Man. Orn., ed. 2, 1, p. 182 — Algéciras, southern Spain.

Breeds in Portugal and southern Spain (southern France, rarely), Balearic Islands, North Africa from Morocco to Algeria, including the desert areas, east to Egypt, northern Sinai Peninsula, Israel, and Jordan; winters south to southern Sahara, southern Sudan, Nile and Red Sea coast. Straggler to Italy, Malta, British Isles, and Heligoland.

Erythropygia galactotes syriaca (Hemprich and Ehrenberg)

Curruca galactodes var. *syriaca* Hemprich and Ehrenberg, 1833, Symb. Phys., fol. bb — Syria [= Beirut, *vide* Hartert, 1909, *op. cit.*, p. 605].

Erythropygia plebeia Reichenow, 1904, Orn. Monatsb., 12, p. 27 — Masinde, East Africa.

Breeds in southern Yugoslavia along the Dalmatian coast, Albania, Greece, Aegean Islands (Samothrace, northern Sporades, Kythera, Mytilene, and Rhodes), southern Bulgaria, Turkey, Syria, and Lebanon; Persian Gulf coast of Saudi Arabia; migrates in winter to Somaliland, Kenya, and Arabia.

Erythropygia galactotes familiaris (Ménétriés)

Sylvia familiaris Ménétriés, 1832, Cat. Rais. Obj. Zoöl. Caucase, p. 32 — banks of the Kura, near Salyany, southeastern Azerbaijan.

Aedon familiaris deserticola Buturlin, 1908, Nasha Okhota, p. 8 — Artyk, Transcaspia [reference not verified].

Aedon familiaris transcaspica Buturlin, 1909, Nasha Okhota, p. 58, new name for preceding [reference not verified].

Aedon familiaris persica Zarudny and Härms, 1911, Journ. f. Orn., 59, p. 238 — Mesopotamia, Zagros, and Baluchistan Districts.

Agrobates galactodes iranica "Zarudny" = Ticehurst, 1922, Ibis, p. 548 — Zagros Mountains and Baluchistan. *In errore.*

Breeds in the southern Caucasus, Iran, Iraq, Afghanistan, Kirghizstan, Tadzhikistan, and West Pakistan, in Baluchistan and North West Frontier Province; wintering in southern Arabia, Iran, Iraq, southern West Pakistan, in Baluchistan, West Punjab, and Sind, and in India, in Kutch, Saurashtra, and Rajasthan.

Erythropygia galactotes minor (Cabanis)

A.(edon) minor Cabanis, 1850, Mus. Hein., 1, p. 39 — Abyssinia.

Sylvia oliviae Alexander, 1908, Bull. Brit. Orn. Club, 23, p. 15 — vicinity of Lake Chad.

Senegal, French Sahara, from Ahaggar and Aïr, south to northern Nigeria and northern Cameroon, east to Sudan, Red Sea hills, Kordofan and Darfur, Ethiopia, Eritrea, and northern Somaliland.

Erythropygia galactotes hamertoni Ogilvie-Grant

Erythropygia hamertoni Ogilvie-Grant, 1906, Bull. Brit. Orn. Club, 19, p. 24 — Beira and Wagar Mountains, Somaliland.

Known only from type locality in Northern Somalia.

ERYTHROPYGIA PAENA

Erythropygia paena benguellensis Hartert

Erythropygia paena benguellensis Hartert, 1907, Bull. Brit. Orn. Club, 19, p. 96 — Huxe, Benguella.

Angola, in Benguella Province.

Erythropygia paena paena Smith

Erythropygia paena Smith, 1836, Rep. Exped. Centr. Africa, p. 46 — between Latakoo and the Tropic [= north of Kuruman, northern Cape Province].

Erythropygia paena damarensis Hartert, 1907, Bull. Brit. Orn. Club, 19, p. 96 — Omaruru, Damaraland.

South West Africa in Namaland, Kaokoveld, Damaraland and southern Great Namaqualand, southwestern Southern Rhodesia, Bechuanaland, and South Africa, in western Transvaal and northern Cape Province.

Erythropygia paena oriens Clancey

Erythropygia paena oriens Clancey, 1957, Durban Mus. Novit., 5, p. 45 — Glen, on Modder River, north of Bloemfontein, Orange Free State, South Africa.

West Orange Free State, west to the Kimberley area, southern Transvaal, and northern Cape Province, where it intergrades with the preceding form.

ERYTHROPYGIA LEUCOSTICTA

Erythropygia leucosticta leucosticta (Sharpe)

Cossypha leucosticta Sharpe, 1883, Cat. Birds Brit. Mus., 7, p. 44, pl. 1 — Accra.

Sierra Leone, Liberia, and Ghana.

Erythropygia leucosticta collsi Alexander

Erythropygia collsi Alexander, 1907, Bull. Brit. Orn. Club, 19, p. 46 — Libokwa, Welle River.

Heavy forests of the northeastern Congo region from

Stanleyville and the Lower Uelle to the vicinity of Irumu and Beni.

Erythropygia leucosticta reichenowi Hartert

Erythropygia reichenowi Hartert, 1907, Bull. Brit. Orn. Club, 19, p. 95 — Canhoca.

Angola, from the type locality south along the escarpment to northern Huila.

ERYTHROPYGIA QUADRIVIRGATA

Erythropygia quadrivirgata erlangeri Reichenow

Erythropygia quadrivirgata erlangeri Reichenow, 1905, Vög. Afr., 3, p. 770 — lower course of the Juba River, between Bardera and Umfudu, Southern Somaliland.

Juba River, Somalia.

Erythropygia quadrivirgata quadrivirgata (Reichenow)

Thamnobia quadrivirgata Reichenow, 1879, Orn. Centralbl., p. 114 — Kapini [= Kipini], lower Tana River [reference not verified].

Erythropygia quadrivirgata rovumae Grote, 1921, Orn. Monatsb., 29, p. 109 — Mbarangandu River, Upper Rovuma.

Coastal districts of Kenya and Tanganyika south to Mozambique, as far south as the districts of Tete, Sofala, Manica, and northern Sul do Save (south to just north of Delagoa Bay, and extending up the Zambesi River to the region between Feira and Chirundu where intergrades occur with the subspecies *interna*), Southern Rhodesia, in eastern Mashonaland, south to Nuanetsi, and northern and eastern Transvaal.

Erythropygia quadrivirgata interna Clancey

Erythropygia quadrivirgata interna Clancey, 1962, Durban Mus. Novit., 6, p. 156 — Zambesi River Valley, 15 mi. west of Victoria Falls, northwestern Southern Rhodesia.

Eastern Caprivi Strip, adjacent Bechuanaland to northwest Southern Rhodesia (mainly along the Zambesi and tributaries), east to Mozambique (Chirundu and Feira see above), Northern Rhodesia, in southern Barotseland and the central and southern provinces, and adjacent southeast Angola.

Erythropygia quadrivirgata wilsoni (Roberts)

Tychaedon barbata wilsoni Roberts, 1936, Ann. Transvaal
Mus., 18, pp. 209 — Mosie Store near Maputa, northeastern Zululand.

Natal, in Zululand from Lake St. Lucia north to Mozambique in the Maputo district of Sul do Save, southeast
Swaziland, and southeast Transvaal.

Erythropygia quadrivirgata greenwayi Moreau

Erythropygia barbata greenwayi Moreau, 1938, Bull.
Brit. Orn. Club, 58, p. 64 — Mafia Island, Tanganyika
Territory.

Mafia Island and Zanzibar.

ERYTHROPYGIA BARBATA

Erythropygia barbata (Hartlaub and Finsch)

Cossypha barbata Hartlaub and Finsch, 1870, in Finsch
and Hartlaub, Vög. Ost.-Afr., p. 864 — Benguella
[= Caconda, Benguella, *vide* Barboza du Bocage, 1881,
Orn. Angola, p. 261].

Angola, Northern Rhodesia and central and northern
Nyasaland.

ERYTHROPYGIA SIGNATA

Erythropygia signata signata (Sundevall)

Cossypha signata Sundevall, 1850, öfv. K. Sv. Vet.-Akad.
Förh., 7, p. 101 — Caffraria inferiore; type from Umh-
langa, Natal.

South Africa in eastern Cape Province and Natal, including southern Zululand.

Erythropygia signata tongensis (Roberts)

Tychaedon signata tongensis Roberts, 1931, Ann. Transvaal Mus., 14, p. 242 — Mangusi Forest, North Zululand.

South Africa, in Natal in the coastal forests of northeastern Zululand.

Erythropygia signata oatleyi Clancey

Erythropygia signata oatleyi Clancey, 1956, Durban Mus.
Novit., 4, p. 251 — Woodbush Forest Reserve, 5,000 ft.,
Pietermaritzburg District, northern Transvaal.

Known only from the type locality in northern Transvaal.

Genus NAMIBORNIS Bradfield

Namibornis Bradfield, 1935, Description of new races of Kalahari birds and mammals; a leaflet reprinted, 1936, Auk, 53, p. 131. Type, by monotypy, *Bradornis Herero* De Schawensee (*sic*).

NAMIBORNIS HERERO

Namibornis herero (de Schauensee)

Bradornis herero de Schauensee, 1931, Proc. Acad. Sci. Philadelphia, 83, p. 449 — Karibib, Damaraland, South West Africa.

Western Damaraland, South West Africa.

Genus CERCOTRICHAS Boie

Cercotrichas Boie, 1831, Isis, p. 542. Type, by subsequent designation (Finsch and Hartlaub, 1870, Vög. Ost.-Afr., p. 249), *Turdus erythropterus* Gmelin = *Turdus podobe* Müller.

CERCOTRICHAS PODOBE

Cercotrichas podobe podobe (Müller)

Turdus podobe P. L. S. Müller, 1776, Syst. Nat., Suppl., p. 145 — Senegal.

Senegal; Gambia (?); east along the thorn savannah belt through French Sudan and Niger, northern Nigeria, south to Maiduguri, northern Cameroun, Lake Chad area, to Sudan, Darfur, Kordofan, south to lat. 10° N., east to Red Sea Province and northern British Somaliland.

Cercotrichas podobe melanoptera (Hemprich and Ehrenberg)

Sphenura erythroptera var. *melanoptera* Hemprich and Ehrenberg, 1833, Symb. Phys., fol. dd — Arabia [= Kunfuda, Yemen].

Western Saudi Arabia from 100 miles north of Jidda south to Yemen, the Aden Protectorate, and east to the Hadramaut.

Genus PINARORNIS Sharpe

Pinarornis Sharpe, 1876, in Layard, Birds South Africa, p. 230. Type, by monotypy, *Pinarornis plumosus* Sharpe.

PINARORNIS PLUMOSUS

Pinarornis plumosus Sharpe

Pinarornis plumosus Sharpe, 1876, in Layard, Birds
South Africa, p. 230 — Victoria Falls [errore = Mato-
pos Hills, Matabeleland, vide Irwin, 1957, Bull. Brit.
Orn. Club, 77, pp. 9-10].
Pinarornis rhodesiae Chubb, 1908, Bull. Brit. Orn. Club,
21, p. 110 — Manzinyama, Gambo Kraal.
Southern Rhodesia from southern Matabeleland (not west
of Bulawayo) patchily distributed to northeastern Mashonaland, below 6,000 feet; adjacent Bechuanaland in the area
between Francistown and the southern Rhodesia border.

GENUS CHAETOPS SWAINSON

Chaëtops Swainson, 1832, in Swainson and Richardson,
Fauna Bor.-Amer., (1831), p. 486. Type, by original
designation, Ch.(aëtops) Burchelli Swainson = Malurus
frenatus Temminck.

CHAETOPS FRENATUS

Chaetops frenatus frenatus (Temminck)

Malurus frenatus Temminck, 1826, Pl. Col., livr. 65, pl.
385 — South Africa [= River Zonde Einde Mountains,
Cape Province, vide Vincent, 1952, Check List Birds
South Africa, p. 64].
Western Cape Province, South Africa, from the Cedar-
burg Mountains, near Clanwilliam, eastward to Knysna.

Chaetops frenatus aurantius Layard

Chaetops aurantius Layard, 1867, Birds South Africa, p.
126 — near Graaff Reinet, Cape Province.
Eastern Cape Province, in Graaff Reinet Mountains, Basutoland, and Natal.

GENUS DRYMODES GOULD

Drymodes Gould, 1840, Proc. Zool. Soc. London, p. 170.
Type, by monotypy, Drymodes brunneopygia Gould.
Drymodina Iredale, 1956, Birds New Guinea, 2, p. 83.
Type, by original designation, Drymodes brevirostris
De Vis.

DRYMODES BRUNNEOPYGIA

Drymodes brunneopygia brunneopygia Gould / ᴜ ⋏ ⋁

Drymodes brunneopygia Gould, 1840, Proc. Zool. Soc. London, p. 170 — belts of the Murray in South Australia.

Drymodes brunneopygia victoriae Mathews, 1912, Novit. Zool., 18, p. 332 — Victoria.

Australia in interior of New South Wales, Victoria, and eastern South Australia.

Drymodes brunneopygia pallidus (Sharpe)

Drymaoedus pallidus Sharpe, 1883, Cat. Birds Brit. Mus., 7, p. 344 — western Australia; restricted to Shark's Bay, West Australia, by Mathews, 1921, Birds Australia, 9, p. 206.

Drymodes brunneopygia intermedia Mathews, 1921, Birds Australia, 9, p. 214 — South Australia.

Mallee country of southern and western South Australia, northwest to the Peron Peninsula.

DRYMODES SUPERCILIARIS

Drymodes superciliaris beccarii Salvadori

Drymoedus beccarii Salvadori, 1875, Ann. Mus. Civ. Genova, 7, p. 965 — Arfak Mountains.

New Guinea in the Arfak and Wandammen Mountains.

Drymodes superciliaris nigriceps Rand

Drymodes superciliaris nigriceps Rand, 1940, Amer. Mus. Novit., no. 1074, p. 1 — Bernhard Camp, Idenburg River, Netherlands New Guinea.

New Guinea in the Cyclops Mountains and northern slope of the Oranje Range.

Drymodes superciliaris brevirostris (De Vis)

Drymaoedus brevirostris De Vis, 1897, Ibis, p. 386 — British New Guinea; restricted to Boirave, Orangerie Bay, by Mayr, 1941, List New Guinea Birds, p. 110.

Drymodes beccarii adjacens Mathews, 1921, Birds Australia, 9, p. 218 — Aru Islands.

Southeastern and southern New Guinea (Oriomo) and Aru Islands.

Drymodes superciliaris colcloughi Mathews

Drymodes superciliaris colcloughi Mathews, 1914, Austral Avian Rec., 2, p. 97 — Roper River, Northern Territory. Northern Territory, Australia.

Drymodes superciliaris superciliaris Gould

Drymodes superciliaris Gould, 1850, in Jardine, Contrib. Orn., p. 105 — Cape York, Queensland.
Cape York Peninsula, north Queensland, south to the Coen River, Australia.

GENUS POGONOCICHLA CABANIS

Pogonocichla Cabanis, 1847, Arch. f. Naturg., 13, pt. 1, p. 314. Type, by original designation, *Muscicapa stellata* Vieillot.
Swynnertonia Roberts, 1922, Ann. Transvaal Mus., 8, p. 232. Type, by monotypy, *Erythracus swynnertoni* Shelley.

cf. Moreau, 1951, Ibis, 93, pp. 383-401 (geographic variation and plumage sequence).

POGONOCICHLA STELLATA

Pogonocichla stellata ruwenzorii (Ogilvie-Grant)

Tarsiger ruwenzorii Ogilvie-Grant, 1906, Bull. Brit. Orn. Club, 19, p. 33 — East Ruwenzori, 6,000-9,000 ft., [= Mubuku Valley, western Uganda, *vide* Sclater, 1930, Syst. Av. Aethiop., 2, p. 487].
Tarsiger eurydesmus Reichenow, 1908, Orn. Monatsb., 16, p. 48 — Rugege Forest.
Northeastern Congo in the mountains from Ruwenzori and the highlands northwest of Lake Edward, through Kivu District to the mountains northwest of Lake Tanganyika; extreme western Uganda.

Pogonocichla stellata elgonensis (Ogilvie-Grant)

Tarsiger elgonensis Ogilvie-Grant, 1911, Bull. Brit. Orn. Club, 27, p. 56 — Mount Elgon.
Higher slopes of Mount Elgon, Uganda-Kenya boundary.

Pogonocichla stellata guttifer (Reichenow and Neumann)

Tarsiger guttifer Reichenow and Neumann, 1895, Orn. Monatsb., 3, p. 76 — Kifinika, 3,000 m., Kilimanjaro.

Pogonocichla intensa Sharpe, 1901, Bull. Brit. Orn. Club, 11, p. 67 — Ntebi [= Entebbe], 3,800 ft., Uganda, *errore* = Mau, Kenya, *vide* Grant and Mackworth-Praed, 1941, Bull. Brit. Orn. Club, 61, p. 38.

Pogonocichla stellata keniensis Mearns, 1911, Smiths. Misc. Coll., 56 (20), p. 9 — Mount Kenya.

Pogonocichla cucullata helleri Mearns, 1913, Smiths. Misc. Coll., 61 (10), p. 1 — Mount Mbololo summit, 4,400 ft., British East Africa.

Southern Sudan mountains, south in Kenya highlands at Gargues, Nyiro, and Kulal; does not occur on Elgon, Marsabit, and Chyulu ranges; northeastern Tanganyika highlands as far as Mount Kilimanjaro.

Pogonocichla stellata macarthuri van Someren

Pogonocichla stellata macarthuri van Someren, 1939, Journ. East Africa Uganda Nat. Hist. Soc., 14, p. 83 — Chyulu Range.

Chyulu Range, southeastern Kenya.

Pogonocichla stellata orientalis (Fischer and Reichenow)

Tarsiger orientalis Fischer and Reichenow, 1884, Journ. f. Orn., 32, p. 57 — Pangani (Küstengebiet).

Pogonocichla johnstoni Shelley, 1893, Ibis, p. 18 — Milanji Plateau.

Tarsiger olivaceus Reichenow, 1900, Orn. Monatsb., 8, p. 100 — Ukinga.

Tarsiger johnstoni montanus Reichenow, 1906, Orn. Monatsb., 14, p. 172 — Usambara.

Tanganyika lowlands (Usambara, Uluguru, Pugu, and Kilosa), Kiboriani Mountain, mountains north and east of Lake Nyasa and on eastern side of Lake Tanganyika; mountains of Nyasaland; Northern Rhodesia (Nyika and Mukutu Mountains); Mozambique (Unangu and Namuli Mountains).

Pogonocichla stellata transvaalensis (Roberts)

Tarsiger stellatus transvaalensis Roberts, 1912, Journ. S. African Orn. Un., 8, p. 21 — Woodbush, Transvaal.

Tarsiger stellatus chirindensis Roberts, 1914, Ann. Transvaal Mus., 4, p. 175 — Chirinda, Mashonaland.

Eastern border of Southern Rhodesia, northeast Transvaal, and Mozambique in Gazaland south to Lourenço Marques.

Pogonocichla stellata lebombo Roberts

Pogonocichla stellata lebombo Roberts, 1935, Ann. Transvaal Mus., 18, p. 208 — Ingwavuma Forest, Lebombo Mountains, Zululand.

Lebombo Mountains, Zululand-Swaziland border, South Africa.

Pogonocichla stellata stellata (Vieillot)

Muscicapa stellata Vieillot, 1818, Nouv. Dict. Hist. Nat., nouv. éd., 21, p. 468, ex Levaillant, 1805, Oiseaux Afrique, 4, pl. 157, "L'Étoile" — Plettenberg Bay, Cape Province.

Pogonocichla margaritata Sundevall, 1850, öfv. K. Sv. Vet.-Akad. Förh., 7, p. 104 — "Caffraria inferiore s. propria;" type from Pietermaritzburg, Natal, *vide* Gyldenstolpe, 1926, Ark. f. Zool., **19A**, p. 54.

Cape Province, South Africa, from Knysna north and east to Orange Free State (Bosch), Natal, and Zululand (north to Eshowe and Richards Bay).

POGONOCICHLA SWYNNERTONI

Pogonocichla swynnertoni (Shelley)

Erythracus swynnertoni Shelley, 1906, Bull. Brit. Orn. Club, 16, p. 125 — Chirinda Forest, Mashonaland.

Eastern border of Southern Rhodesia, in the Chirinda Mountains from Mount Selinda to the Vumba, in evergreen forest.

GENUS **ERITHACUS** CUVIER

Erithacus Cuvier, 1800, Leçons Anat. Comp., 1, tab. 2. Type, by monotypy, *Motacilla Rubecula* Linnaeus.

Luscinia Forster, 1817, Synop. Cat. Brit. Birds, p. 14. Type, by monotypy, "*Sylvia luscinia*" = *Luscinia megarhynchos* Brehm.

Aedon Forster, 1817, Synop. Cat. Brit. Birds, pp. 14; 53. Type, by monotypy, *Aedon luscinia* Forster = *Motacilla luscinia* Linnaeus.

Cyanosylvia Brehm, 1828, Isis, col. 920. Type, by monotypy, *Motacilla sinensis* Linnaeus.

Cyanecula Brehm, 1828, Handl. Vög. Deutschl., p. 349
Type, by monotypy, *Motacilla suecica* Linnaeus.
Calliope Gould, 1836, Birds Europe, pt. 2, pl. 118, text.
Type, by monotypy, *Calliope Lathamii* Gould = *Motacilla calliope* Pallas.
Larvivora Hodgson, 1837, Journ. Asiat. Soc. Bengal, 6, p. 102. Type, by original designation, *Motacilla cyane* Pallas.
Nemura Hodgson, 1844, in Gray, Zool. Misc., p. 83, *nom. nud. Idem*, 1845, Proc. Zool. Soc. London, p. 27. Type, by subsequent designation (Blyth, 1847, Journ. Asiat. Soc. Bengal, 16, p. 132), *Nemura rufilata* Hodgson.
Tarsiger Hodgson, 1845, Proc. Zool. Soc. London, p. 28, ex Hodgson, in Gray, 1844, Zool. Misc., p. 83, *nom. nud.* Type, by monotypy, *Tarsiger chrysaeus* Hodgson.
Ianthia Blyth, 1847, Journ. Asiat. Soc. Bengal, 16, pt. 1, p. 132, new name for *Nemura* Hodgson, *nec Nemura* Latreille, 1798.
Stiphrornis Hartlaub, 1855, Journ. f. Orn., 3, p. 355. Type, by original designation, *Stiphrornis erythrothorax* Hartlaub (ex Temminck MS).
Icoturus Stejneger, 1886, Proc. U. S. Nat. Mus., 9, p. 643. Type, by original designation, *Icoturus namiyei* Stejneger.
Sheppardia Haagner, 1909, Ann. Transvaal Mus., 1, p. 180. Type, by monotypy, *Sheppardia gunningi* Haagner.
Pseudaëdon Buturlin, 1910, Mess. Orn., ann. 1 (2), pp. 136; 139. Type, by monotypy, *Larvivora sibilans* Swinhoe.
Vibrissosylvia Neumann, 1920, Journ. f. Orn., 68, p. 78. Type, by monotypy, *Callene cyornithopsis* Sharpe.

cf. Witherby *et al.*, 1938, Handbook Brit. Birds, 2, pp. 187; 199 (*megarhyncha, rubecula*).
Macdonald, 1940, Ibis, pp. 663-671 (*sharpei, gunningi*).
Prigogine, 1954, Rev. Zool. Bot. Africa, 50, pp. 1-12 (*aequatorialis* and *erythrothorax*).
Vaurie, 1955, Amer. Mus. Novit., no. 1731, pp. 1-14 (*Erithacus, Luscinia, Tarsiger*).
Ripley, 1958, Postilla, Yale Univ., no. 37, pp. 1-3 (*obscurus* and *pectardens*).
Hall, 1961, Bull. Brit. Orn. Club, 81, pp. 45-46 (*gabela*).

ERITHACUS GABELA

Erithacus gabela (Rand)

Muscicapa gabela Rand, 1957, Fieldiana: Zool. [Chicago], 39, p. 41 — 15 kilometers south of Gabela, Angola.
Known only from the type locality.

ERITHACUS CYORNITHOPSIS

Erithacus cyornithopsis houghtoni (Bannerman)

Sheppardia cyornithopsis houghtoni Bannerman, 1931, Bull. Brit. Orn. Club, 51, p. 128 — Sugarloaf Forest, near Freetown, Sierra Leone.
Evergreen forest in Sierra Leone and Liberia.

Erithacus cyornithopsis cyornithopsis (Sharpe)

Callene cyornithopsis Sharpe, 1901, Bull. Brit. Orn. Club, 12, p. 4 — Efulen, Kamerun.
Lowland forest of southern Cameroons.

Erithacus cyornithopsis lopezi (Alexander)

Callene lopezi Alexander, 1907, Bull. Brit. Orn. Club, 19, p. 46 — Libokwa, Welle River.
Forests of northeastern Congo, ranging to central Congo and Uganda (Manzira Forest near Masaka).

Erithacus cyornithopsis acholiensis (Macdonald)

Sheppardia aequatorialis acholiensis Macdonald, 1940, Ibis, p. 670 — Kitibol, Acholi Hills.
Imatong Mountains, southern Sudan.

ERITHACUS AEQUATORIALIS

Erithacus aequatorialis (Jackson)

Callene aequatorialis Jackson, 1906, Bull. Brit. Orn. Club, 16, p. 46 — Kericho, Lumbwa.
Highlands of the eastern Congo from Ituri, Kivu District, to the base of Mount Elgon; Ruanda; Uganda; and western Kenya to western base of the Mau Plateau.

ERITHACUS ERYTHROTHORAX

Erithacus erythrothorax erythrothorax (Hartlaub)

Stiphrornis erythrothorax Hartlaub, 1855, Journ. f. Orn., 3, p. 355 — Dabocrom.

Sierra Leone, Liberia, Ghana (Gold Coast), and southern Nigeria north to lat. 9° N.

Erithacus erythrothorax gabonensis (Sharpe)

Stiphrornis gabonensis Sharpe, 1883, Cat. Birds Brit. Mus., 7, p. 174, pl. 6, fig. 2 — Gabon.

Fernando Po Island and coastal region of Cameroons and Gabon.

Erithacus erythrothorax xanthogaster (Sharpe)

Stiphrornis xanthogaster Sharpe, 1903, Bull. Brit. Orn. Club, 14, p. 19 — River Ja.

Southern and southeast Cameroons and Congo along the Middle Congo River.

Erithacus erythrothorax mabirae (Jackson)

Stiphrornis mabirae Jackson, 1910, Bull. Brit. Orn. Club, 25, p. 85 — Mabira Forest, Uganda.

Congo, in the eastern half of the upper Congo forest, and Uganda, in forest patches east to the Mabira.

ERITHACUS SHARPEI

Erithacus sharpei usambarae (Macdonald)

Sheppardia sharpei usambarae Macdonald, 1940, Ibis, p. 669 — Amani Forest, Usambara Mountains, Tanganyika Territory.

Forests of Usambara and Ngura Mountains, Tanganyika.

Erithacus sharpei sharpei (Shelley)

Callene sharpei Shelley, 1903, Bull. Brit. Orn. Club, 13, p. 60 — Masisi Hill.

Sheppardia cyornithopsis bangsi Friedmann, 1930, Occ. Papers Boston Soc. Nat. Hist., 5, p. 323 — Uluguru, Tanganyika Territory.

Mountain forests of southwest Tanganyika and northern Nyasaland.

ERITHACUS GUNNINGI

Erithacus gunningi sokokensis (van Someren)

Callene sokokensis van Someren, 1921, Bull. Brit. Orn. Club, 41, p. 125 — Sokoke Forest, coast of British East Africa.

Coastal areas of Kenya and Tanganyika, from Malindi to the Pugu Hills.

Erithacus gunningi bensoni (Kinnear)

Sheppardia bensoni Kinnear, 1938, Bull. Brit. Orn. Club, 58, p. 138 — near Nkata Bay, Chinteche District, West Nyasa District, Nyasaland.
Rainforest in northwestern Nyasaland.

Erithacus gunningi gunningi (Haagner)

Sheppardia gunningi Haagner, 1909, Ann. Transvaal Mus., 1, p. 180 — Mzimbiti, near Beira.
Forests of central Mozambique.

ERITHACUS RUBECULA

Erithacus rubecula melophilus Hartert

Erithacus rubecula melophilus Hartert, 1901, Novit. Zool., 8, p. 317 — British Isles; restricted to Hertfordshire by Lack, 1946, Bull. Brit. Orn. Club, 66, p. 62.
Erithacus rubecula hebridium Horniman, 1940, Prelim. Descr. New Birds, p. 2 — Skye, Inner Hebrides; *idem* Grant, 1956, Ann. Mag. Nat. Hist., ser. 12, 9, p. 366.[1]

Breeds in the British Isles and on adjacent coast, where intergrades with the nominate form in the region of the North Sea and English Channel; winters on the home range; some individuals migrating to Portugal, Italy (?), and North Africa (Morocco).

Erithacus rubecula rubecula (Linnaeus)

Motacilla Rubecula Linnaeus, 1758, Syst. Nat., ed. 10, 1, p. 188 — Europe [= Sweden, *vide* Linnaeus, 1746, Fauna Svecica, no. 232].
Erithacus rubecula maior Prazák, 1897, Journ. f. Orn., 45, p. 249 — Stryj, East Galicia.
E. (rithacus) r. (ubecula) microrhynchos Reichenow, 1906, Journ. f. Orn., 54, p. 153 — Madeira.
Erithacus rubecula monnardi Kleinschmidt, 1916, Falco, 12, p. 14 — northeast Germany and Rhine.

---

[1] Under Opinion 480 (1957, Ops. Decls. Internat. Comm. Zool. Nomencl., 16, pp. 417-454) this name is suppressed for purposes of Law of Priority, but not for those of Law of Homonymy.

Erithacus rubecula atlas Lynes, 1919, Bull. Brit. Orn. Club, 40, p. 32 — Middle Atlas of Morocco.

Erithacus rubeculus armoricanus Lebeurier and Rapine, 1936, Oiseau Rev. Franç. Orn., 6, p. 252 — Primel, Plougasnou (Finistère).

Erithacus rubecula hispaniae von Jordans, 1950, Sylleg. Biol. Leipzig, Festchr. O. Kleinschmidt, p. 174 — Linares de Ríofrio, Salamanca.

Breeds in western Europe from central coastal Norway, central coastal Sweden, and central Finland east to the Urals (intergrading with *melophilus* on the English Channel islands and southern North Sea coast, and in color approaching *witherbyi* in southern Spain and Portugal), south in Europe to Italy and in northwest Morocco, and the Azores, Madeira, and western Canary Islands; winters from western France, England (scarce), and northwest Germany south as far as eastern North Africa, Greece, and Egypt.

Erithacus rubecula superbus Koenig

Erithacus superbus Koenig, 1889, Journ. f. Orn., 37, p. 183 — higher region of Teneriffe.

Mountains of Teneriffe and Grand Canary, central Canary Islands.

Erithacus rubecula witherbyi Hartert

Erithacus rubecula witherbyi Hartert, 1910, Vög. pal. Fauna, 1, p. 753 — Hamman R'Hira, north Algeria.

Erithacus rubecula lavaudeni Bannerman, 1926, Bull. Brit. Orn. Club, 47, p. 24 — Les Sources, 2,200 ft., near Ain Draham.

Breeds in eastern Algeria and Tunisia. Color variants of *rubecula* approach *witherbyi* in southern Spain and Italy (see Lack, 1947, Bull. Brit. Orn. Club, 67, pp. 52-54).

Erithacus rubecula sardus Kleinschmidt

Erithacus Dandalus sardus Kleinschmidt, 1906, Falco, p. 71 — Sardinia.

Mountains of Corsica and Sardinia, possibly intergrading into the breeding form of Italy.

Erithacus rubecula balcanicus Watson

Erithacus rubecula balcanicus Watson, 1961, Postilla, Yale Univ., no. 52, p. 6 — Boz Dag, 4,200 ft., Odemis, Izmir, western Turkey.

Breeds from northern Yugoslavia south in Balkans through Greece, Bulgaria, and northern and western Turkey; winters at lower altitudes, on the Aegean islands, and in southern Turkey.

Erithacus rubecula hyrcanus Blanford

Erithacus hyrcanus Blanford, 1874, Ibis, p. 79 — Ghilan Province, Persia.

Erithacus rubecula caucasicus Buturlin, 1907, Orn. Monatsb., 15, p. 9 — Caucasus and Transcaucasus.

Erithacus xanthothorax Salvadori and Festa, 1913, Boll. Mus. Zool. Anat. Comp. Torino, 28, no. 673, p. 15 — Rhodes.[1]

Erithacus rubeculus ciscaucasicus Buturlin, 1929, Syst. Notes Birds N. Caucasus, p. 22 — Wladikawkaz.

E.(rithacus) r.(ubecula) valens Portenko, 1954, Birds U.S.S.R., 3, p. 193 — Crimea.

Intergrades with *rubecula* in the Caucasus; breeds in eastern Turkey, Armenia, southeastern Russia, and Transcaspia; winters south to Iraq, Iran, and Lebanon (subspecies ?), west to Greek islands (Rhodes) and to Fao on the Persian Gulf.

Erithacus rubecula tataricus Grote

Erithacus rubecula tataricus Grote, 1928, Orn. Monatsb., 36, p. 52 — Orenbourg District, Urals.

Breeds in Russia in western Siberia from the Ural Mountains north to lat. 64° 20′ N. on the Pechora River, east to Semipalatinsk (rare); vagrant to Tomsk and northwest Mongolia, Kobdo, south to the lower Volga steppes; winters south to Iran.

ERITHACUS AKAHIGE

Erithacus akahige akahige (Temminck)

Sylvia akahige Temminck, 1835, Pl. Col., livr. 96, p. 571 — Riu Kius, south of Japan; corrected to Hondo by Kuroda, 1923, Bull. Brit. Orn. Club, 43, p. 106.

Southern Kurile Islands, Sakhalin, Hokkaido, Honshu, Shikoku; winter visitor to Kyushu, Formosa, and southern China coast.

---

[1] This name based on a winter migrant.

Erithacus akahige tanensis Kuroda

Erithacus akahige tanensis Kuroda, 1923 (March), Bull. Brit. Orn. Club, 43, p. 106 — Nishino-omote, Tanegashima, south of Kiu Siu (*sic*).

Erithacus akahige sgectatoris (misprint for *spectatoris*) Momiyama, 1923 (Dec.), Dobuts. Zasshi, 35, p. 403 — Hachijô (Island).

Luscinia akahige kobayashii Momiyama, 1940, Dobuts. Zasshi, 52, p. 463 — Yakushima.

Izu Islands, Tanegashima Island, and Yakushima Island, southern Japan.

ERITHACUS KOMADORI

Erithacus komadori komadori (Temminck)

Sylvia komadori Temminck, 1835, Pl. Col., livr. 96, pl. 570 — Korea; corrected to northern Riu Kius and Tanegashima by Kuroda, 1923, *op. cit.*, p. 106.

Tanegashima; Amami-oshima and Tokunoshima, northern Riu Kiu Islands.

Erithacus komadori namiyei (Stejneger)

Icoturus namiyei Stejneger, 1886, Proc. U. S. Nat. Mus., 9, p. 644 — Okinawa, Riu Kiu Islands.

Okinawa, central Riu Kiu Islands.

Erithacus komadori subrufus (Kuroda)

Icoturus komadori subrufus Kuroda, 1923, Bull. Brit. Orn. Club, 43, p. 106 — Yonakuni Island, southernmost island of Yaeyama Islands, southern Riu Kiu group.

Ishigaki, Iriomote, and Yonakuni, southern Riu Kiu Islands.

ERITHACUS SIBILANS

Erithacus sibilans (Swinhoe)

Larvivora sibilans Swinhoe, 1863, Proc. Zool. Soc. London, p. 292 — Macao, southeastern China.

P.(seudaëdon) s.(ibilans) swistun Portenko, 1954, Birds U.S.S.R., 3, p. 196 — Lake Mazharskoye, east of Minusinsk.

Breeds in southern east Siberia from Lake Teletskoe, Altai, along Yenisei River to Eloguya, lower Tunguska, Olekmi, mouth of the Aldan, south to the foothills of Altai, Sayan

Mountains, southern Transbaikalia, crossing northeast China to Vladivostock, Kamchatka and Sakhalin Island; winters in China in southeast Yunnan, Kwangtung, northwest Fokhien, and northern Laos; migrant in Shaweishan, northeast Chihli, Korea; straggler to Japan (Honshu and Shikoku).

ERITHACUS LUSCINIA

Erithacus luscinia (Linnaeus)

> *Motacilla Luscinia* Linnaeus, 1758, Syst. Nat., ed. 10, 1, p. 184 — "Europae frondosis"; restricted to Sweden by Hartert, 1910, Vög. pal. Fauna, 1, p. 736.

Breeds in coastal eastern Denmark, southern Sweden, northeast Germany, southern Finland, Russia from the Baltic east to west Siberia north to Kargopol, upper Vichegda, Tobolsk to the western Altai (Krasnoyarsk, vagrant); Karatau range; south to Karkaralinsk, middle Ural region, northern Caucasus, Crimea west to southeast Europe, Austria, Yugoslavia, Hungary, and Rumania; wintering in Northern and Southern Rhodesia, Nyasaland, Kenya, Sudan, Somaliland, and southern Arabia; migrant in Israel, Lebanon, Egypt, and Iraq.

ERITHACUS MEGARHYNCHOS

Erithacus megarhynchos megarhynchos (Brehm)

> *Luscinia megarhynchos* Brehm, 1831, Naturg. Vög. Deutsch., p. 356 — Germany.
> *Luscinia megarhyncha corsa* Parrot, 1910, Orn. Monatsb., 18, p. 155 — Ajaccio, Corsica.
> *Luscinia megarhynchos luscinioides* von Jordans, 1923, Falco, 19, sonderheft, p. 3 — Arta, Mallorca.
> *Luscinia megarhynchos caligiformis* Clancey and von Jordans, 1950, Auk, 67, p. 361 — Martlesham, Woodbridge, East Suffolk, southeastern England.
> *L.(uscinia) m.(egarhyncha) tauridae* Portenko, 1954, Birds U.S.S.R., 3, p. 207 — Simferopol, Crimea.

Breeds in southern England, western Europe north to Germany and Poland (except the Baltic coast), and southwestern Ukraine, south to southern Spain, Portugal, Italy, the Balearic Islands, Corsica, Sardinia, Greece, Crete, Cyprus, northern Middle East and western Transcaucasia, and northwest Africa from Morocco to eastern Libya; migrates through Lebanon; winters in tropical Africa from Sierra Leone and Ghana (Gold Coast), Nigeria, and the northern Congo to Uganda and Kenya; vagrant to Denmark, Canary Islands, and Madeira.

Erithacus megarhynchos africanus (Fischer and Reichenow)

Lusciola africana Fischer and Reichenow, 1884, Journ. f. Orn., 32, p. 182 — Lower Arusha, near Kilimanjaro.
Philomela transcaucasia Buturlin, 1910, Mess. Orn., 1, p. 140 — Aresh District, Elisabetpolsk.

Breeds in Armenia, Syria, Iran (from southern Caucasus to Shiraz), and Iraq; winters in Africa in northern Tanganyika and Kenya; on passage in southern Arabia (Taif), Yemen, and Aden.

— **Erithacus megarhynchos hafizi** (Severtzov)

Luscinia Hafizi Severtzov, 1873, Vertikal. . . Turkest. Zhivotn., (1872), p. 120 — Turkestan.
Luscinia Golzii Cabanis, 1873, Journ. f. Orn., 21, p. 79 — Turkestan.

Breeds in Asia from west Kazakhstan east through Turgai, Syr-Darya, Kirghiz Mountains, Semirechye, Irtysh River, and Zaissan basin, south to the Tian Shan range and Afghanistan; winters in coastal east Africa from Somaliland to Kenya and northeastern Tanganyika; on passage in Iraq and Arabia; occasional in West Pakistan.

ERITHACUS CALLIOPE

— **Erithacus calliope** (Pallas)

Motacilla Calliope Pallas, 1776, Reise versch. Prov. Russ. Reichs, 3, p. 697 — between the Yenisei and the Lena Rivers.
Turdus camtschatkensis Gmelin, 1789, Syst. Nat., 1 (2), p. 817 — Kamchatka.
Calliope calliope sachalinensis Portenko, 1937, Mitt. Zool. Mus. Berlin, 22, p. 223 — Aleksandrovsk, Sakhalin Island.
Luscinia calliope beicki Meise, 1937, Journ. f. Orn., 85, p. 562 — Sin-tien-pu, north Kansu.
Calliope calliope natio *anadyrensis* Portenko, 1939, Tamzhe, p. 128 — Anadyr [reference not verified].

Breeds in Siberia from the middle Ural Mountains north

to tree line, along the Ob and Yenisei Rivers, ranging farther north in eastern Asia, as far as the Anadyr range, Kamchatka, Commander, and Kurile Islands, south in the southern taiga belt from Tobolsk to Novosibirsk and Barnaul, the Altai and northern Mongolia, northern China, Manchuria and Hokkaido Island, breeding south in China in the Nan Shan Mountains of Kansu and Szechwan. Winters from eastern India and East Pakistan (irregularly), northern Burma, Thailand, Laos, and Tonkin, east in southern China, Formosa, Philippine Islands, and Japan; on passage in Korea; straggler to western Aleutian Islands.

ERITHACUS SVECICUS

Erithacus svecicus svecicus (Linnaeus)

Motacilla svecica Linnaeus, 1758, Syst. Nat., ed. 10, 1, p. 187 — in Europae alpinis [= Sweden and Lappland, *vide* Hartert, 1910, Vög. pal. Fauna, 1, p. 745].

Erithacus gaetkei Kleinschmidt, 1904, Journ. f. Orn., 52, p. 302 — Helgoland.

Cyanecula svecica robusta Buturlin, 1907, Psov. Rush. Okh., no. 6 — Kolyma Delta [reference not verified].

Luscinia svecica weigoldi Kleinschmidt, 1924, Abh. Ber. Mus. Dresden, 16 (2), p. 43 — Bago, 90 km. east of Jehol, northern Chihli (Hopeh).

Breeding from Norway, Sweden, and Finland across Russia (reaching lat. 72° N.), northern Alaska from Wales to Point Barrow, south to north central Russia and China in Sinkiang and northern Mongolia; winters in northeast Africa, Sudan, Ethiopia, Israel, Iraq, and Arabia (irregular) east to Iran, Afghanistan, West Pakistan, Nepal, India, Ceylon, Burma, Thailand, and southern China; on passage in middle Europe west to British Isles, southwest Asia, and central China.

Erithacus svecicus cyaneculus (Meisner)

Sylvia Cyanecula Meisner, 1804, Syst. Verz. Vögel Schweiz, p. 30 — France, ex Buffon, 1783, Pl. en., 6, pl. 361; restricted to Ardennes [restriction not verified].

Luscinia svecica namnetum Mayaud, 1934, Bull. Brit. Orn. Club, 54, p. 179 — Noirmoutier, Vendée.

Breeds in middle Europe from France north to Germany, Denmark, Poland, and Baltic Russia (east to Smolensk and central Kiev districts) south to mildle Spain and Yugoslavia; winters in northwest Africa east to Egypt, Sudan, and Lebanon (subspecies).

—Erithacus svecicus volgae Kleinschmidt

[*Cyanecula succica*] *occidentalis* Zarudny, 1892, Mater. Kenntn. Faun. Flor. Russ. Reichs., Zool., (1), p. 146 — Ortchik, Ukraine [reference not verified], *nec Luscinia occidentalis* Severtzov, 1872.

Erithacus volgae Kleinschmidt, 1907, Falco, 3, p. 47 — lower Volga; restricted to Sarpa by Hartert, 1910, Vög. pal. Fauna, 1, p. 749.

Luscinia succica grotei Dementiev, 1932, Alauda, 4, p. 8 — Vaskin Potok, Ivanovo, central Russia.

Breeds in south-central European Russia in the Dnieper and Don River area; winter range not worked out.

Erithacus svecicus luristanicus Ripley

Cyanecula wolfi magna Zarudny and Loudon, 1904, Orn. Jahrb., p. 225 — Bidesar, Arabistan, southwest Persia, *nec Philomela magna* Blyth, 1833 (Aug.), Rennie's Field Nat., 1, p. 355, substitute name for *Sylvia Philomela* Bechstein (Temminck MS), 1802; see, also, Blyth, 1833 (May), Rennie's Field Nat., 1, p. 200, here called *Philomela major*.

Erithacus svecicus luristanicus Ripley, 1952, 1954, Postilla, Yale Univ., no. 13, p. 23, *et ad.* p. 1, *nom. nov.* for *Cyanecula wolfi magna* Zarudny and Loudon—Luristan.

Breeds in Armenia west to Artvin, south to southwest Iran, from Arabistan to Mazanderan; winters to Iraq, Arabia (one record), and Sudan.

—Erithacus svecicus pallidogularis (Zarudny)

C.(yanecula) succica var. *pallidogularis* Zarudny, 1897, Zapiski Mem. Imp. Acad. Sci. St. Petersberg, suppl. Mater. Kenntn. Flor. Faun. Russ. Reichs., Zool., (3), p. 186 — Orenburg.

Cyanecula discessa Madarász, 1902, Termész. Füzetek, 25, p. 535 — Transcaspia [reference not verified].

Cyanecula succica aralensis Zarudny, 1916, Izvest. Turkest. Otd. Russk. Geogr. Obsht., 12. p. 71 — delta of Syr-Darya [references not verified].

Cyanosylvia suecica kobdensis Tugarinov, 1929, Ann. Mus.
Zool. Acad. Sci. (U.S.S.R.; Leningrad), 29 (1928), p. 9
— Uliassutai.
Cyanosylvia suecica kaschgariensis Tugarinov, 1929, *ibid.*,
p. 10 — Yarkand-Darya.
Cyanosylvia suecica caucasica Buturlin, 1929, Syst. Notes
Birds North Caucasus, p. 21 — Kotlarevskaia on the
Terek.

Breeds in the southern Urals from Irtysh to the Volga
east to Transcaspia, Tadzhikistan, and Kirgizstan, and the
Tian Shan range below 5,000 feet; winters in Afghanistan,
West Pakistan, and India.

Erithacus svecicus abbotti (Richmond)

Cyanecula abbotti Richmond, 1896, Proc. U. S. Nat. Mus.,
18, p. 484 — Nubra Valley, Ladakh.

Breeds in West Pakistan and India in Gilgit, Baltistan,
Ladakh, and Zanskar; winters in foothills and adjacent
plains of West Pakistan, Kashmir, and India.

Erithacus svecicus saturatior (Sushkin)

Cyanecula svecica saturatior Sushkin, 1925, List Distrib.
Birds Russian Altai, p. 77 — moor Djoievo, near Minu-
sinsk; moor near Bijsk, lake Dzhagatai-kul, Uriankh-
land.
Cyanecula svecica altaica Sushkin, 1925, *ibid.*, p. 77 —
near Kosh-agach, Topolevka-Karaghem, tributary of
Arkhyt, plateau of Chulyshman.
Cyanosylvia suecica tianshanica Tugarinov, 1929, Ann.
Mus. Zool. Acad. Sci. (U.S.S.R.; Leningrad), 29 (1928),
p. 9 — Baingol, Tian Shan.
Cyanosylvia suecica przevalskii Tugarinov, 1929, *ibid.*, p.
11 — Zagan-bulyk, Ala Shan range.

Breeds in the Tian Shan and Pamirs, above 5,000 feet,
east in the mountains of the lower Turguska basin, Altai,
Sayan, Angara, northern Mongolia, and eastern Tibet; winters in adjacent foothills south to Afghanistan and eastern
Tibet; southeast Tibet (?).

ERITHACUS PECTORALIS

Erithacus pectoralis pectoralis (Gould)

Calliope pectoralis Gould, 1837, Icones Avium, pl. 4 and

text — Himalaya Mountains [= western Himalayas, *vide* Hartert, 1910, Vög. pal. Fauna, 1, p. 739].

Luscinia Ballioni Severtzov, 1873, Vertikal . . . Turkest. Zhivotn., (1872), p. 122 — Tian Shan.

Breeds in mountains of Tadzhikistan and southern Kirgizstan south to West Pakistan, Gilgit, Baltistan, Kaghan Valley, Kashmir, and India east along the Himalayas to Kumaon; winters in foothills east to Nepal; on passage in Chitral and Himalayan foothills from Kashmir to Simla.

Erithacus pectoralis confusus (Hartert)

Luscinia pectoralis confusa Hartert, 1910, Vög. pal. Fauna, 1, p. 740 — Sikkim.

Breeds in mountains of Nepal, Darjeeling, Sikkim, and Bhutan; winters in adjacent plains of Nepal, eastern India, and East Pakistan.

Erithacus pectoralis tschebaiewi (Przevalski)

Calliope Tschebaiewi Przevalski, 1876, Mongol. i Strana Tangut., 2, p. 44, pl. 9, fig. 1 — Kansu.

Breeds from extreme east Ladakh east to Kansu and Tsinghai in northwest China, southeast Tibet, and extreme north Burma; winters from Sikkim and Bhutan duars east in Assam, East Pakistan, and Burma; on passage in Kashmir.

ERITHACUS RUFICEPS

Erithacus ruficeps (Hartert)

Larvivora ruficeps Hartert, 1907, Bull. Brit. Orn. Club, 19, p. 50 — Tai pai Shan, Tsinling Mountains.

Known only from type locality in Tsinling Mountains, southwest Shensi, west China, and one migrant record from Mount Brinchang, 6,500 feet, Malaya (1963).

ERITHACUS OBSCURUS

Erithacus obscurus (Berezowsky and Bianchi)

Larvivora obscura Berezowsky and Bianchi, 1891, Ptitzi Kansus. Puteshest, G. N. Potanina, p. 97 — Kansu, *nec Cyanecula obscura* Brehm, 1831, Vög. Deutschl., p. 353.

Erithacus hachisukae Ripley, 1952, 1954, Postilla, Yale Univ., no. 13, p. 24, *et ad.*, p. 1, *nom. nov.* for *Larvivora obscura* Berezowski and Bianchi — Kansu.[1]

Southeast Kansu and southwest Shensi, west China.

ERITHACUS PECTARDENS

Erithacus pectardens (David)

Calliope pectardens David, 1871, Nouv. Arch. Mus. Hist. Nat. [Paris], no. 167, *nom. nud.*

Calliope pectardens David, 1877, in David and Oustalet, Oiseaux Chine, 1, p. 236 — Moupin [= Paohing, eastern Sikang].

Luscinia Davidi Oustalet, 1892, Bull. Mus. Hist. Nat. Paris, p. 222 — Tatsien-lu.

Luscinia davidi gloriosa Sushkin, 1926, Auk, 43, p. 181 — Li Kiang Mountains, China.

Luscinia daulias Koelz, 1954, Contrib. Inst. Regional Explor., no. 1, p. 12 — Phulbari, Garo Hills.

Breeds in the mountains of southeast Tibet, southeast Sikang, and Yunnan (one specimen from southwest Shensi; post-breeding vagrant ?) ; winters in lower hills, and south occasionally to the Himalayas in Sikkim and Assam and in northern Burma (one record, Bhamo district).

ERITHACUS BRUNNEUS

Erithacus brunneus brunneus (Hodgson)

Larvivora brunnea Hodgson, 1837, Journ. Asiat. Soc. Bengal, 6, p. 102 — Nepal.

Larvivora brunnea angamea Koelz, 1952, Journ. Zool. Soc. India, 4, p. 41 — Tekhubama, Naga Hills.

Breeds in the Himalayas from the Afghanistan-West Pakistan boundary (Safed Koh) east to Kashmir, Nepal, Sikkim, and Bhutan; winters in lower Himalayas and south on Indian Peninsula to Ceylon; presumably winters in East Pakistan.

[1] By Opinion 444 (1957, Ops. Decls. Comm. Zool. Nomencl., 15, pp. 175-190) *Larvivora obscura* Berezowsky and Bianchi, 1891, is placed on the Official List of Specific Names in Zoology, and *Cyanecula obscura* Brehm, 1831, and *Erithacus hachisukae* Ripley, 1952, are placed on the Official Index of Rejected and Invalid Specific Names in Zoology.

Erithacus brunneus wickhami (Baker)

Larvivora wickhami Baker, 1916, Novit. Zool., 23, p. 298 — Chin Hills.

Chin Hills, Burma.

ERITHACUS CYANE

Erithacus cyane cyane (Pallas)

Motacilla Cyane Pallas, 1776, Reise versch. Prov. Russ. Reichs, 3, p. 697 — Dauria, between the Onon and Argu Rivers [southeast Transbaicalia].

Breeds in Russia in the western Altai from about Lake Teletskoye north to Krasnoyarsk, east across Lake Baikal to Chita and north to central Lena River, about lat. 60° N.; winters south to Burma, Malaya, Sumatra, Borneo, northern Thailand, and Laos; straggler to Bhutan duars.

Erithacus cyane bochaiensis (Shulpin)

Larvivora cyane bochaiensis Shulpin, 1928, Ann. Mus. Zool. Acad. Sci. (U.S.S.R.; Leningrad), 27 (1927), p. 404 — Fansa Station, Suchan railroad, southern Ussuria [reference not verified].

Breeds in Ussuria, Korea (rarely), and Japan (southern Sakhalin, Hokkaido, Honshu, Shikoku and Kyushu; eggs found only on Honshu) ; winters south in the Indochinese subregion, Malaya, and Borneo.

ERITHACUS CYANURUS

Erithacus cyanurus cyanurus (Pallas)

Motacilla Cyanurus Pallas, 1773, Reise versch. Prov. Russ. Reichs, 2, p. 709 — Yenisei.

Tarsiger cyanurus ussuriensis Stegmann (ex Sushkin MS), 1929, Ann. Mus. Zool. Acad. Sci. U.R.S.S. [Leningrad], 29 (1928), p. 229 — Ussuria.

Breeds in northern Lapland (and probably Finland) and Russia, from sources of Pechora River east to Stanovoi Mountains and Okhotsk, reaching north to lat. 65° N.; Kamchatka, Commander Islands, Sakhalin, and northern Japan (Hokkaido and Honshu) ; winters in southern Korea, southern Japanese islands, southern China, and Formosa, reaching upper Burma; vagrant to Assam; on passage in north China, north of the Yangtse River.

Erithacus cyanurus pallidior (Baker)

Ianthia cyanura pallidiora Baker, 1924, Fauna Brit. Ind.,
Birds, ed. 2, 1, p. 101 — Simla.

Breeds in the hills from the Afghanistan-West Pakistan
boundary (Safed Koh), Gilgit, Astor, and Kashmir (excluding Ladakh), east to Garhwal and perhaps extreme west
Nepal; winters in adjacent foothills.

Erithacus cyanurus rufilatus (Hodgson)

Nemura rufilata Hodgson, 1845, Proc. Zool. Soc. London,
p. 27 — central and northern regions of hills, Nepal.
Ianthia practica Bangs and Phillips, 1914, Bull. Mus.
Comp. Zool., 58, p. 292 — Loukouchai, southern Yunnan.
Tarsiger cyanurus albocaeruleus Meise, 1937, Journ. f.
Orn., 85, p. 550 — Tschau-tou on the Tetung-ho, north-
ern Kansu [= northeastern Tsinghai].

Nepal east to Assam, southern and eastern Tibet, extreme
north Burma (breeding ?), and west China, in Kansu, Shensi, Szechwan, and Yunnan; winters in adjacent foothills
south in Burma, northern Thailand (uncommon), Laos, and
southern China, south of the Yangtse.

ERITHACUS CHRYSAEUS

Erithacus chrysaeus whistleri (Ticehurst)

Tarsiger chrysaeus whistleri Ticehurst, 1922, Bull. Brit.
Orn. Club, 42, p. 121 — Simla, Punjab, Himalayas.

West Pakistan in Hazara and Murree; India from Kashmir (Sind Valley) east to Duala Dhar, and from Simla to
Garhwal, intergrading with the nominate form in Kumaon,
winters at lower elevations in the same hills.

Erithacus chrysaeus chrysaeus (Hodgson)

Tarsiger chrysaeus Hodgson, 1845, Proc. Zool. Soc. London, p. 28 — Nepal.
Tarsiger chrysaeus vitellinus Stresemann, 1923, Journ. f.
Orn., 71, p. 365 — base of the Was Shan, Szechwan
[= eastern Sikang].

From Kumaon, where intergrades with the preceding
form, east through Nepal, to Assam, southeast Tibet, north
Burma (breeding ?), and west China in Yunnan, Sikang,
western Szechwan, Kansu, and Shensi; winters in the adjacent hills and south to Tonkin.

ERITHACUS INDICUS

Erithacus indicus indicus (Vieillot)

> *Sylvia indica* Vieillot, 1817, Nouv. Dict. Hist. Nat., nouv.
> éd., 11, p. 267 — India, ex Sonnerat, 1817, Voy. Ind., 2,
> p. 208; restricted to Darjeeling by Baker, 1921, Journ.
> Bombay Nat. Hist. Soc., 27, p. 74.

Breeds above 6,600 feet in Garhwal, Nepal, Darjeeling, Sikkim, Bhutan, southeast Tibet, and Assam; winters in foothills.

Erithacus indicus yunnanensis (Rothschild)

> *Tarsiger indicus yunnanensis* Rothschild, 1922, Bull. Brit.
> Orn. Club, 43, p. 10 — Lichiang Range, 10,000 feet,
> northern Yunnan.

Breeds in Yunnan and probably in Sikang and Szechwan; winters south to extreme north Burma and northern Tonkin.

Erithacus indicus formosanus (Hartert)

> *Tarsiger indicus formosanus* Hartert, 1909, Bull. Brit.
> Orn. Club, 25, p. 32 — Mount Arizan.

Mountains of Formosa, above 6,000 feet.

ERITHACUS HYPERYTHRUS

Erithacus hyperythrus (Blyth)

> *Ianthia hyperythra* Blyth, 1847, Journ. Asiat. Soc. Bengal,
> 16, p. 132 — Darjeeling.

Breeds above 5,000 feet in Nepal, Darjeeling, Sikkim, southeast Tibet, Assam, and north Burma.

ERITHACUS JOHNSTONIAE

Erithacus johnstoniae (Ogilvie-Grant)

> *Ianthia johnstoniae* Ogilvie-Grant, 1906, Bull. Brit. Orn,
> Club, 16, p. 118 — Mount Morrison, 8,000 feet.
> *Erithacus taiwan* Hachisuka, 1953, Bull. Brit. Orn. Club,
> 73, p. 33, *nom. nov.* for *Ianthia johnstoniae* Ogilvie-
> Grant, *nec Pogonocichla johnstoni* Shelley, 1893.[1]

Mountains of Formosa, above 6,000 feet.

---

[1] Under the International Rules, as these names differ by one letter or more, *johnstoniae* Ogilvie-Grant is not pre-occupied by *johnstoni* Shelley, even should the genera be combined.

GENUS COSSYPHA VIGORS

Cossypha Vigors, 1825, Zool. Journ., 2, p. 396. Type, by original designation. *Turdus vociferans* Swainson = *Muscicapa dichroa* Gmelin.
Dessonornis (sic) A. Smith, 1836, Rep. Exped. Centr. Africa, p. 46. Type, by monotypy, *Dessonornis humeralis* Smith; corrected to *Bessonornis* by A. Smith, 1840, Illustr. Zool. South Africa, Aves, pl. 48.
Xenocopsychus Hartert, 1907, Bull. Brit. Orn. Club, 19, p. 81. Type, by monotypy, *Xenocopsychus ansorgei* Hartert.
Caffrornis Roberts, 1922, Ann. Transvaal Mus., 8, p. 232. Type, by monotypy, *Cossypha caffra* (Linnaeus).
Hyloaedon Roberts, 1922, Ann. Transvaal. Mus., 8, p. 232. Type, by original designation, *Cossypha dichroa* (Gmelin).
Cossyphicula Grote, 1934, Anz. Orn. Ges. Bayern, 2, p. 311. Type, by monotypy, *Callene roberti* Alexander.

cf. Chapin, 1948, Auk, 65, pp. 292-293 (*ansorgei*).
  Chapin, 1953, Bull. Amer. Mus. Nat. Hist., 75A, pp. 517-534 (Congo).
  Moreau and Benson, 1956, Bull. Brit. Orn. Club, 76, pp. 62-63 (*bocagei*).

COSSYPHA ROBERTI

Cossypha roberti roberti (Alexander)

Callene roberti Alexander, 1903, Bull. Brit. Orn. Club, 13, p. 37 — Bakaki, Fernando Po.
Hills of Fernando Po and Kumba Division, Cameroons.

Cossypha roberti rufescentior Hartert

Cossypha roberti rufescentior Hartert, 1908, Bull. Brit. Orn. Club, 23, p. 9 — forest west of Lake Albert Edward.
Eastern Congo to the west of Lake Edward and the Ruzizi Valley.

COSSYPHA BOCAGEI

Cossypha bocagei insulana Grote

Callene poensis Alexander, 1903, Bull. Brit. Orn. Club, 13, p. 9 — Bilelipi, *nec Cossypha poensis* Strickland, 1844.

Cossypha insulana Grote, 1935, Orn. Monatsb., 43, p. 95, *nom. nov.* for *Callene poensis* Alexander.
Fernando Po, 4,000 feet and above.

Cossypha bocagei granti Serle

Cossypha insulana granti Serle, 1949, Bull. Brit. Orn. Club, 69, p. 53 — Kupé Mountain (lat. 40° 50′ N., long. 9° 40′ E.), 4,500 ft., British Cameroons.
Known only from Mount Kupé, southern Cameroons.

Cossypha bocagei kungwensis Moreau

Cossypha polioptera kungwensis Moreau, 1941, Bull. Brit. Orn. Club, 61, p. 60 — Ujamba forest, Kungwe Mountain (7,900 feet), Kigoma District, western Tanganyika.
Known only from the Kungwe-Mahare Mountains and the Nyamansi River area, western Tanganyika.

Cossypha bocagei schoutedeni Prigogine

Cossypha insulana schoutedeni Prigogine, 1952, Rev. Zool. Bot. Africa, 46, p. 409 — Lutunguru (lat. 0° 28′ S., long. 28° 49′ E.), *ca.* 1,500 m., Congo.
Cossypha bocagei kaboboensis Prigogine, 1955, Rev. Zool. Bot. Africa, 52, p. 181 — Mount Kabobo (lat. 5° 8′ S., long. 29° 2′ E.), 1,670 m., Kivu, Congo.
Congo in Kivu and mountains west of Lake Edward.

Cossypha bocagei chapini Benson

Cossypha bocagei chapini Benson, 1955, Bull. Brit. Orn. Club, 75, p. 104 — Mporokoso, Northern Rhodesia.
Northwestern Northern Rhodesia in Mwinilunga District, Ndola, and Northern Province east to Abercorn.

Cossypha bocagei bocagei Finsch and Hartlaub

Cossypha bocagei Finsch and Hartlaub, 1870, Vög. Ost-Afr., p. 284 — Mossamedes Province [= Biballa, *vide* Barboza du Bocage, 1881, Orn. Angola, p. 259].
Southern Congo, in Katanga, and in western highlands of Angola.

COSSYPHA POLIOPTERA

Cossypha polioptera polioptera Reichenow

Cossypha polioptera Reichenow, 1892, Journ. f. Orn., 40, p. 59 — Bukoba, Victoria Nyanza.

Southern Sudan in Imatong Mountains; Uganda southeast to Mount Elgon, Kisumu, and Bukoba; northern Angola (Ndala Tando).

Cossypha polioptera nigriceps Reichenow

Cossypha nigriceps Reichenow, 1910, Orn. Monatsb., 18, p. 7 — Genderu Mountains, Cameroons.

Highland savannah of Sierra Leone (Tingi Mountains), northern Nigeria, and western Cameroons.

Cossypha polioptera tessmanni Reichenow

Cossypha tessmanni Reichenow, 1921, Journ. f. Orn., 69, p. 49 — Upper Kadei River, Cameroons.

Grasslands of eastern Cameroons (upper Kadei River).

Cossypha polioptera grimwoodi White

Cossypha polioptera grimwoodi White, 1954, Bull. Brit. Orn. Club, 74, p. 88 — source of the Zambesi River, Mwinilunga District, Northern Rhodesia.

Known only from type locality in Northern Rhodesia.

COSSYPHA ARCHERI

Cossypha archeri Sharpe

Cossypha archeri Sharpe, 1902, Bull. Brit. Orn. Club, 13, p. 9 — Ruwenzori.

Cossypha bocagei albimentalis Sassi, 1914, Anz. Akad. Wiss. Wien, math.-naturwiss., 51, p. 311 — forest west of Lake Tanganyika.

Cossypha archeri kimbutui Prigogine, 1955, Rev. Zool. Bot. Africa, 51, p. 33 — Mount Kabobo (lat. 5° 8' S., long. 29° 2' E.).

Congo, in higher mountains from Ruwenzori through Kivu District to region northwest of Lake Tanganyika and west of Ruzizi Valley.

COSSYPHA ISABELLAE

Cossypha isabellae batesi (Bannerman)

Callene batesi Bannerman, 1922, Bull. Brit. Orn. Club, 42, p. 130 — Banso Mountains, north of Kumbo, 6,000 ft., highlands of Nigerian-Cameroons boundary.

Banso Mountains, Manenguba Mountain, and Oku, eastern Nigeria, about 6,000 feet.

Cossypha isabellae isabellae Gray

Cossypha Isabellae Gray, 1862, Ann. Mag. Nat. Hist., ser. 3, **10**, p. 443 — Cameroon Mountain, 7,000 feet.

Mount Cameroon, from 3,000 to 7,500 feet.

COSSYPHA NATALENSIS[1]

Cossypha natalensis intensa Mearns

Cossypha natalensis intensa Mearns, 1913, Smiths. Misc. Coll., **61** (20), p. 2 — Taveta, British East Africa.

Cossypha natalensis hylophona Clancey, 1952, Durban Mus. Novit., 4, p. 15 — Chinteche, 1,700 ft., Nyasaland.

Cossypha natalensis egregior Clancey, 1956, Bull. Brit. Orn. Club, **76**, p. 118 — near Manhiça, Sul do Save, southern Portuguese East Africa.

Southern Sudan in Zande District, Imatong and Didinga Mountains, southern Ethiopia, Somalia, south in Kenya, eastern Uganda, southern Tanganyika west to French Cameroons, lower and middle Congo region, eastern Angola, Northern Rhodesia, eastern Southern Rhodesia, Nyasaland, and the adjacent highlands and lowlands of Mozambique, north of Lourenço Marques to the coast of Beira and Mzimbiti, and eastern Transvaal from the lowlands to 7,000 feet.

Cossypha natalensis larischi Meise

Cossypha natalensis larischi Meise, 1958, Abh. Verh. Naturwiss. Ver. Hamburg, 2 (1957), p. 73 — Canzêle. Northern Angola.

Cossypha natalensis garguensis Mearns

Cossypha natalensis garguensis Mearns, 1913, Smiths. Misc. Coll., **61** (20), p. 2 — Mount Gargues (South Creek, alt. 3,600 ft.), north-central British East Africa.

Known only from Mount Gargues (Uraguess), Matthews Range, north of the Uaso Nyiro, Kenya.

Cossypha natalensis tennenti Williams

Cossypha natalensis tennenti Williams, 1962, Bull. Brit. Orn. Club, **82**, p. 137 — mist forest, Mount Endau, 3,400 ft., Kitui District, Kenya.

Known only from Mount Endau, Kitui District, Kenya.

[1] This species is in need of further revision.

Cossypha natalensis natalensis Smith

Cossypha natalensis A. Smith, 1840, Ill. Zool. South Africa, Aves., pl. 60 (text) — neighborhood of Port Natal [= Durban].

Coastal Cape Province from Pondoland to Natal, Zululand, and extreme southern Mozambique, in the Lebombo Mountains and adjacent littoral, Lourenço Marques.

COSSYPHA DICHROA

Cossypha dichroa (Gmelin)

Muscicapa dichroa Gmelin, 1789, Syst. Nat., 1 (2), p. 949 — South Africa.

Cossypha haagneri Gunning, 1901, Ann. Transvaal Mus., 1, p. 174, pl. 3 — Mgqeleni, western Pondoland.

Southern Cape Province south to Knysna Forest, east to Natal, Zululand, and eastern Transvaal.

COSSYPHA SEMIRUFA

Cossypha semirufa semirufa (Rüppell)

Petrocincla semirufa Rüppell, 1840, Neue Wirbelt., Vögel, p. 81 — Abyssinia.

Cossypha semirufa saturatior Neumann, 1906, Orn. Monatsb., 14, p. 7 — Bolagoschana, Doko, southwestern Abyssinia.

Boma Hills, southeast Sudan, Eritrea, central, western, and southeastern Ethiopia, south to Moyale and Marsabit, Kenya.

Cossypha semirufa donaldsoni Sharpe

Cossypha donaldsoni Sharpe, 1895, Bull. Brit. Orn. Club, 4, p. 28 — no locality [= Sheik Husein (Arussi, south-central Ethiopia), *vide* Sharpe, 1895, Proc. Zool. Soc. London, p. 484].

Eastern and southeastern Ethiopia in the Harrar area and eastern Gallaland.

Cossypha semirufa intercedens (Cabanis)

Bessonornis intercedens Cabanis, 1878, Journ. f. Orn., 26, pp. 205; 219 — Kitui, Ukamba.

Kenya in the south-central highlands, Ukamba, and Kikuyu to Mount Kenya, Aberdare Mountains, northern Tanganyika (Mount Kilimanjaro).

COSSYPHA HEUGLINI

Cossypha heuglini pallidior Berlioz and Gillet

Cossypha heuglini pallidior Berlioz and Gillet, 1956, Oiseau
Rev. Franç. Orn., 26, p. 137 — Fort-Lamy, Chad.
Chad (Chari Valley).

Cossypha heuglini heuglini Hartlaub

Cossypha heuglini Hartlaub, 1866, Journ. f. Orn., 14, p. 36
— "Keren" [error = Wau, Bahr el Ghazal, *fide* Heuglin,
1869, Orn. Nordost. Afr., 1, p. 375].
Cossypha heuglini occidentalis Reichenow, 1909, Journ. f.
Orn., 57, p. 108 — Lufuku, West Tanganyika.
Cossypha heuglini mwinilunga Horniman, 1940, Prelim.
Descr. New Birds, p. 2 — Mwinilunga, Northern Rho-
desia; *idem* Grant, 1956, Ann. Mag. Nat. Hist., ser. 12,
9, p. 366.[1]

Southeastern Ethiopia, southern Sudan in Equatoria,
Bahr el Ghazal and Darfur, south Uganda, eastern Congo
in Kivu, Kasai, and Katanga, western Kenya, interior Tanganyika south to Northern Rhodesia and Nyasaland, up to
7,000 feet.

Cossypha heuglini subrufescens Bocage

Cossypha subrufescens Barboza du Bocage, 1869, Proc.
Zool. Soc. London, p. 436 — Caconda.
Congo and Gabon south to lower Congo and Angola, at
least to Capelongo.

Cossypha heuglini intermedia (Cabanis)

Bessornis intermedia Cabanis, 1868, Journ. f. Orn., 16, p.
412 — inner East Africa.
Coastal areas of Somalia, Kenya, and Tanganyika, at least
to Rovuma.

Cossypha heuglini euronota Friedmann

Cossypha heuglini euronota Friedmann, 1930, Occ. Papers
Boston Soc. Nat. Hist., 5, p. 327 — Lumbo, Mozambique.
Southern Rhodesia, Mozambique, eastern Transvaal, and
northern Zululand in the Lebombo Mountains up to 5,000
feet.

---

[1] Under Opinion 480 (1957, Ops. decls. Internat. Comm. zool. Nomencl., 16, pp. 417-454), this name is suppressed for purposes of the
Law of Priority, but not for those of the Law of Homonymy.

COSSYPHA CYANOCAMPTER

Cossypha cyanocampter cyanocampter (Bonaparte)

B. (*essonornis*) *cyanocampter* Bonaparte (ex Cabanis MS), 1850, Consp. Av., 1, p. 301 — "Patria ignota" [= Dabokrom, Gold Coast, *apud* Hartlaub, 1857, Syst. Orn. Westafr., p. 76].
Cossypha periculosa Sharpe, 1883, Cat. Birds Brit. Mus., 7, p. 40 — River Danger, Gabon.
Sierra Leone, Liberia, and Ghana through forested Nigeria to Cameroons and Gabon.

Cossypha cyanocampter bartteloti Shelley

Cossypha bartteloti Shelley, 1890, Ibis, p. 159, pl. 5 — Yambuya.
Northeastern Congo, Uganda, and Kenya in Kakamega.

COSSYPHA CAFFRA

Cossypha caffra iolaema Reichenow

Cossypha caffra iolaema Reichenow, 1900, Orn. Monatsb., 8, p. 5 — Kilimanjaro.
Cossypha caffra mawensis Neumann, 1900, Journ. f. Orn., 48, p. 309 — Mau Mountains.
Mountains of extreme southern Sudan, Uganda, western highlands of Kenya, Tanganyika and Nyasaland, northern Mozambique, southeastern highlands of the Congo, Nyika, Northern Rhodesia, eastern highlands of Southern Rhodesia, and adjacent southern Mozambique (also on Mount Gorongoza).

Cossypha caffra kivuensis Schouteden

Cossypha caffra kivuensis Schouteden, 1937, Rev. Bot. Africa, 30, p. 165 — Kivu.
From the mountains northwest of Baraka on Lake Tanganyika across the Kivu Highland to Ankole, Congo.

Cossypha caffra drakensbergi (Roberts)

Caffrornis caffra drakensbergi Roberts, 1936, Ostrich, 7, p. 110 — Wakkerstroom.
South Africa in Natal-Transvaal border country to central and eastern Transvaal. Birds of eastern Southern Rhodesia are intergrades between this form and *iolaema*.

Cossypha caffra namaquensis Sclater

Cossypha caffra namaquensis Sclater, 1911, Ibis, p. 415
— Klipfontein.
Southern South West Africa and Orange River valley east
to western Orange Free State and western Transvaal, where
intergrades with *drakensbergi* and *caffra* occur.

Cossypha caffra caffra (Linnaeus)

Motacilla caffra Linnaeus, 1771, Mantissa Plant., p. 527 —
Cape of Good Hope.
South Africa in the southern part of Cape Province north
to Transvaal (see above), and east to Natal and Zululand,
north to Ingwzvuma, southern Lebombo Mountains (Swazi-
land-Zululand border); Basutoland ?

COSSYPHA ANOMALA

Cossypha anomala mbuluensis (Grant and Mackworth-
Praed)

Bessonornis macclounii mbuluensis Grant and Mackworth-
Praed, 1936, Bull. Brit. Orn. Club, 57, p. 80 — Nou
Forest, 7,000 ft., Mbulu District, Tanganyika.
Mbulu District, northern Tanganyika.

Cossypha anomala albigularis (Reichenow)

Callene albigularis Reichenow, 1895, Orn. Monatsb., 3,
pp. 87; 96 — east Ulugura (*sic*) [= Morogoro District,
Tanganyika].
Bessonornis grotei Reichenow, 1932, Verh. Orn. Ges. Bay-
ern, p. 584 — Uluguru, Morogoro District; new name
for *C. albigularis* Reichenow *nec Bessornis albigularis*
Tristam, 1867, = *Irania gutturalis.*
Alethe macclouniei njombe Benson, 1936, Bull. Brit. Orn.
Club, 56, p. 100 — Njombe, southern Tanganyika.
Morogoro District to Njombe and Songea Districts, Tan-
ganyika.

Cossypha anomala macclounii (Shelley)

Callene macclounii Shelley, 1903, Bull. Brit. Orn. Club,
13, p. 61 — Nwenembe.
Bessonornis anomala porotensis Bangs and Loveridge,
1931, Proc. New England Zool. Club, 12, p. 94 — Igale,
Poroto Mountains, southwestern Tanganyika.

Southwestern Tanganyika, in Tukuyu District and northern Nyasaland in Vipya, Nyankhowa, and Nyika, above 6,000 feet.

Cossypha anomala anomala (Shelley)

Callene anomala Shelley, 1893, Ibis, p. 14 — Milanji Plateau.

Milanje, Nyasaland, above 4,000 feet.

Cossypha anomala gurue (Vincent)

Alethe anomala gurué Vincent, 1933, Bull. Brit. Orn. Club, 53, p. 138 — Namuli Mountain, Quelimane Province, Portuguese East Africa.

Namuli Mountains, northern Mozambique.

COSSYPHA HUMERALIS

Cossypha humeralis humeralis (Smith)

Dessonornis (sic) *humeralis* A. Smith, 1836, Rep. Exped. Centr. Africa, p. 46 — banks of the Marikwa [= Marico River, western Transvaal].

Southern Rhodesia, except in the extreme southeast in the Sabi-Lundi confluence and Birchenough Bridge on the Sabi River; perhaps absent in northern and western Mashonaland, eastern Bechuanaland, and Transvaal, except in the eastern Transvaal lowveld where intergrades with the following subspecies.

Cossypha humeralis crepuscula Clancey

Cossypha humeralis crepuscula Clancey, 1962, Durban Mus. Novit., 6, p. 155 — Panda, Inhambane District, southern Portuguese East Africa.

Extreme southeast Southern Rhodesia (see under preceding subspecies), eastern Transvaal lowveld where intergrades with *humeralis*, Natal, including Zululand and Swaziland, and Sul do Save, southern Mozambique; Bazaruto Island.

COSSYPHA ANSORGEI

Cossypha ansorgei (Hartert)

Xenocopsychus ansorgei Hartert, 1907, Bull. Brit. Orn. Club, 19, p. 82 — Lobango, Mossamedes, Angola.

Western Angola (Vila Salazar and Lubango), near limestone caves.

COSSYPHA NIVEICAPILLA

Cossypha niveicapilla (Lafresnaye)

Turdus niveicapilla Lafresnaye, 1838, Essai Nouv. Manière Grouper Genres Espèces Ordre Passereaux, p. 16 — Senegal.

Bessornis melanonota Cabanis,[1] 1875, Journ. f. Orn., 23, p. 235 — Chinchoncho on the Loango Coast [= Chinchoxo, Portuguese Congo].

Senegal, Gambia, Sierra Leone, Ghana, Nigeria, and Cameroons east in the Congo and Uganda to the Sudan, as far north as Darfur and Sennar; southwest Ethiopia south to extreme western Kenya in the Elgon area, northeast side of Lake Tanganyika, and northern Angola.

COSSYPHA HEINRICHI

Cossypha heinrichi Rand

Cossypha heinrichi Rand, 1955, Fieldiana: Zool. [Chicago], 34, p. 327 — about 30 km. northeast of Duque de Braganza, Angola.

Known only from gallery forest northeast of Duque de Braganza, northern Angola.

COSSYPHA ALBICAPILLA

Cossypha albicapilla albicapilla (Vieillot)

Turdus albicapillus Vieillot, 1818, Nouv. Dict. Hist. Nat., nouv. éd., 20, p. 254 — Senegal.

Senegal, Gambia, Casamance, and Portuguese Guinea; Sierra Leone (?).

Cossypha albicapilla giffardi Hartert

Cossypha giffardi Hartert, 1899, Bull. Brit. Orn. Club, 10, p. 5 — Gambaga, Gold Coast hinterland.

Cossypha albicapilla genderuensis Reichenow, 1910, Orn. Monatsb., 18, p. 176 — Genderu Mountains, Cameroons.

Ghana east through northern Nigeria and the northern Cameroons highlands of Nigeria to northern Cameroun (Bamingui River).

[1] Blackish specimens (*melanonota*?) appear to occur discontinuously in Cameroons, Uganda, northern Tanganyika, and southwest Ethiopia, perhaps correlated with certain forest types. For the time being, it seems better to merge these two phenotypes.

Cossypha albicapilla omoensis Sharpe

Cossypha omoensis Sharpe, 1900, Bull. Brit. Orn. Club,
11, p. 28 — Omo River, Equatorial Africa.
Extreme southeast Sudan and southwest Ethiopia, north
of Lake Rudolf in the Omo River area.

GENUS **MODULATRIX** RIPLEY

Modulatrix Ripley, 1952, Postilla, Yale Univ., no. 12, p. 2.
Type, by monotypy, *Turdinus stictigula* Reichenow.

MODULATRIX STICTIGULA

Modulatrix stictigula stictigula (Reichenow)

Turdinus stictigula Reichenow, 1906, Orn. Monatsb., 14,
p. 10 — Mbaramo, Usambara.
Tanganyika in the Usambara and Nguru Mountains.

Modulatrix stictigula pressa (Bangs and Loveridge)

Illadopsis stictigula pressa Bangs and Loveridge, 1931,
Proc. New England Zool. Club, 12, p. 94 — Nkuka Forest, Rungwe Mountain, southwestern Tanganyika Territory.
Southwestern Tanganyika in Uzungwe and Ukinga Mountains, Rungwe Mountain, and extreme northern Nyasaland
in the Masuku Mountains.

GENUS **CICHLADUSA** PETERS

Cichladusa Peters, 1863, Monatsb. Kön. Akad. Wiss. Berlin, p. 134. Type, by original designation, *Cichladusa
arquata* Peters.

CICHLADUSA GUTTATA

Cichladusa guttata guttata (Heuglin)

Crateropus guttatus Heuglin, 1862, Journ. f. Orn., 10, p.
300 — Bahr el Abiad.
Southern Sudan in Bahr el Ghazal and Equatoria, southwestern Ethiopia, south to Uganda, central and southeastern
Kenya, Congo on shores of Lake Albert, and Tanganyika
south to the Central Railway Line, but not east of Mombo
and Dodoma.

Cichladusa guttata rufipennis Sharpe

Cichladusa rufipennis Sharpe, 1901, Bull. Brit. Orn. Club, 12, p. 35 — coast region of East Africa (Lamu, etc.).
Cichladusa guttata mülleri Zedlitz, 1916, Journ. f. Orn., 64, p. 108 — Afgoi, southern Somaliland.

Southeastern Ethiopia, southern Somalia, south in coastal Kenya, west to the Orr Valley, and eastern Tanganyika, east of Mombo and Dodoma.

CICHLADUSA ARQUATA

Cichladusa arquata Peters

Cichladusa arquata Peters, 1863, Monatsb. Kön. Akad. Wiss. Berlin, p. 134 — Sena, near the Zambezi, Mozambique [reference not verified].

Southern Kenya, southwestern Uganda, southeastern Congo, lowlands round Lake Tanganyika and Lualaba and Luapula rivers, Northern Rhodesia, Southern Rhodesia, along the Sabi and Zambezi rivers, Nyasaland, and Mozambique, primarily in lowlands.

CICHLADUSA RUFICAUDA

Cichladusa ruficauda (Hartlaub)

Bradyornis ruficauda Hartlaub, 1857, Syst. Orn. Westafr., p. 66 — Gabon.

Gabon and Congo, from Stanley Pool along Congo River north to Coquilhatville, south to Angola.

GENUS ALETHE CASSIN

Alethe Cassin, 1859, Proc. Acad. Nat. Sci. Philadelphia, 11, p. 43. Type, by monotypy, *Napothera castanea* Cassin.
Chamaetylas Heine, 1859, Journ. f. Orn., 7, p. 425. Type, by monotypy, *Geocichla compsonota* Cassin = *Napothera castanea* Cassin.

ALETHE DIADEMATA

Alethe diademata diademata (Bonaparte)

Bessonornis (*Turdus*) *diadematus* Bonaparte (ex Temminck MS), 1851, Consp. Av., 1 (1850), p. 302 — Guinea.

Portuguese Guinea, Sierra Leone, Liberia, Ghana, and Togo.

Alethe diademata castanea (Cassin)

Napothera castanea Cassin, 1856, Proc. Acad. Nat. Sci. Philadelphia, 8, p. 158 — Moonda River, western Africa.

Southern Nigeria, Cameroon Mountains, Cameroun, and Gabon, south to Congo in the Lower Congo and Mayombe District; Fernando Po Island; northern Angola (?).

Alethe diademata woosnami Ogilvie-Grant

Alethe woosnami Ogilvie-Grant, 1906, Bull. Brit. Orn. Club, 19, p. 24 — forest near Urumu, 3,000 feet, northwest of Ruwenzori.

Forests of the Upper Congo and of Uganda east to Mabira.

ALETHE POLIOPHRYS

Alethe poliophrys Sharpe

Alethe poliophrys Sharpe, 1902, Bull. Brit. Orn. Club, 13, p. 10 — Ruwenzori.

Alethe poliophrys kaboboensis Prigogine, 1957, Rev. Zool. Bot. Africa, 55, p. 42 — Mount Kabobo (lat. 5° 8' S., long. 29° 3' E.), 1,600 m.

Congo-Uganda border from Ruwenzori and the highland west of Lake Edward south to the Kivu Volcanoes and the highlands northwest of Lake Tanganyika; Mount Kabobo (near Albertville), Congo.

ALETHE FUELLEBORNI

Alethe fuelleborni usambarae Reichenow

Alethe fülleborni usambarae Reichenow, 1905, Orn. Monatsb., 13, p. 182 — Mlalo, Usambara.

Eastern Tanganyika from the Usambara Mountains to the Uluguru Mountains and Mahenge.

Alethe fuelleborni fuelleborni Reichenow

Alethe fülleborni Reichenow, 1900, Orn. Monatsb., 8, p. 99 — Peroto-Ngosi, Tandalla.

South-central and southwestern Tanganyika from Njombe to the Tukuyu District south to northern Nyasaland, Nyika, and Masuku, in mountain forests.

ALETHE MONTANA

Alethe montana Reichenow

Alethe montana Reichenow, 1907, Orn. Monatsb., **15**, p. 30 — Usambara.

Usambara Mountains, northeast Tanganyika.

ALETHE LOWEI

Alethe lowei Grant and Mackworth-Praed

Alethe lowei Grant and Mackworth-Praed, 1941, Bull. Brit. Orn. Club, **61**, p. 61 — Njombe area, southern Tanganyika.

Njombe area, Iringa Province, southern Tanganyika, and Njombe, northern Nyasaland.

ALETHE POLIOCEPHALA

Alethe poliocephala castanonota Sharpe

Alethe castanonota Sharpe, 1871, Cat. African Birds, p. 20 — Fantee (Fanti), Gold Coast.

Sierra Leone, Liberia, and Ghana.

Alethe poliocephala poliocephala (Bonaparte)

Trichophorus (Criniger) poliocephalus Bonaparte (ex Temminck MS), 1851, Consp. Av., **1** (1850), p. 262 — Africa [= Fernando Po Island].

Alethe alexandri Sharpe, 1901, Bull. Brit. Orn. Club, **12**, p. 4 — Efulen, Kamerun.

Fernando Po Island; southern Cameroons; Gabon; northwestern Angola, Quicolungo; probably in lower Congo.

Alethe poliocephala hallae Traylor

Alethe poliocephala hallae Traylor, 1961, Bull. Brit. Orn. Club, **81**, pp. 44 — 15 km. south of Gabela, Cuanza Sul, Angola.

Known only from region of Gabela, on the escarpment zone in Cuanza Sul, Angola.

Alethe poliocephala carruthersi Ogilvie-Grant

Alethe carruthersi Ogilvie-Grant, 1906, Bull. Brit. Orn. Club, **19**, p. 25 — 150 miles west of Entebbe, 5,000 ft.

Alethe uellensis Reichenow, 1912, Journ. f. Orn., **60**, p. 321 — Angu on Uelle River.

Extreme southern Sudan, western Imatong mountains, northeastern Congo, and forest of Uganda east to Mount Elgon.

Alethe poliocephala akeleyae Dearborn

Alethe akeleyae Dearborn, 1909, Field Mus. Nat. Hist. Publ., Orn. Ser., 1, p. 170 — Mount Kenya.
Alethe kikuyuensis Jackson, 1910, Bull. Brit. Orn. Club, 27, p. 7 — Kikuyu Forest, 5,400 ft., British East Africa.
Kenya from Mount Kenya to Kikuyu and Nairobi.

Alethe poliocephala kungwensis Moreau

Alethe poliocephala kungwensis Moreau, 1941, Bull. Brit. Orn. Club, 61, p. 46 — forest above Ujamba, 6,900 ft., Kungwe Mountain.
Kungwe-Mahare Mountains to headwaters of the Nyamansi River, western Tanganyika.

Alethe poliocephala ufipae Moreau

Alethe poliocephala ufipae Moreau, 1942, Bull. Brit. Orn. Club, 62, p. 54 — Mbisi Forest, *ca.* 8,000 ft.
Ufipa Plateau, southwestern Tanganyika.

ALETHE CHOLOENSIS

Alethe choloensis choloensis Sclater

Alethe choloensis Sclater, 1927, Bull. Brit. Orn. Club, 47, p. 86 — Cholo Mountain, Nyasaland.
Eastern and southern Nyasaland, east of the Rift.

Alethe choloensis namuli Vincent

Alethe choloensis namuli Vincent, 1933, Bull. Brit. Orn. Club, 53, p. 138 — Namuli Mountain, Quelimane Province, Portuguese East Africa.
Known only from the Namuli massif, Mozambique.

GENUS **COPSYCHUS** WAGLER

Copsychus Wagler, 1827, Syst. Av., note to art. *Gracula*, p. 306. Type, by subsequent designation (Gray, 1840, List Gen. Birds, ed. 1, p. 21), *Gracula saularis* Linnaeus.
Notodela Lesson, 1831, Traité Orn. (1830 ?), p. 374. Type, by subsequent designation (Baker, 1930, Fauna Brit.

India, Birds, ed. 2, 7, p. 112; 8, p. 622), *Gracula saularis* Linnaeus.

Kittacincla Gould, 1836, Proc. Zool. Soc. London, p. 7. Type, by monotypy, *Turdus macrourus* Gmelin = *Muscicapa malabarica* Scopoli.

Trichixos Lesson, 1839, Rev. Zool. [Paris], p. 167. Type, by monotypy, *Trichixos pyrropyga* Lesson.

Gervaisia Bonaparte, 1854, Compt. Rend. Acad. Sci. Paris, 38. Type, by monotypy, *Turdus albospecularis* Eydoux and Gervais.

Shama Hachisuka, 1934, Tori, 8, p. 222. Type, by monotypy, *Cittocincla cebuensis* Steere.

Kurodornis Hachisuka, 1941, Tori, 11, p. 86. Type, by original designation, *Turdus luzoniensis* Kittlitz.

COPSYCHUS SAULARIS

Copsychus saularis saularis (Linnaeus)

Gracula Saularis Linnaeus, 1758, Syst. Nat., ed. 10, 1, p. 165 — Asia; restricted to Bengal by Baker, 1921, Journ. Bombay Nat. Hist. Soc., 27, p. 714.

Low country of West Pakistan and India, except the desert regions, intergrading with *ceylonensis* in Mysore and western Madras and with *erimelas* in East Pakistan.

Copsychus saularis ceylonensis Sclater

Copsychus ceylonensis Sclater, 1861, Proc. Zool. Soc. London, p. 186 — Ceylon.

From the Wynaad, Mysore, and southern Madras, where intergrades with the preceding form, south Kerala and Ceylon; in low country, reaching 6,000 feet in Ceylon.

Copsychus saularis erimelas Oberholser

Copsychus saularis erimelas Oberholser, 1923, Smiths. Misc. Coll., 76 (6), p. 1 — Kaukarit, Houndraw Branch, Tenasserim.

Copsychus saularis haliblectus Oberholser, 1923, Smiths. Misc. Coll., 76 (6), p. 2 — Domel Island, Mergui Archipelago.

Intergrades with *saularis* in western East Pakistan, Bhutan Duars, and Assam, thence east in Burma, including Mergui Archipelago, Thailand, except the Peninsula, and Indochina (North and South Vietnam and Laos).

Copsychus saularis andamanensis Hume

Copsychus andamanensis Hume, 1874, Stray Feathers, 2, p. 231 — Andamans.
Andaman Islands.

Copsychus saularis prosthopellus Oberholser

Copsychus saularis prosthopellus Oberholser, 1923, Smiths. Misc. Coll., 76 (6), p. 1 — Deep Bay, Hong Kong, China.
Southern and southeastern China in Yunnan, southeast Sikang, southwest Szechwan, Kweichow, Hunan, Kwungtung, Fukien, Kiangsi, Chekiang, Hupeh, southern Anhwei, and southern Kiangsu; Hainan Island.

Copsychus saularis musicus (Raffles)

Lanius musicus Raffles, 1822, Trans. Linn. Soc. London, 13, p. 307 — Bencoolen, West Sumatra.
Copsychus saularis ephalus Oberholser, 1923, Smiths. Misc. Coll., 76 (6), p. 2 — Tarussan Bay, northwestern Sumatra.
Copsychus saularis nesiotes Oberholser, 1923, Smiths. Misc. Coll., 76 (6), p. 3 — Tanjong Bedaan, Banka Island, southeastern Sumatra.
Peninsular Thailand, Malaya, and Sumatra and offlying islands, Singapore, Tioman, Rhio Archipelago, Billiton and Bangka; up to 5,000 feet.

Copsychus saularis zacnecus Oberholser

Copsychus saularis zacnecus Oberholser, 1912, Smiths. Misc. Coll., 60 (7), p. 12 — Simalur Island.
Simalur Island, west Sumatra islands.

Copsychus saularis nesiarchus Oberholser

Copsychus saularis nesiarchus Oberholser, 1923, Smiths. Misc. Coll., 76 (6), p. 3 — Lafau, Nias Island, western Sumatra.
Nias Island, west Sumatra islands.

Copsychus saularis masculus Ripley

Copsychus saularis masculus Ripley, 1943, Notulae Naturae, no. 114, p. 1 — Tana Massa Island, Batu Islands.
Batu Islands, west Sumatra islands: Pini, Tello, Tana Massa.

Copsychus saularis pagiensis Richmond

Copsychus saularis pagiensis Richmond, 1912, Proc. Biol.
Soc. Washington, 25, p. 105 — North Pagi Island, west
Sumatra.

Siberut, Sipora, and North Pagi Islands, west Sumatra
islands.

Copsychus saularis javensis Chasen and Kloss

Copsychus saularis javensis Chasen and Kloss, 1930, Bull.
Raffles Mus., 4, pp. 87; 89 — Wynkoops Bay, southwest
Java.

Western Java, intergrading with the following form in
mid-Java.

Copsychus saularis amoenus (Horsfield)

Turdus amoenus Horsfield, 1821, Trans. Linn. Soc. London, 13, p. 147 — Java [= East Java, *vide* Sharpe, 1883,
Cat. Birds Brit. Mus., 7, p. 63].

Eastern Java, intergrading with the preceding form in
mid Java; Bali Island.

Copsychus saularis problematicus Sharpe

Copsychus problematicus Sharpe, 1876, Ibis, p. 36 —
Sibu.

Southwestern (Kapuas River basin) and western Borneo
to the Baram River in eastern Sarawak, where intergrades
with *adamsi*.

Copsychus saularis adamsi Elliott

Copsychus niger Wardlaw Ramsay, 1886, Proc. Zool. Soc.
London, p. 123 — Sandakan, North Borneo, *nec Kittacincla nigra* Sharpe, 1877.

Copsychus adamsi Elliott, 1890, Auk, 7, p. 348 — Sandakan.

Copsychus saularis ater Delacour, 1945, Zoologica, 30,
p. 112, new name for *C. niger* Wardlaw Ramsay.

Brunei, eastern Sarawak (where intergrades with *prob-
lematicus*), North Borneo, Banguey, Balembangan, Malle-
wallé, and Sibatik Island.

Copsychus saularis pluto Bonaparte

Copsychus (Turdus) pluto Bonaparte, 1851, Consp. Av.,
1 (1850), p. 267 — ex Borneo [= neighborhood of Sa-

marinda, *vide* Chasen and Kloss, 1930, Bull. Raffles
Mus., 4, p. 90].

Eastern Borneo; Maratua Island; intergrades with *prob-lematicus* in Sampit District, southeastern Borneo (Kali-mantan).

Copsychus saularis deuteronymus Parkes

Copsychus saularis deuteronymus Parkes, 1963, Bull. Brit.
Orn. Club, 83, p. 50, new name for *Copsychus saularis
heterogynus* Parkes, 1962, *nec Kittacincla malabarica
heterogyna* Oberholser, 1917, preoccupied in *Copsychus*.
Copsychus saularis heterogynus Parkes, 1962, Postilla,
Yale Univ., no. 67, p. 3 — Pangil, Laguna Province,
Luzon, Philippines.

Luzon; may occur on Polillo, Catanduanes, and Marin-duque Islands.

Copsychus saularis mindanensis (Boddaert)

Turdus mindanensis Boddaert, 1783, Tabl. Pl. enlum., p.
38 — Philippine Islands; restricted to Mindanao by
Parkes, 1962, *ibid.*, p. 3.

Philippine Islands of Sibuyan, Mindoro, Samar, Cebu,
Negros, Mindanao, Basilan, and Sulu archipelago.

COPSYCHUS SECHELLARUM

Copsychus sechellarum Newton

Copsychus sechellarum Newton, 1865, Ibis, p. 332, pl. 8
— on some islands of the Seychelles.

Seychelles Archipelago: Frigate Island; extinct on Mari-anne, Ladigue, Aride and Praslin; Alphonse Island (intro-duced; extinct ?).

COPSYCHUS ALBOSPECULARIS

Copsychus albospecularis albospecularis (Eydoux and Ger-vais)

Turdus albo-specularis Eydoux and Gervais, 1836, Mag.
Zool. [Paris], p. 9, pls. 64, 65 — Madagascar; restricted
to Maroantsetra, northeastern Madagascar, by Dela-cour, 1931, Ois. Rev. Franç. Orn., 1, p. 623.

Northern Madagascar, intergrading with the following
form in the Fanovana region.

Copsychus albospecularis inexpectatus Richmond

Copsychus inexpectatus Richmond, 1897, Proc. U. S. Nat. Mus., 19, p. 688 — mouth of River Fanantra, east coast of Madagascar.

Eastern Madagascar, south to Manombo; intergrading with the preceding form in the Fanovana and Sianka areas.

Copsychus albospecularis pica Pelzeln

Copsychus (Turdus) pica Pelzeln, 1858, Sitzungsb. K. Akad. Wiss. Wien., Math.-Naturwiss. Cl., 31, p. 323 — Bontebok Bay, northwestern Madagascar.

Wooded areas of western Madagascar and the extreme north as far as Vohemar, and south at least to Ampotaka; intergrades with *inexpectatus* in Ivohibé area in the southeast.

COPSYCHUS MALABARICUS

Copsychus malabaricus malabaricus (Scopoli)

Muscicapa malabarica Scopoli, 1788, Del. Flor. Fauna Insubr., fasc. 2, p. 96 — Mahé, Malabar.

India from Gujarat (Surat Dangs) south through the Ghats to Mysore, western Madras, and Kerala; plains to 2,000 feet.

Copsychus malabaricus leggei (Whistler)

Kittacincla malabarica leggei Whistler, 1941, Ibis, p. 319 — Uragaha, Ceylon.

Ceylon, in the low country to nearly 3,000 feet.

Copsychus malabaricus indicus (Baker)

Kittacincla malabarica indica Baker, 1924, Fauna Brit. Ind., Birds, ed. 2, 2, p. 118 — Bhutan Duars.

From Nepal east along the Himalyan foothills through Darjeeling, Sikkim, Bhutan Duars, and Assam, south in Uttar Pradesh, Bihar, eastern Madhya Pradesh (Rajmahal Hills), Orissa, and northern Andhra Pradesh; from the plains to 2,000 feet.

Copsychus malabaricus albiventris (Blyth)

Kittacincla albiventris Blyth, 1859, Journ. Asiat. Soc. Bengal, 27, p. 269 — Andamans.

Andaman Islands.

Copsychus malabaricus interpositus (Robinson and Kloss)

Kittacincla malabarica interposita Robinson and Kloss, 1922, Journ. Fed. Malay States Mus., 10, p. 262 — Daban, South Annan.

Kittacincla malabarica pellogyna Oberholser, 1923, Smiths. Misc. Coll., 76 (6), p. 4 — Bok Pyim, Tennasserim.

Kittacincla malabarica lamprogyna Oberholser, 1923, Smiths. Misc. Coll., 76 (6), p. 5 — St. Luke Island, Mergui Archipelago.

Burma, Thailand, North and South Vietnam, and Laos.

Copsychus malabaricus minor (Swinhoe)

Cittacincla macrura minor Swinhoe, 1870, Ibis, p. 244 — Hainan Island.

Cittacincla brevicauda Ogilvie-Grant, 1899, Ibis, p. 584 — interior of Hainan.

Hainan Island.

Copsychus malabaricus mallopercnus (Oberholser)

Kittacincla malabarica mallopercna Oberholser, 1923, Smiths. Misc. Coll., 76 (6), p. 5 — Sing Kep Island, Berhala Strait, off southeastern Sumatra.

Malay Peninsula, Langkawi group, Penang, Singapore; Tioman, Rhio, and Lingga Archipelagos; intergrades with *interposita* in the northern peninsula.

Copsychus malabaricus tricolor (Vieillot)

Turdus tricolor Vieillot, 1818, Nouv. Dict. Hist. Nat., nouv. éd., 30, p. 291 — islands of the South Sea [= Bantam, western Java, *vide* Robinson and Kloss, 1921, Journ. Fed. Malay States Mus., 10, p. 210].

Kittacincla malabarica abbotti Oberholser, 1923, Smiths. Misc. Coll., 76 (6), p. 5 — Tanjong Bedaan, Banka Island, southeastern Sumatra.

Sumatra, Banka, Billiton, and Karimata Islands, Riouw Archipelago, and Java, in the western districts east at least to Cape Indramaju.

Copsychus malabaricus mirabilis Hoogerwerf

Copsychus malabaricus mirabilis Hoogerwerf, 1962, Ardea, 50, p. 184 — Tjiharashas, Prinsen Island, West Java.

Prinsen Island, southern Sunda Strait, Indonesia.

Copsychus malabaricus melanurus (Salvadori)

Cittacincla melanura Salvadori, 1887, Ann. Mus. Civ.
Genova, 4, p. 549, pl. 8, fig. 1 — Nias Island.
Kittacincla melanura hypoliza Oberholser, 1912, Smiths.
Misc. Coll., 60 (7), 13 — Simalur Island.
Kittacincla melanura opisthochra Oberholser, 1912,
Smiths. Misc. Coll., 60 (7), p. 13 — Pulo Lasia.
Kittacincla melanura pagiensis Oberholser, 1923, Smiths.
Misc. Col., 76 (6), p. 3 — North Pagi Island, western
Sumatra.
Simalur, Lasia, Babi, Nias, Siberut, Sipora, and Pagi
Islands, west Sumatra Islands.

Copsychus malabaricus opisthopelus (Oberholser)

Kittacincla malabarica opisthopela Oberholser, 1912,
Smiths. Misc. Coll., 60 (7), p. 13 — Tana Bala Island,
Batu Islands.
Kittacincla malabarica opisthisa Oberholser, 1912, Smiths.
Misc. Coll., 60 (7), p. 13 — Pulo Tuanku, Banjak
Islands.
Tunagku, Bangkaru, Tello, Tana Massa, Tana Bala, and
Banyak Islands, west Sumatra Islands.

Copsychus malabaricus javanus (Kloss)

Kittacincla malabarica javana Kloss, 1921, Journ. Fed.
Malay States Mus., 10, p. 210 — Karangbolang, south
coast of mid-Java (not Karangboland of Noesa Kam-
bangan Island).
Western and central Java.

Copsychus malabaricus omissus (Hartert)

Kittacincla macrurus omissa Hartert, 1902, Novit. Zool.,
9, p. 572 — Lawang, East Java.
Eastern Java.

Copsychus malabaricus ochroptilus (Oberholser)

Kittacincla malabarica ochroptila Oberholser, 1917, Bull.
U. S. Nat. Mus., 98, p. 51 — Pulo Siantan, Anamba
Islands.
Kittacincla malabarica heterogyna Oberholser, 1917, Bull.
U. S. Nat. Mus., 98, p. 53 — Pulo Riabu, Anamba
Islands.
Anamba Islands.

Copsychus malabaricus eumesus (Oberholser)

> *Kittacincla malabarica eumesa* Oberholser, 1932, Bull.
> U. S. Nat. Mus., 159, p. 81 — Bunguran Island, Natuna
> Islands.

Natuna Islands.

Copsychus malabaricus suavis Sclater

> *Copsychus suavis* Sclater, 1861, Proc. Zool. Soc. London,
> p. 185 — Banjermassing, southern Borneo.
> *Kittacincla malabarica zaphotina* Oberholser, 1923,
> Smiths. Misc. Coll., 76 (6), p. 6 — central Borneo.

Borneo, except North Borneo.

Copsychus malabaricus nigricauda (Vorderman)

> *Cittacincla nigricauda* Vorderman, 1893, Nat. Tijds.
> Nederl. Ind., 42, p. 197 — Kangean.

Kangean Islands and Mata Siri Island, Java Sea.

COPSYCHUS STRICKLANDII

Copsychus stricklandii stricklandii Motley and Dillwyn

> *Copsychus Stricklandii* Motley and Dillwyn, 1855, Nat.
> Hist. Labuan, p. 20, pl. 4 — Labuan Island [reference
> not verified].

Labuan, Balambangan, and Banggi (Banguey) Islands,
North Borneo south to Lawas District, Sarawak, and east
to northeastern Borneo (Kalimantan).

Copsychus stricklandii barbouri (Bangs and Peters)

> *Kittacincla barbouri* Bangs and Peters, 1927, Occ. Papers
> Boston Soc. Nat. Hist., 5, p. 239 — Maratua Island.

Maratua Island, eastern Borneo (Kalimantan).

COPSYCHUS LUZONIENSIS

Copsychus luzoniensis luzoniensis (Kittlitz)

> *Turdus luzoniensis* Kittlitz, 1832, Kupfertafeln Natur.
> Vögel, p. 7, pl. 11, fig. 2 — Luzon [reference not veri-
> fied].

Luzon, Cantaduanes, and Marinduque Islands, Philippines.

Copsychus luzoniensis parvimaculatus (McGregor)

Kittacincla parvimaculata McGregor, 1910, Philippine Journ. Sci., 5, no. 2, sect. D, p. 112 — Polillo, Polillo Island, Philippine Islands.
Polillo Island, Philippines.

Copsychus luzoniensis superciliaris (Bourns and Worcester)

Cittocincla superciliaris Bourns and Worcester, 1894, Occ. Papers Minnesota Acad. Nat. Sci., 1, p. 23 — Masbate, Negros and Ticao Islands.
Negros, Panay, Masbate, and Ticao Islands, Philippines.

COPSYCHUS NIGER

Copsychus niger niger (Sharpe)

Cittocincla nigra Sharpe, 1877, Trans. Linn. Soc. London, 1, p. 335, pl. 52 — Palawan.
Balabac, Calamianes, and Palawan Islands, Philippines.

Copsychus niger cebuensis (Steere)

Cittocincla cebuensis Steere, 1890, List. Birds Mamm. Steere Exped., p. 20 — no locality; Cebu Island designated by Bourns and Worcester, 1894, Occ. Papers Minnesota Acad. Nat. Sci., 1, p. 58.
Cebu Island, Philippines.

COPSYCHUS PYRROPYGUS

Copsychus pyrropygus (Lesson)

Trichixos pyrropyga Lesson, 1839, Rev. Mag. Zool. [Paris], p. 167 — Sumatra.
Malaya, as far north as Wellesley, Sumatra, and Borneo; locally up to 3,000 feet.

GENUS IRANIA DE FILIPPI

Irania de Filippi, 1863, Arch. Zool. Anat. Fisiol. Genova, 2, p. 380. Type, by monotypy, *Irania Finoti* de Filippi = *Cossypha gutturalis* Guérin-Méneville.

IRANIA GUTTURALIS

Irania gutturalis (Guérin-Méneville)

Cossypha gutturalis Guérin-Méneville, 1843, Rev. Zool., p. 162 — Abyssinia.

Breeds from Turkey, Armenia and Lebanon, east through northern Iran to northern Afghanistan, Tadzhikistan, western Kirghizstan, and southern Kazakhstan, in the Kara Tau Mountains; on passage in Syria and Israel; winters in southwest Saudi Arabia, Yemen, Aden, Eritrea, Somalia, and southern Ethiopia south to the Taveta District of Kenya and to central Tanganyika.

GENUS **PHOENICURUS** FORSTER

Phoenicurus Forster, 1817, Synopt. Cat. Brit. Birds, **16**, p. 53. Type, by tautonymy, *Motacilla phoenicurus* Linnaeus.

Adelura Bonaparte, 1854, Compt. Rend. Acad. Sci. Paris, **38**, p. 8, footnote. Type, by original designation, *Phoenicura caeruleocephala* Vigors.

Diplootocus Hartert, 1902, Novit. Zool., **9**, p. 325. Type by monotypy, *Erythacus Moussieri* Olphe-Galliard.

cf. Stegmann, 1928, Journ. f. Orn., **76**, pp. 496-503 (*ochruros; phoenicurus*).

Portenko, 1954, Birds U.S.S.R., **3**, pp. 180-189 (genus).

Vaurie, 1955, Amer. Mus. Novit., no. 1731, pp. 14-20 (*ochruros; auroreus; fuliginosus*).

PHOENICURUS ALASCHANICUS

Phoenicurus alaschanicus (Przevalski)

Rutirilla (sic) *alaschanica* Przevalski, 1876, Mongol. Strana Tangut., 2, p. 40, pl. 9, fig. 2 — Ala Shan [reference not verified].

Mountains of western China in Ningsia, western Kansu, and northeastern Tsinghai.

PHOENICURUS ERYTHRONOTUS

Phoenicurus erythronotus (Eversmann)

Sylvia Erythronota Eversmann, 1841, Add. Pallas Zoogr. Rosso-Asiat., 2, p. 11 — Altai.

Breeds from the mountains of Tadzhikistan northeast across Middle Asia to Altai, Targabatai, Tian Shan, Sayan, and northwest Mongolia; Ala Shan mountains, Ningsia. Winters south in Kirghizstan and Tadzhikistan, Turkmenia,

Iraq (rare), Iran, Afghanistan, West Pakistan, northern India, and west-central Nepal. Vagrant to eastern Siberia.

PHOENICURUS CAERULEOCEPHALUS

Phoenicurus caeruleocephalus Vigors

> Phoenicura caeruleocephala Vigors, 1831, Proc. Zool. Soc. London, p. 35 — Himalayas.

Breeds from Tadzhikistan northeast to Altai and Targabatai, south to eastern Afghanistan, West Pakistan from the Safed Koh to Kashmir, and along Himalayas to Garhwal; found east to Nepal, Darjeeling, Sikkim, and Bhutan (no breeding records).

PHOENICURUS OCHRUROS

Phoenicurus ochruros gibraltariensis (Gmelin)

> Motacilla gibraltariensis Gmelin, 1789, Syst. Nat., 1(2), p. 987 — Gibraltar.
> Erithacus Domesticus Kleinschmidt, 1903, Journ. f. Orn., 51, p. 357 — no locality.
> Phoenicurus ochruros aterrimus von Jordans, 1923, Falco, Sonderheft, p. 8 — Bellas, Portugal.

Southern Norway, and Sweden, Denmark, Netherlands, Belgium, southern British Isles (range extending), south in France, Portugal, and Spain to Atlas Mountains, Morocco; east in Europe to the Crimea and Greece; wintering in the southern parts of the range, Mediterranean Islands, North Africa from Morocco and Tangier east to Egyptian Delta, and Israel.

Phoenicurus ochruros ochruros (Gmelin)

> Motacilla Ochruros S. G. Gmelin, 1774, Reise Russl., 3, p. 101, pl. 19, fig. 3 — Mountains of northern Gilan, Iran [reference not verified].

Breeds in the Caucasus, eastern Turkey, Armenia, and northern Iran; on migration in Syria, Israel, and Iraq; Arabia, two winter records.

Phoenicurus ochruros semirufus (Hemprich and Ehrenberg)

> Sylvia semirufa Hemprich and Ehrenberg, 1833, Symb. Phys., Av., fol. bb. — Egypt [errore = Syria, vide Stresemann, 1954, Abh. Deutschen Akad. Wiss., Math.-Naturwiss., Berlin, no. 1, p. 175].

Breeds in the hills of Syria and Lebanon; winters south to Israel and Sinai.

Phoenicurus ochruros phoenicuroides (Moore)

Ruticilla phoenicuroides Moore, 1854, in Horsfield and Moore, Cat. Birds Mus. East India Co., 1, p. 301 — Shikarpur, Sind.

Ruticilla rufiventris var. *paradoxa* Zarudny, 1896, Mater. Kpoznan. Faun. Flor. Ross. Imp., p. 75 — Transcaspia [reference not verified].

Phoenicurus phoenicuroides alexandrovi Zarudny, 1908, Izvest. Zakasp. Muz., 1, p. 13 — Great Balkhan, western Transcaspia [reference not verified].

Breeds in northeast Iran, Afghanistan, and Kirghizstan, and Tadzhikistan north to Altai, Tian Shan, and western Sayan Mountains; mountains north of Lake Balkash, south to northern Baluchistan, mountains of West Pakistan, Kashmir east to Ladakh, Lahul, and Spiti; winters south of breeding range to eastern Sudan, Ethiopia, Somaliland, Arabia, Iraq, southern Iran, West Pakistan, and Indian plains.

Phoenicurus ochruros rufiventris (Vieillot)

Oenanthe rufiventris Vieillot, 1818, Nouv. Dict. Hist. Nat., nouv. éd., 21, p. 431 — southern Africa, *errore*; restricted to Gyantse, Tibet by Baker, 1921, Jour. Bombay Nat. Hist. Soc., 27, p. 712.

Phoenicurus ochruros xerophilus Stegmann, 1928, Journ. f. Orn., 76, p. 501 — Russki Mountains.

In Russki, Nan Shan, and Humboldt Mountains of Ningsia and Tsinghai, northwestern China, east to western Szechwan and south to Kansu, Sikang, and Garhwal-Tibet border, higher Himalayas east to Nepal and Sikkim; on passage, Mount Everest, 20,000 feet; wintering in India and northern Burma.

PHOENICURUS PHOENICURUS

Phoenicurus phoenicurus phoenicurus (Linnaeus)

Motacilla Phoenicurus Linnaeus, 1758, Syst. Nat., ed. 10, 1, p. 187 — "in Europa"; restricted to Sweden by Hartert, 1910, Vög. pal. Fauna, 1, p. 718.

Erithacus algeriensis Kleinschmidt, 1904, Orn. Monatsb., 12, p. 197 — Lambèse, Algeria.

Phoenicurus phoenicurus turkestanicus Zarudny, 1910, Orn. Monatsb., 18, p. 189 — Bukhara.

Phoenicurus phoenicurus caesitergum Clancey, 1947, Bull. Brit. Orn. Club, 67, p. 77 — the wooded policies of Gawthorpe Estate, near Burnley, Lancashire, England.

British Isles, Scandanavia and across Russia north to the tree line, east to Yenisei River at about lat. 64° N. and Lake Baikal, south to Kirghistan and the Altai range; south in Europe to Algeria and the Mediterranean and Black Sea; on passage through Morocco, the Nile Valley, Lebanon, Israel, and Iraq; winters in tropical west Africa south to Equatorial Africa east as far as Tanganyika, western Arabia, and northwest West Pakistan.

Phoenicurus phoenicurus samamisicus (Hablizl)

Motacilla samamisica Hablizl, 1783, Neue Nord. Beyträge, 4, p. 60 — Gilan Alps, Iran [reference not verified].

Sylvia mesoleuca Ehrenberg, 1833, in Hemprich and Ehrenberg, Symb. Phys. Av., fol. ee — Jidda, Arabia.

Ruticilla semenowi Zarudny, 1904, Orn. Jahrb., 15, p. 213 — Luristan, Arabistan, Mazanderan [, Iran].

Phoenicurus mesoleuca incognita Zarudny, 1910, Orn. Monatsb., 18, p. 189 — Luristan, Gilan, Mazanderan [, Iran].

Phoenicurus mesoleuca bucharensis Zarudny, 1910, Orn. Monatsb., 18, p. 189 — Bukhara.

Crimea, Caucasus, eastern Turkey, northern and western Iran, western Afghanistan, southern Turkmenistan, and Tadzhikistan south to Syria, Cyprus, and Iraq; winters in southern part of breeding range south to Lebanon (breeding ?), Arabia, Yemen, eastern Sudan, Eritrea, and Ethiopia.

PHOENICURUS HODGSONI

Phoenicurus hodgsoni (Moore)

Ruticilla Hodgsoni Moore, 1854, in Horsfield and Moore, Cat. Birds Mus. East India Co., 1, p. 303 — Bootan.

Western China in the mountains of Ningsia, Kansu, Tsinghai, western Szechwan, and Sikang south to southeast Tibet; winters in Szechwan, Yunnan, northern Burma, northeastern India, Bhutan, and Nepal.

PHOENICURUS FRONTALIS

Phoenicurus frontalis Vigors

> *Phoenicura frontalis* Vigors, 1832, Proc. Zool. Soc. London, p. 172 — Himalayas; restricted to Garhwal by Baker, 1924, Fauna Brit. India, Birds, ed. 2, 2, p. 69.
>
> *Phoenicurus frontalis sinae* Hartert, 1918, Bull. Brit. Orn. Club, 38, p. 78 — Kansu.
>
> *Phoenicurus frontalis perates* Koelz, 1954, Contrib. Inst. Regional Explor., no. 1, p. 13 — Karong, Manipur.

Breeding in Himalayas from Chitral, in West Pakistan, east through Kashmir, Nepal, Sikkim, Bhutan, and southern and southeast Tibet north in west China in Sikang, Szechwan, Tsinghai, Kansu, and Ningsia; winters at lower altitudes and south in north Burma, Yunnan, and North Vietnam (Tonkin).

PHOENICURUS SCHISTICEPS

Phoenicurus schisticeps (Gray)

> *Ruticilla schisticeps* Gray, 1846, Cat. Mamm. Birds Nepal, pp. 69, 153 — Nepal.
>
> *Phoenicurus schisticeps beicki* Stresemann, 1927, Orn. Monatsb., 35, p. 134 — Lan-hu-kou (Kansu).

Nepal east to southeast Tibet and west China, in northern Yunnan, Sikang, Tsinghai, Szechwan, western Shensi, and Kansu; winters at lower altitudes and south to Assam and northern Burma.

PHOENICURUS AUROREUS

Phoenicurus auroreus leucopterus Blyth

> *Phoenicura leucoptera* Blyth, 1843, Journ. Asiat. Soc. Bengal, 12, p. 962 — Malay Peninsula.
>
> *Ruticilla rufiventris pleskei* Schalow, 1901, Journ. f. Orn., 49, p. 454 —Nan Shan.

Probably breeds in west China in Ningsia, western Shensi, Kansu, and Sikang; breeds in southeast Tibet and Yunnan; winters in breeding area and south to India (Himalayas from Darjeeling east to Assam), northern Burma, southern China, northern Thailand, Laos, and northern Vietnam.

Phoenicurus auroreus auroreus (Pallas)

Motacilla aurorea Pallas, 1776, Reise Russ. Reichs, 3, p. 695 — Selenga River.

Erithacus auroreus Filchneri Parrot, 1907, in Filchner, Exped. China-Tibet, Zool.-bot. Ergeb., p. 130 — Kintschou and Ping-liang, north China [reference not verified].

Phoenicurus auroreus orientalis Domaniewski, 1933, Acta Orn. Mus. Zool. Polonici, 1, p. 81 — Sidemi, near Vladivostok.

Breeds in Siberia, northern Mongolia east to Amur, Manchuria, Korea, and China in northern Hopeh; winters in the southern parts of breeding range and south into China, Hainan Island, Ryu Kyu Islands, and central and southern Japan (rare in Hokkaido).

PHOENICURUS MOUSSIERI

Phoenicurus moussieri (Olphe-Galliard)

Erythacus Moussieri Olphe-Galliard, 1852, Ann. Soc. Agric. Lyon, ser. 2, 4, p. 101, pl. 2 — Oran Province, Algeria [reference not verified].

Breeds in Atlas Mountains and associated high ranges in Morocco, Algeria, and Tunisia; in winter wanders to adjacent foothills.

PHOENICURUS ERYTHROGASTER

Phoenicurus erythrogaster erythrogaster (Güldenstädt)

Motacilla erythrogastra Güldenstädt, 1775, Nov. Comm. Acad. Petrop., 19, p. 469, pls. 16, 17 — Caucasus.

Caucasus and southern Caspian region of Iran.

Phoenicurus erythrogaster grandis (Gould)

Ruticilla grandis Gould, 1850, Proc. Zool. Soc. London, (1849), p. 112 — Afghanistan and Tibet.

Phoenicurus erythrogaster maximus Kleinschmidt, 1924, Abh. Ber. Mus. Dresden, 16, p. 42 — Janeti, Rombatsa, Chuwo, eastern Tibet [= Sikang].

Breeds from Ferghana east to Altai and Tarbagatai; Transbaicalia? south into the Pamirs, Tian Shan, hills of West Pakistan and India east to Nepal, Darjeeling, Sikkim, southeast Tibet, Sikang, Tsinghai, and Kansu; winters at lower elevations in adjacent hills and in Hopeh and Jehol, northeastern China.

GENUS **RHYACORNIS** BLANFORD

Rhyacornis Blanford, 1872, Journ. Asiat. Soc. Bengal, 41, p. 51. Type, by monotypy, *Phoenicura fuliginosa* Vigors.
Kawabitakia Hachisuka, 1934, Tori, 8, no. 38, p. 222. Type, by monotypy, *Chaimarrornis bicolor* Ogilvie-Grant.

RHYACORNIS BICOLOR

Rhyacornis bicolor (Olgilvie-Grant)
Chimarrhornis (sic) *bicolor* Ogilvie-Grant, 1894, Bull. Brit. Orn. Club, 3, p. 49 — northern Luzon.
Mountains of northern Luzon, Philippines.

RHYACORNIS FULIGINOSUS

Rhyacornis fuliginosus fuliginosus (Vigors)
Phoenicura fuliginosa Vigors, 1831, Proc. Zool. Soc. London, p. 35 — Himalayas; restricted to Simla-Almora district by Ticehurst and Whistler, 1924, Ibis, p. 471.
Ch[a]imarronis fuliginosa tenuirostris Stresemann, 1923, Journ. f. Orn., 71, p. 364 — Siuhang, Kwangtung.
West Pakistan (Safed Koh), Indian Himalayas, Nepal, Sikkim, and Bhutan, east through southeast Tibet, Sikang, Yunnan, Szechwan, and Kansu to Hopeh, thence south throughout to Burma, northern Thailand, northern Vietnam, and Hainan Island.

Rhyacornis fuliginosus affinis (Ogilvie-Grant)
Xanthopygia affinis Ogilvie-Grant, 1906, Bull. Brit. Orn. Club, **16**, p. 118 — Mount Morrison, 6,000 ft., central Formosa.
Formosa.

GENUS **HODGSONIUS** BONAPARTE

Hodgsonius Bonaparte, 1851, Consp. Av., 1 (1850), p. 300. Type, by monotypy, *Brachypterus phaenicuroides* "Hodgs." = *Bradypterus phaenicuroides* Gray.

HODGSONIUS PHAENICUROIDES

Hodgsonius phaenicuroides phaenicuroides (Gray)

Bradypterus phaenicuroides Gray, 1846, Cat. Mamm. Birds Nepal Thibet, pp. 70, 153 — Nepal.

West Pakistan and Indian Himalayas, Nepal, Sikkim, and Bhutan east through southeast Tibet, Sikang, Yunnan, and north Burma.

Hodgsonius phaenicuroides ichangensis Baker

Hodgsonius phoenicuroides ichangensis Baker, 1922, Bull. Brit. Orn. Club, 43, p. 18 — Ichang, Upper Yangtse Valley.

China in Szechwan, Kansu, east Kuku Nor, and Shensi; south in winter to northern Vietnam and Laos.

GENUS CINCLIDIUM BLYTH

Cinclidium Blyth, 1842, Journ. Asiat. Soc. Bengal, 11, p. 181. Type, by montypy, *C. frontale* Blyth.

Muscisylvia Hodgson, 1845, Proc. Zool. Soc. London, p. 27. Type, by original designation, *M. leucura* Hodgson, *nec Musicisylvia* (sic) Agassiz, 1841.

Myiomela Gray, 1846, Gen. Birds, 1, p. 172. New name for *Muscisylvia* Hodgson, preoccupied.

Callene Blyth, 1847, Journ. Asiat. Soc. Bengal, 16, p. 136. New name for *Cinclidium* Blyth, 1842, preoccupied in botany.

cf. Mayr, 1938, Ibis, p. 293 (names *Muscisylvia, Myiomela*).

CINCLIDIUM LEUCURUM

Cinclidium leucurum leucurum (Hodgson)

M.(uscisylvia) leucura Hodgson, 1845, Proc. Zool. Soc. London, p. 27 — Nepal.

Notodela leucura rhipidura Koelz, 1952, Journ. Zool. Soc. India, 4, p. 41 — Tekhubama, Naga Hills.

Nepal, northeastern Pakistan and India, northern Vietnam (Tonkin), Laos, and southern Vietnam; Malaya (Perak); Burma, Yunnan, and northern Thailand; foothills to 8,000 feet.

Cinclidium leucurum cambodianum (Delacour and Jabouille)
Notodela cambodiana Delacour and Jabouille, 1928, Bull.
Brit. Orn. Club, 48, p. 132 — Bokor, 1,000 m., southern
Cambodia.
Cambodia, in the Chaine de l'Eléphant.

CINCLIDIUM DIANA

Cinclidium diana sumatranum (Robinson and Kloss)
Notodela diana sumatrana Robinson and Kloss, 1918,
Journ. Fed. Malay States Mus., 8, p. 215 — Sungei,
4,600 ft., Kumbang, Korinchi, Sumatra.
Mountains of northern and west-central Sumatra.

Cinclidium diana diana (Lesson)
Lanius (*Notodela*) *diana* Lesson, 1834, in Bélanger, Voy.
Ind. Zool. Ois., p. 246, pl. 3 — Pegu [*errore* = Java, *vide*
Sharpe, 1883, Cat. Birds Brit. Mus., 7, p. 25].
Mountains of Java.

CINCLIDIUM FRONTALE

Cinclidium frontale frontale Blyth
Cinclidium frontale Blyth, 1842, Journ. Asiat. Soc. Ben-
gal, 11, p. 181 — Sikkim.
In the hills of Nepal, Darjeeling District, and Sikkim.

Cinclidium frontale orientale Delacour and Jabouille
Callene frontalis orientalis Delacour and Jabouille, 1930,
Oiseau Rev. Franç. Orn., 11, p. 397 — Chapa, Tonkin.
North Vietnam, in Tonkin; and Laos.

GENUS **GRANDALA** HODGSON

Grandala Hodgson, 1843, Journ. Asiat. Soc. Bengal, 12,
p. 447. Type, by monotypy, *Grandala côelicolor* Hodgson.

GRANDALA COELICOLOR

Grandala coelicolor Hodgson
Gr.(*andala*) *côelicolar* (sic) Hodgson, 1843, Journ. Asiat.
Soc. Bengal, 12, p. 447 — Nepal.
Grandala coelicolor florentes Bangs, 1926, Proc. New

England Zool. Club, 9, p. 78 — Tatsienlu [=Kangting],
Szechwan.

From Kashmir east along the Himalayas, Nepal, Sikkim,
Bhutan, southeast Tibet, north Burma (Adung Valley, win-
ter), Sikang, western Szechwan, Tsinghai, and adjacent
Kansu.

GENUS SIALIA SWAINSON

Sialia Swainson, 1827, Phil. Mag., n.s., 1 (5), p. 369.
Type, by monotypy, *Sialia azurea* Swainson = *Motacilla
sialis* Linnaeus.

SIALIA SIALIS

Sialia sialis sialis (Linnaeus)

Motacilla Sialis Linnaeus, 1758, Syst. Nat., ed. 10, 1, p.
187 — *in Bermudis & America calidiore* [= South
Carolina].

Southern central and eastern Canada from Saskatchewan
to Nova Scotia, south in the United States from the eastern
slope of the Rockies in Montana and the Dakotas south to
central Texas and east through eastern United States, except peninsular Florida; Bermuda; wintering in the central
and southern latitudes of its range south to Mexico (Nuevo
León) and western Cuba (rarely).

Sialia sialis grata Bangs

Sialia sialis grata Bangs, 1898, Auk, 15, p. 182 — Miami,
Dade County, Florida.

Peninsular Florida from Lake County south.

Sialia sialis episcopus Oberholser

Sialia sialis episcopus Oberholser, 1917, Proc. Biol. Soc.
Washington, 30, p. 27 — Santa Engracia, Tamaulipas,
Mexico.

Southern coastal Texas (Rockport) south to southern
Tamaulipas, Mexico.

Sialia sialis fulva Brewster

Sialia sialis fulva Brewster, 1885, Auk, 2, p. 85 — Santa
Rita Mountains, Arizona.

Mountains of south-central Arizona south through the

Mexican tableland to Guerrero and México; wintering to Veracruz, Chiapas, and Guatemala.

Sialia sialis guatemalae Ridgway

Sialia sialis guatemalae Ridgway, 1882, Proc. U. S. Nat. Mus., 5, p. 13 — highlands of Guatemala and Honduras.

Mountains of eastern and southeastern Mexico from southern Tamaulipas to Chiapas and Oaxaca; and Guatemala.

Sialia sialis meridionalis Dickey and van Rossem

Sialia sialis meridionalis Dickey and van Rossem, 1930, Condor, 32, p. 69 — Los Esesmiles, Chalatenango, El Salvador.

Mountains of El Salvador and Nicaragua; Honduras (?).

SIALIA MEXICANA

Sialia mexicana occidentalis Townsend

S. (ialia) occidentalis Townsend, 1837, Journ. Acad. Nat. Sci. Philadelphia, 7, p. 188 — Columbia River [= Fort Vancouver, Washington].

Southern British Columbia, Montana, eastern Oregon, Idaho, and Wyoming south in the mountains to southern California and western Nevada; wintering at lower altitudes; wandering to southeast California, Santa Catalina, and San Clemente Islands.

Sialia mexicana bairdi Ridgway

Sialia mexicana bairdi Ridgway, 1894, Auk, 11, pp. 151, 157 — Camp 110, New Mexico [= Cactus Pass, 20 miles east of Kingman, Mohave County, Arizona].

Southern Nevada, central Utah, and Colorado south to Arizona, western Texas, northern Sonora, and northern Chihuahua; wintering at lower altitudes in same areas and as far as southeast California.

Sialia mexicana anabelae Anthony

Sialia mexicana anabelae Anthony, 1889, Proc. California Acad. Sci., ser. 2, 2, p. 79 — San Pedro Mountain, Lower California.

Northern Baja California in Sierra Juarez and San Pedro Martir; wintering at lower altitudes and on Todos Santos Island.

Sialia mexicana amabilis Moore

Sialia mexicana amabile Moore, 1939, Proc. Biol. Soc. Washington, 52, p. 125 — Nievero, 4 miles west of Ciudad Durango, Mexico.

Sierra Madre Occidental from southern Chihuahua to Zacatecas.

Sialia mexicana mexicana Swainson

Sialia Mexicana Swainson, 1832, in Swainson and Richardson, Fauna Bor.-Amer., 2, p. 202 — tableland of Mexico.

Northeastern central plateau in Coahuila, San Luis Potosí, Nuevo León, and southwest Tamaulipas.

Sialia mexicana australis Nelson

Sialia mexicana australis Nelson, 1903, Proc. Biol. Soc. Washington, 16, p. 159 — Mount Tancítaro, Michoacán.

Southern plateau in Jalisco, Michoacán, Guanajuato, México, Morelos, Puebla, and western Veracruz.

SIALIA CURRUCOIDES

Sialia currucoides (Bechstein)

Motacilla s. Sylvia Currucoides Bechstein (ex Borkh MS), 1798, in Latham, Allgem. Uebers. Vögel, 3, pt. 2, p. 546, pl. 121 — Virginien [= western America].

Breeds from central Alaska, southern Yukon, southern Mackenzie, and southwest Manitoba south in mountains to southern California, northern Arizona, southern New Mexico, and in plains of Dakotas; winters from southern British Columbia south to Baja California, Sinaloa, Michoacán, Guanajuato, Nuevo León, and southern Texas.

GENUS **ENICURUS** TEMMINCK

Enicurus Temminck, 1822, Pl. Col., livr. 19, pl. 113. Type, by monotypy, *Enicurus coronatus* Temminck = *Turdus Leschenaulti* Vieillot.

Microcichla Sharpe, 1883, Cat. Birds Brit. Mus., 7, pp. 312, 322. Type, by monotypy, *Enicurus scouleri* Vigors.

ENICURUS SCOULERI

Enicurus scouleri scouleri Vigors

Enicurus Scouleri Vigors, 1832, Proc. Zool. Soc. London, (1831), p. 174 — Himalayas; restricted to Simla by Baker, 1924, Fauna Brit. India, Birds, ed. 2, 2, p. 65.

Southeast Russia in Tadzhikistan and Kirghizstan, northeast Afghanistan, West Pakistan and India from Chitral along the Himalayas, Nepal, Sikang, Yunnan, and Szechwan northeast to the hills of Shensi, and south to southwest China and Vietnam (Tonkin) ; winters in foothills and south to northern Burma (resident ?).

Enicurus scouleri fortis (Hartert)

Microcichla scouleri fortis Hartert, 1910, Vög. pal. Fauna, 1, p. 761 — Tapposha, Formosa.

Hills of Formosa (Taiwan).

ENICURUS VELATUS

Enicurus velatus sumatranus (Robinson and Kloss)

Henicurus velatus sumatranus Robinson and Kloss, 1923, Journ. Fed. Malay States Mus., 11, p. 56 — Siolak Dras, Korinchi Valley, Sumatra.

Sumatra.

Enicurus velatus velatus Temminck

Enicurus velatus Temminck, 1822, Pl. Col., livr. 27, pl. 160 — Java.

Java.

ENICURUS RUFICAPILLUS

Enicurus ruficapillus Temminck

Enicurus ruficapillus Temminck, 1823, Pl. Col., livr. 90, pl. 534 — Palembang, Sumatra.

Malay Peninsula from Burma (southern Tenasserim) and Thailand (Chumphon) south; Sumatra; Borneo.

ENICURUS IMMACULATUS

Enicurus immaculatus (Hodgson)

Motacilla (Enicurus) Immaculatus Hodgson, 1836, Asiat. Res., 19, p. 190 — Nepal.

Himalayas of India from Garhwal east, Nepal, East Pakistan, Burma (except Tenasserim), and Thailand.

ENICURUS SCHISTACEUS

Enicurus schistaceus (Hodgson)

Motacilla (Enicurus) schistaceus Hodgson, 1836, Asiat. Res., 19, p. 189 — Nepal.

Himalayas of India from Kumaon east, Nepal, East Pakistan (?), Burma, northern Thailand, Laos, Vietnam in Annam and Tonkin, and China in Yunnan, Kwangtung, and Fukien.

ENICURUS LESCHENAULTI

Enicurus leschenaulti indicus Hartert

Enicurus leschenaulti indicus Hartert, 1909, Vög. pal. Fauna, 1, p. 760 — Margherita, Upper Assam.

Himalayas of eastern India from Darjeeling eastward, East Pakistan, Burma, northern Thailand, Laos, and Vietnam in northern Annam and Tonkin.

Enicurus leschenaulti sinensis Gould

Enicurus sinensis Gould, 1865, Proc. Zool. Soc. London, p. 665 — China [= Shanghai; *vide* Sharpe, 1883, Cat. Birds Brit. Mus., 7, p. 314].

China from Sikang, Yunnan, Szechwan, southern Kansu, and Shensi to Kwangsi, Fukien, Hupeh, and Chekiang; Hainan Island.

Enicurus leschenaulti frontalis Blyth

Enicurus frontalis Blyth, 1847, Journ. Asiat. Soc. Bengal, 16, p. 156 — Malayan peninsula; restricted to Malacca by Chasen, 1935, Bull. Raffles Mus., 11, p. 233.

Malaya, Sumatra, Nias Island (race?), and Borneo in the lowlands.

Enicurus leschenaulti chaseni de Schauensee

Enicurus leschenaulti chaseni de Schauensee, 1940, Proc. Acad. Nat. Sci. Philadelphia, 92, p. 38 — Tana Massa Island, Batu Islands.

Tana Massa, Batu Islands, west Sumatra.

Enicurus leschenaulti leschenaulti (Vieillot)

Turdus Leschenaulti Vieillot, 1818, Nouv. Dict. Hist. Nat., nouv. éd., 20, p. 269 — Java.

Java and Bali.

Enicurus leschenaulti borneensis (Sharpe)

Henicurus borneensis Sharpe, 1889, Ibis, p. 277 — Kinabalu.

Borneo from Mount Kinabalu south to Mahakam drainage and Mount Liang Kubung, upper Kapuas; Mount Dulit.

ENICURUS MACULATUS

Enicurus maculatus maculatus Vigors

Enicurus maculatus Vigors, 1831, Proc. Zool. Soc. London, p. 9 — Himalaya; restricted to Simla by Baker, 1921, Journ. Bombay Nat. Hist. Soc., **27**, p. 711.

Himalayas of West Pakistan and India from Chitral and Kashmir east to hills of central Nepal.

Enicurus maculatus guttatus Gould

Enicurus guttatus Gould, 1865, Proc. Zool. Soc. London, p. 664 — Sikkim; restricted to Darjeeling by Baker, 1921, Journ. Bombay Nat. Hist. Soc., **27**, p. 711.

Himalayas from extreme eastern Nepal east in India, Sikkim, Bhutan, East Pakistan (?), southwest Sikang, northern Yunnan, and Burma to the southern Shan States.

Enicurus maculatus bacatus Bangs and Phillips

Enicurus guttatus bacatus Bangs and Phillips, 1914, Bull. Mus. Comp. Zool., **58**, p. 292 — Loukouchai, southern Yunnan.

Enicurus maculatus omissus Rothschild, 1921, Novit. Zool., **28**, p. 26 — Fokien.

Mountains of southeastern Yunnan, North Vietnam in northeast Tonkin (Chapa), and northwest Fukien.

Enicurus maculatus robinsoni Baker

Enicurus maculatus robinsoni Baker, 1922, Bull. Brit. Orn. Club, **43**, p. 19 — Langham [= Langbiang] peaks, southern Annam.

Mountains of Da Lat district, South Vietnam.

GENUS COCHOA HODGSON

Cochoa Hodgson, 1836, Journ. Asiat. Soc. Bengal, **5**, p. 359. Type, by original designation, *Cochoa purpurea* Hodgson.

COCHOA PURPUREA

Cochoa purpurea Hodgson

Cochoa *purpurea* Hodgson, 1836, Journ. Asiat. Soc. Bengal, 5, p. 359 — Nepal.

Indian Himalayas from Almora to Assam, Nepal, Burma, Yunnan, northern Thailand, and North Vietnam in Tonkin.

COCHOA VIRIDIS

Cochoa viridis Hodgson

Co.(*choa*) *Viridis* Hodgson, 1836, Journ. Asiat. Soc. Bengal, 5, pp. 359, 360 — Nepal.

Cochoa *rothschildi* Baker, 1924, Fauna Brit. India, Birds, 2, p. 186 — Sikkim.

Indian Himalayas from Kumaon to Assam, Nepal, northern Burma, northern Thailand, Laos, North Vietnam in Tonkin, South Vietnam in Annam, and Fukien.

COCHOA AZUREA

Cochoa azurea beccarii Salvadori

Cochoa *beccarii* Salvadori, 1879, Ann. Mus. Civ. Genova, 14, p. 228 — Padang highlands, Sumatra.

West Sumatra in the Padang highlands and Korinchi.

Cochoa azurea azurea (Temminck)

Turdus *azureus* Temminck, 1824, Pl. Col., livr. 46, pl. 274 — Java.

Highlands of central and western Java.

GENUS **MYADESTES** SWAINSON

Myadestes Swainson, 1838, Nat. Libr. Flycatchers, p. 132. Type, by monotypy, *Myidestes* (sic) *genibarbis* Swainson.

Cichlopsis Cabanis, 1851, Mus. Hein., 1 (1850), p. 54. Type, by monotypy, C. *leucogenys* Cabanis.

MYADESTES TOWNSENDI

Myadestes townsendi townsendi (Audubon)

Ptiliogony's (sic) *Townsendi* Audubon, 1838, Birds Amer. (folio), 4, pl. 419, fig. 2 — Columbia River [=

near Astoria, Oregon, *vide* Amer. Orn. Union, 1910,
Check-list North Amer. Birds, ed. 3, p. 359].

Central and southeastern Alaska, Yukon, southwest Mackenzie, southwest Alberta, and central British Columbia
south through the western mountains of the United States
from Oregon east to South Dakota and Nebraska, south to
northeastern Arizona, New Mexico, and northern Chihuahua; winters at lower altitudes east to Kansas and central
Texas, south to Baja California, Guadalupe Island, northeast
Sonora, and Coahuila.

Myadestes townsendi calophonus Moore

Myadestes townsendi calophonus Moore, 1937, Proc. Biol.
Soc. Washington, 50, p. 201 — within 1,000 feet of summit of Mt. Mohinora, southwest Chihuahua, Mexico.

Southern Chihuahua and Durango; southeast Sonora, Jalisco, and Zacatecas (race?).

MYADESTES OBSCURUS

Myadestes obscurus obscurus Lafresnaye

Myadestes obscurus Lafresnaye, 1839, Rev. Zool., 2,
p. 98 — Mexico; probably Veracruz.

Mountains of eastern Mexico from southern Tamaulipas
to Nuevo León, western Veracruz, eastern Puebla, Hidalgo,
San Luis Potosí, eastern México, Distrito Federal, eastern
Guanajuato, and eastern Oaxaca.

Myadestes obscurus cinereus Nelson

Myadestes obscurus cinereus Nelson, 1899, Proc. Biol. Soc.
Washington, 13, p. 30 — mountains near Alamos, Sonora, Mexico.

Southeastern Sonora, Sinaloa, southern Chihuahua, and
Durango.

Myadestes obscurus occidentalis Stejneger

Myadestes obscurus var. *occidentalis* Stejneger, 1882,
Proc. U. S. Nat. Mus., 4, pp. 371-372 — Tonila, Jalisco.

Western mountains from Nayarit, Jalisco, Michoacán,
Guerrero, western Oaxaca, western Guanajuato, western
México, and Morelos; intergrading with preceding forms in
north and east of the range.

Myadestes obscurus insularis Stejneger

Myadestes obscurus var. *insularis* Stejneger, 1882, Proc.
U. S. Nat. Mus., 4, pp. 371 ; 373 — Tres Marías Islands.
Tres Marías Islands, off Nayarit, Mexico.

Myadestes obscurus oberholseri Dickey and van Rossem

Myadestes obscurus oberholseri Dickey and van Rossem,
1925, Proc. Biol. Soc. Washington, 38, p. 133 — Volcán
de San Rafael, El Salvador.
Mountains of southern Mexico (Chiapas), Guatemala,
and El Salvador.

MYADESTES ELISABETH

Myadestes elisabeth elisabeth (Lembeye)

Muscicapa elisabeth Lembeye, 1850, Aves Isla Cuba, p.
39, pl. 5, fig. 3 — Cuba.
Mountains of western and eastern Cuba.

Myadestes elisabeth retrusus Bangs and Zappey

Myadestes elizabeth (sic) *retrusus* Bangs and Zappey,
1905, Amer. Nat., 39, p. 208 — Pasadita, Isle of Pines.
Isle of Pines, Cuba.

MYADESTES GENIBARBIS

Myadestes genibarbis solitarius Baird

Myiadestes solitarius Baird, 1866, Rev. Amer. Birds, 1,
p. 421 — Jamaica [*viz.* Port Royal Mountains, *vide*
Hellmayr, 1934, Field Mus. Nat. Hist. Publ., Zool. Ser.,
13, pt. 7, p. 437].
Jamaica.

Myadestes genibarbis montanus Cory

Myiadestes montanus Cory, 1881, Bull. Nuttall Orn. Club,
6, p. 130 — Haiti [= Massif de la Selle].
Hispaniola.

Myadestes genibarbis dominicanus Stejneger

Myiadestes dominicanus Stejneger, 1882, Proc. U. S. Nat.
Mus., 5, p. 22, pl. 2, fig. 5 — Dominica, Lesser Antilles.
Dominica.

Myadestes genibarbis genibarbis Swainson

Myidestes (sic) *genibarbis* Swainson, 1838, Nat. Libr. Flycatchers, p. 134, pl. 13 — Africa or India [= Martinique, *vide* Sclater, 1871, Proc. Zool. Soc. London, p. 270].

Martinique, Lesser Antilles.

Myadestes genibarbis sanctaeluciae Stejneger

Myadestes sanctae-luciae Stejneger, 1882, Proc. U.S. Nat. Mus., 5, p. 20, pl. 2, fig. 4 — Santa Lucia.

St. Lucia, Lesser Antilles.

Myadestes genibarbis sibilans Lawrence

Myiadestes sibilans Lawrence, 1878, Ann. New York Acad. Sci., 1, p. 147 — St. Vincent.

St. Vincent, Lesser Antilles.

MYADESTES RALLOIDES

Myadestes ralloides melanops Salvin

Myiadestes melanops Salvin, 1865, Proc. Zool. Soc. London, (1864), p. 580, pl. 36 — Tucurrique, Costa Rica.

Costa Rica and western Panama.

Myadestes ralloides coloratus Nelson

Myadestes coloratus Nelson, 1912, Smiths. Misc. Coll., 60, no. 3, p. 23 — Mount Pirri, near head of Río Limon, eastern Panama.

Eastern Panama.

Myadestes ralloides plumbeiceps Hellmayr

Myadestes ralloides plumbeiceps Hellmayr, 1921, Anz. Orn. Ges. Bayern, no. 4, p. 27 — Siató, Río Siató, near Pueblo Rico, Chocó, western Colombia.

Western and central Andes of Colombia and western Ecuador.

Myadestes ralloides candelae de Schauensee

Myadestes ralloides candelae de Schauensee, 1947, Proc. Acad. Nat. Sci. Philadelphia, 99, p. 117 — La Candela, Huila, Colombia.

Head of the Magdalena Valley, Colombia.

— Myadestes ralloides venezuelensis Sclater

> Myiadestes venezuelensis Sclater, 1856, Ann. Mag. Nat. Hist., ser. 2, 17, p. 468 — vicinity of Caracas, Venezuela.

Eastern Andes of Colombia, northern Venezuela, eastern Ecuador, and extreme northern Peru (Dept. Cajamarca).

Myadestes ralloides ralloides (d'Orbigny)

> Muscipeta ralloides d'Orbigny, 1840, Voy. Amér. Mérid., 4, pt. 3, Ois., p. 322 — Chulumani, Prov. Yungas, east side of the Cordillera, Bolivia.

Western Bolivia and subtropical zone of Peru, north to Depts. Libertad (Utcubamba) and Huánuco (Chinchao).

MYADESTES UNICOLOR

— Myadestes unicolor unicolor Sclater

> Myiadestes unicolor Sclater, 1857, Proc. Zool. Soc. London, p. 299 — Córdova, Veracruz, Mexico.

Southern Mexico in Hidalgo, Puebla, Veracruz, Oaxaca, and Chiapas.

Myadestes unicolor veraepacis Griscom

> Myadestes unicolor veraepacis Griscom, 1930, Amer. Mus. Novit., no. 438, p. 6 — Finca Sepacuite, fifty miles east of Coban, Alta Vera Paz, Guatemala.

Highlands of Guatemala and northern Honduras.

— Myadestes unicolor pallens Miller and Griscom

> Myadestes unicolor pallens Miller and Griscom, 1925, Amer. Mus. Novit., no. 183, p. 5 — San Rafael del Norte, Nicaragua.

Nicaragua.

MYADESTES LEUCOGENYS

Myadestes leucogenys gularis (Salvin and Godman)

> Cichlopsis gularis Salvin and Godman, 1882, Ibis, p. 76 — Merumé Mountains, British Guiana.

British Guiana.

Myadestes leucogenys chubbi (Chapman)

> Cichlopsis chubbi Chapman, 1924, Amer. Mus. Novit., no. 138, p. 15 — Mindo, Huila, western Ecuador.

Western Ecuador.

Myadestes leucogenys peruvianus (Hellmayr)

Cichlopsis leucogenys peruvianus Hellmayr, 1930, Novit.
Zool., 35, p. 265 — Perené, Dept. Junin, Peru.
Tropical zone of central Peru.

Myadestes leucogenys leucogenys (Cabanis)

C.(ichlopsis) leucogenys Cabanis, 1851, Mus. Hein., 1
(1850), p. 54 — Brazil.
Espirito Santo, southeastern Brazil.

GENUS ENTOMODESTES STEJNEGER

Entomodestes Stejneger, 1883, Proc. U. S. Nat. Mus., 5
(1882), p. 456. Type, by monotypy, *Entomodestes leucotis* (Tschudi).

ENTOMODESTES LEUCOTIS

Entomedestes leucotis (Tschudi)

Ptilogonys leucotis Tschudi, 1844, Arch. f. Naturg., 10, p.
270 — Peru.
Subtropics of Peru, in Libertad, Junín, Cuzco, and Puno,
and in Bolivia, in La Paz and Cochabamba.

ENTOMODESTES CORACINUS

Entomodestes coracinus (Berlepsch)

Myiadestes coracinus Berlepsch, 1897, Orn. Monatsb.,
5, p. 175 — near San Pablo, Prov. Túquerres, southwestern Colombia.
Western Colombia and western Ecuador.

GENUS STIZORHINA OBERHOLSER

Stizorhina Oberholser, 1899, Proc. Acad. Nat. Sci. Philadelphia, 51, p. 213. Type, by original designation, *Muscicapa fraseri* Strickland.

STIZORHINA FRASERI

Stizorhina fraseri fraseri (Strickland)

Muscicapa Fraseri Strickland, 1844, Proc. Zool. Soc. London, p. 101 — Fernando Po.
Fernando Po Island.

Stizorhina fraseri rubicunda (Hartlaub)

C.(assinia) rubicunda Hartlaub, 1860, Rev. Zool. [Paris],
p. 82 — Gabon.

Southern Cameroons, vicinity of Mount Cameroon, Cameroun, Gabon, Rio Muni (?), lower Congo, and Kasai, south to northern and western Angola, south to the Cuanza River and Gabela.

Stizorhina fraseri vulpina Reichenow

Stizorhina vulpina Reichenow, 1902, Journ. f. Orn., **50**, p. 125 — Bundeko, Sembiki Valley.
Stizorhina vulpina intermedia Clarke, 1913, Bull. Brit. Orn. Club, 31, p. 107 — Entebbe.

Northern and eastern Congo, Ruanda, and western Uganda in forested areas.

STIZORHINA FINSCHII

Stizorhina finschii (Sharpe)

Cassinia finschii Sharpe, 1870, Ibis, p. 53, pl. 2, fig. 2 — Fantee.

Lowland forest from Sierra Leone, Liberia, Ivory Coast, Ghana, Togo, Dahomey, and Nigeria.

GENUS **NEOCOSSYPHUS** FISCHER AND REICHENOW

Neocossyphus Fischer and Reichenow, 1884, Zeitschr. ges. Orn., 1, p. 301. Type, by original designation, *Pseudocossyphus rufus* Fischer and Reichenow.

NEOCOSSYPHUS RUFUS

Neocossyphus rufus gabunensis Neumann

Neocossyphus rufus gabunensis Neumann, 1908, Bull. Brit. Orn. Club, **21**, p. 77 — Ohumbe, Lake Onange, Ogowe River.
Neocossyphus rufus arrhenii Lönnberg, 1917, Ark. f. Zool., 10, no. 24, p. 30 — Beni, Semliki Valley.

Lowland forest from southern Cameroons, Cameroun, Gabon, and northern Congo to Budonga forest, Uganda.

Neocossyphus rufus rufus (Fischer and Reichenow)

Pseudocossyphus rufus Fischer and Reichenow, 1884, Journ. f. Orn., 32, p. 58 — Pangani.

Tanganyika, Uluguru Mountains, north in Kenya in coastal mountains and forests to the Tana River; Zanzibar.

NEOCOSSYPHUS POENSIS

Neocossyphus poensis poensis (Strickland)

Cossypha poensis Strickland, 1844, Proc. Zool. Soc. London, p. 100 — Clarence, Fernando Po.

Coastal forest of Sierra Leone, Liberia, Ivory Coast, Ghana, Togo, Dahomey, Nigeria, Cameroons, Cameroun, Gabon, Fernando Po, and perhaps Mayombe Forest of Congo.

Neocossyphus poensis pallidigularis Meise

Neocossyphus poensis pallidigularis Meise, 1958, Abh. Verh. Naturwiss. Hamburg, no. 2 (1957), p. 74 — Canzele.

Northern Angola.

Neocossyphus poensis praepectoralis Jackson

Neocossyphus praepectoralis Jackson, 1906, Bull. Brit. Orn. Club, 16, p. 90 — Kibera, Toro.

Neocossyphus granti Alexander, 1908, Bull. Brit. Orn. Club, 23, p. 15 — Beritio, Lower Uelle River.

Forests of northern and eastern Congo east to western Uganda.

Genus **CERCOMELA** Bonaparte

Cercomela Bonaparte, 1856, Compt. Rend. Acad. Sci. Paris, 42, p. 766. Type, by original designation, *Cercomela asthenia* Bonaparte = *Saxicola melanura* Temminck.

Pinarochroa Sundevall, 1872, Tentamen, p. 4. Type, by original designation, *Saxicola sordida* Rüppell.

Emarginata Shelley, 1896, Birds Africa, 1, p. 89. Type, by original designation, *E. sinuata* = *Luscinia sinuata* Sundevall.

Karrucincla Roberts, 1922, Ann. Transvaal Mus., 8, p. 230. Type, by monotypy, *Saxicola pollux* Hartlaub.

Phoenicuroides Roberts, 1922, Ann. Transvaal Mus., 8, p. 231. Type, by monotypy, 'Tractrac' ex Levaillant = *Motacilla tractrac* Wilkes.

Psammocichla Roberts, 1922, Ann. Transvaal Mus., 8, p. 231. Type, by monotypy, *Saxicola albicans* Wahlberg.

cf. Sclater, 1928, Bull. Brit. Orn. Club, 49, pp. 11-19 (African forms).
Macdonald, 1957, Contr. Orn. W. South Africa. Brit. Mus. (Nat. Hist.), pp. 127-128 (*schlegelii* and *pollux*).
Clancey, 1962, Ostrich, 33 (4) :24-28 (South African races of *familiaris*).

CERCOMELA SINUATA

Cercomela sinuata sinuata (Sundevall)

Lusc.(*inia*) *sinuata* Sundevall, 1858, K. Sv. Vet.-Akad., Handl., 2 (3) (1857), p. 44 — Cape Town [= Saldanha Bay, *vide* Gyldenstolpe, 1922, Ark. f. Zool., 19A, p. 55].
Cape Province of South Africa, from coastal areas north of Saldanha Bay east to Humansdorp and Kei Road, in winter rainfall areas.

Cercomela sinuata ensifera Clancey

Cercomela sinuata ensifera Clancey, 1958, Durban Mus. Novit., 5, p. 102 — Rietfontein, 4,300 ft., Griquatown-Niekerkshoop road, Asbestos Mountains, northern Cape Province.
Little Namaqualand and the Calvinia district eastwards through the Karroo to the eastern Cape, Griqualand West, the Orange Free State, and southern Transvaal; intergrading with *sinuata* in the border areas of the range.

Cercomela sinuata hypernephela Clancey

Cercomela sinuata hypernephela Clancey, 1956, Durban Mus. Novit., 4, p. 281 — 40 miles east of Maseru on new mountain road (lat. 29° 28' S., long. 27° 55' E.), *ca.* 8,000 ft., Basutoland.
Highlands of Basutoland.

CERCOMELA FAMILIARIS

Cercomela familiaris falkensteini (Cabanis)

Saxicola Falkensteini Cabanis, 1875, Journ. f. Orn., 23, p. 235 — Chinchoxo, Loango Coast.

Saxicola sennaarensis Seebohm, 1881, Cat. Birds Brit. Mus., 5, p. 391 — Sennaar; emended to Yemen by Sclater, 1930, Syst. Av. Aethiop., 2, p. 459; *errore* = Sennaar, *vide* Meinertzhagen, 1954, Birds Arabia, p. 257.
Bessonornis (? *Cossypha*) *gambagae* Hartert, 1899, Bull. Brit. Orn. Club, 10, p. 5 — near Gambaga, Gold Coast hinterland.
Phoenicurus tessmanni Reichenow, 1921, Journ. f. Orn., 69, p. 49 — Buera, Cameroons.
Cercomela familiaris genderuensis Bannerman, 1922, Bull. Brit. Orn. Club, 43, p. 8 — Genderu, Cameroons highlands.

Ghana, northern Nigeria, southern Cameroons, Lower Congo and Loango Coast, northern Congo in Kivu, Uganda, Kenya, southwestern Sudan, and northern Ethiopia.

Cercomela familiaris angolensis Lynes

C. (ercomela) f. (amiliaris) angolensis Lynes, 1926, Ibis, p. 393 — Huxe [= Uchi], Benguella.
Cercomela familiaris hoeschi Niethammer, 1955, Bonn. Zool. Beitr., 6, p. 189 — Otju in the Hoarusib/Kaokoveld.

Angola from Malanje south to Kaokoveld and Ovamboland in northern South West Africa.

Cercomela familiaris galtoni (Strickland)

Erythropygia galtoni Strickland, 1852, in Jardine, Contrib. Orn., p. 147 — Damaraland.
Saxicola familiaris lübberti Reichenow, 1902, Orn. Monatsb., 10, p. 77 — Windhoek.
Phoenicurus familiaris damarensis Roberts, 1931, Ann. Transvaal Mus., 14, p. 242 — Windhoek.
Cercomela familiaris dodsoni Macdonald, 1953, Ibis, 95, p. 72 — Deelfontein, eastern Cape Province.
Cercomela familiaris richardi Macdonald, 1953, Ibis, 95, p. 73 — Springbok, Little Namaqualand.

South West Africa in Great Namaqualand and Damaraland, northern Little Namaqualand, and northern Cape Province to western districts of Bechuanaland.

Cercomela familiaris familiaris (Stephens)

Saxicola familiaris Stephens, 1826, Gen. Zool., 13, p. 241 — Southern Africa, ex Levaillant; restricted to Table

Mountain, southwestern Cape Province, by Macdonald, 1957, Contrib. Orn. W. South Africa, Brit. Mus. (Nat. Hist.), p. 129.

Southern Cape Province north to northern Karroo and Little Namaqualand, Basutoland, Natal, Zululand, Swaziland, and southern Mozambique.

Cercomela familiaris hellmayri (Reichenow)

Saxicola familiaris hellmayri Reichenow, 1902, Orn. Monatsb., 10, p. 78 — south bank of Limpopo River, north of Pietersburg, Transvaal.

Northern Orange Free State, Transvaal, eastern Bechuanaland, and southern parts of Southern Rhodesia.

Cercomela familiaris modesta (Shelley)

Bessornis modesta Shelley, 1897, Ibis, p. 539, pl. 12, fig. 1 — Karonga, northern Nyasaland.

Low country of Zambezi Valley, Southern Rhodesia, Nyasaland, northern Mozambique, Northern Rhodesia, and northeastern Angola north to Katanga, Tanganyika, and Uganda.

Cercomela familiaris omoensis (Neumann)

Saxicola galtoni omoensis Neumann, 1904, Orn. Monatsb., 12, p. 163 — Baka, Omo Territory, southern Ethiopia.

Boma Hills of southeast Sudan and southwest Ethiopia.

CERCOMELA TRACTRAC

Cercomela tractrac hoeschi (Niethammer)

Oenanthe tractrac hoeschi Niethammer, 1955, Bonn. Zool. Beitr., 6, p. 188 — Koako-Namib, west of Orupembe.

South West Africa in the Kaokoveld and Namib, and the coastal desert of southern Angola.

Cercomela tractrac albicans (Wahlberg)

Saxicola albicans Wahlberg, 1855, öfv. K. Sv. Vet.-Akad. Förh., p. 213 — "In locis arenosis terrae Damararum" [= Walvis Bay].

South West Africa from Cape Cross to about Luderitz in the coastal Namib.

Cercomela tractrac barlowi (Roberts)

Phoenicuroides tractrac barlowi Roberts, 1937, Ostrich, 8, p. 102 — Aus, South West Africa.

South West Africa in western Great Namaqualand.

Cercomela tractrac nebulosa Clancey

Cercomela tractrac nebulosa Clancey, 1962, Durban Mus. Novit. 6, p. 186 — white coastal sand-dunes at Mc-Dougall Bay, south of Port Nolloth, Little Namaqualand, northwestern Cape Province.

South West Africa on the white coastal sand-dunes of the arid coast of Little Namaqualand, from some miles to the south of Port Nolloth to the mouth of the lower Orange River, and on the white coastal dunes of southwestern Great Namaqualand.

Cercomela tractrac tractrac (Wilkes)

Motacilla tractrac Wilkes, 1817, Encycl. Londinensis, 16, p. 89 — Auteniquois country, ex Levaillant [probably Orange River, *vide* Sclater, 1930, Syst. Av. Aethiop., 2, p. 456].

Saxicola layardi Sharpe, 1876, in Layard, Birds S. Afr., ed. 2, p. 236 — South Africa.

Little Namaqualand (except for range of *nebulosa*) and eastern Great Namaqualand south to Olifants River and east to Aliwal North.

CERCOMELA SCHLEGELII

Cercomela schlegelii benguellensis (Sclater)

Karrucincla schlegelii benguellensis Sclater, 1928, Bull. Brit. Orn. Club, 49, p. 14 — Huxe [= Uchi], Benguella.

Southern Angola.

Cercomela schlegelii schlegelii (Wahlberg)

Erithacus schlegelii Wahlberg, 1855, öfv. K. Sv. Vet.-Akad. Förh., 12, p. 213 — Damaraland.

Namib coastal areas east to Erongo Mountains, Otjimbinque, and Onanis, Damaraland, South West Africa.

Cercomela schlegelii namaquensis (Sclater)

Karrucincla schlegelii namaquensis Sclater, 1928, Bull. Brit. Orn. Club, 49, p. 15 — Great Namaqualand; restricted to Bethanie, Great Namaqualand by Macdonald,

1957, Contr. Orn. South Africa. Brit. Mus. (Nat. Hist.), p. 128.
Little Namaqualand, South Africa and southern Great Namaqualand, South West Africa.

Cercomela schlegelii kobosensis (Roberts)

Karrucincla schlegelii kobosensis Roberts, 1937, Ostrich, 8, p. 102 — Kobos, Rehoboth District, South West Africa.
Central Great Namaqualand, South West Africa.

Cercomela schlegelii pollux (Hartlaub)

Saxicola pollux Hartlaub, 1866, Proc. Zool. Soc. London, (1865), p. 747 — Karroo [= Traka, Willowmore Dist., Cape Province, *vide* Sharpe, 1877, in Layard, Birds S. Afr., ed. 2, p. 244].
Cape Province, Karroo areas from Little Namaqualand east to Orange Free State and Griqualand West.

CERCOMELA FUSCA

Cercomela fusca (Blyth)

Saxicola fusca Blyth, 1851, Journ. Asiat. Soc. Bengal, 20, p. 523 — Muttra.
Cercomela fusca ruinarum Koelz, 1939, Proc. Biol. Soc. Washington, 52, p. 66 — Sanchi, Bhopal State.
West Pakistan in West Punjab, India from East Punjab south to Delhi, Rajasthan, Kutch, northern Gujarat, and central India east to western West Bengal.

CERCOMELA DUBIA

Cercomela dubia (Blundell and Lovat)

Myrmecocichla dubia Blundell and Lovat, 1899, Bull. Brit. Orn. Club, 10, p. 22 — Fontaly, Abyssinia.
Cercomela scotocerca enigma Neumann and Zedlitz, 1913, Journ. f. Orn., 61, p. 368 — Dire Dawa, Harar Province, Ethiopia.
Central Ethiopia east to Somalia.

CERCOMELA MELANURA

Cercomela melanura melanura (Temminck)

Saxicola melanura Temminck, 1824, Pl. Col., livr. 43, pl.

257, fig. 2 — Arabia [= Sinai Peninsula, *vide* Hartert, 1922, Vög. pal. Fauna, 3, p. 2165].

From Dead Sea basin of Israel and Jordan through the Egyptian Sinai Peninsula to Riyadh and Medina, Saudi Arabia.

Cercomela melanura neumanni Ripley

C.(ercomela) m.(elanura) erlangeri Neumann and Zedlitz, 1913, Journ. f. Orn., 61, p. 364 — Khareba, southern Arabia.

Cercomela melanura neumanni Ripley, 1952, Postilla, Yale Univ., no. 13, p. 31, new name for *Cercomela sordida erlangeri* Neumann and Zedlitz, 1913, *nec Pinarochroa sordida erlangeri* Reichenow, 1905, in *Cercomela*.

Saudi Arabia in the Taif and Mecca areas, intergrading north of these areas into the preceding form, south to Yemen and Aden and east into the Hadramaut.

Cercomela melanura lypura (Hemprich and Ehrenberg)

Sylvia lypura Hemprich and Ehrenberg, 1833, Symb. Phys. Av., fol. ee — Abyssinia [= eastern Eritrea, *vide* Neumann and Zedlitz, 1913, Journ. f. Orn., 61, p. 365].

Western coastal areas of the Red Sea from Gebel Elba in Egypt south in the Sudan to Eritrea.

Cercomela melanura aussae Thesiger and Meynell

Cercomela melanura aussae Thesiger and Meynell, 1934, Bull. Brit. Orn. Club., 55, p. 79 — Aussa Danokil.

Danokil country of eastern Ethiopia and adajcent French Somaliland.

Cercomela melanura airensis Hartert

Cercomela melanura airensis Hartert, 1921, Novit. Zool., 28, p. 114 — Mount Baguezan, Aïr [= Asben, French West Africa].

Aïr Mountains of southern Sahara in upper Niger east to Ennedi in Chad and Darfur in Sudan.

Cercomela melanura ultima Bates

Cercomela melanura ultima Bates, 1933, Bull. Brit. Orn. Club, 53, p. 175 — Niger, near Gao.

From the Niger River, Bourem, in the eastern Mali Federation east to southern Niger.

CERCOMELA SCOTOCERCA

Cercomela scotocerca scotocerca (Heuglin)

Saxicola scotocerca Heuglin, 1869, Orn. Nordost. Afr., 1, p. 363 — near Keren, Bogosland.

Sudan in Red Sea coastal area south to Eritrea.

Cercomela scotocerca furensis Lynes

C.(ercomela) s.(cotocerca) furensis Lynes, 1926, Ibis, p. 391 — Jebel Marra, 7,300 feet, central Darfur.

Darfur, western Sudan.

Cercomela scotocerca turkana van Someren

Cercomela turkana van Someren, 1920, Bull. Brit. Orn. Club, 40, p. 91 — Turkana country, west of Lake Rudolf.

Southwest Ethiopia and northwest Kenya.

Cercomela scotocerca spectatrix Clarke

Cercomela spectatrix Clarke, 1919, Bull. Brit. Orn. Club, 40, p. 49 — 10 miles inland from Las Khorai in eastern British Somaliland.

Somalia.

CERCOMELA SORDIDA

Cercomela sordida sordida (Rüppell)

Saxicola sordida Rüppell, 1837, Neue Wirbelt., Vögel, p. 75, pl. 26, fig. 2 — Simen, Abyssinia.

Pinarochroa sordida erlangeri Reichenow, 1905, Orn. Monatsb., 13, p. 25 — Gara-Mulata [, near Harrar].

Pinarochroa sordida schoana Neumann, 1905, Orn. Monatsb., 13, p. 78 — Abuje, Gindeberet, Shoa.

Pinarochroa sordida djamdjamensis Neumann, 1905, Orn. Monatsb., 13, p. 79 — Abera, Djamdjam.

Highlands of Ethiopia.

Cercomela sordida rudolfi (Madarász)

Pinarochroa rudolfi Madarász, 1912, Orn. Monatsb., 20, p. 175 — Mount Elgon.

Mount Elgon, Kenya.

Cercomela sordida ernesti (Sharpe)

Pinarochroa ernesti Sharpe, 1900, Bull. Brit. Orn. Club, 10, p. 36 — Mount Kenya.

Mount Kenya, Aberdare range and Kinangop Mountains, Kenya.

Cercomela sordida olimotiensis (Elliott)

Pinarochroa sordida olimotiensis Elliott, 1945, Bull. Brit. Orn. Club, 66, p. 19 — Olimoti Mountain, 9,000 feet, north of Ngorongoro, northeastern Tanganyika.

Highlands of northern Tanganyika between long. 35° E. and long. 36° E., north of Lakes Eyasi and Manyara.

Cercomela sordida hypospodia (Shelley)

Pinarochroa hypospodia Shelley, 1885, Proc. Zool. Soc. London, p. 226, pl. 13 — Mount Kilimanjaro.

Mount Kilimanjaro, Tanganyika.

GENUS SAXICOLA BECHSTEIN

Saxicola Bechstein, 1803, Orn. Taschenb., (1802), p. 216. Type, by subsequent designation (Swainson, 1827, Zool. Journ., 3, p. 172), *Motacilla rubicola* Linnaeus.

Pratincola Koch, 1816, Syst. baier. Zool., 1, p. 190. Type, by original designation, *Motacilla rubicola* Linnaeus, nec *Pratincola* Forster, 1795.

Rhodophila Jerdon, 1863, Birds India, 2, p. 128. Type, by monotypy, *Rhodophila melanoleuca* Jerdon = *Oreicola jerdoni* Blyth.

cf. Stegmann, 1935, Doklady Akad. Nauk. S.S.R., n.s., 3, pp. 45-48 (*torquata*).

Vaurie, 1955, Amer. Mus. Novit., no. 1731, pp. 24-27 (*rubetra; ferrea*).

SAXICOLA RUBETRA

Saxicola rubetra (Linnaeus)

Motacilla Rubetra Linnaeus, 1758, Syst. Nat., ed. 10, 1, p. 186 — Europa [= Sweden].

Pratincola rubetra spatzi Erlanger, 1900, Journ. f. Orn., 48, p. 101 — Gafsa Oasis, Tunisia.

Pratincola rubetra noskae Tschudi, 1902, Orn. Jahrb., p. 234 — Labathal, northern Caucasus.

Pratincola rubetra margaretae Johansen, 1903, Orn. Jahrb., p. 232 — Tomsk, western Siberia.

Saxicola rubetra incerta Trischitta, 1939, Arti grafiche "Solunto" Bagheria, p. 1 — Sardinia and Sicily.

Saxicola rubetra hesperophila Clancey, 1949, Limosa, 22, p. 370 — Newton Mearns, East Renfrewshire, southwestern Scotland.

Breeds in Scandinavia, Finland, Russia to lat. 68° N., east in Siberia to about Krasnoyarsk, Yenisei River, and south in Great Britain, northern Ireland and France to northern Spain and Portugal, North Africa in Morocco (?) and Algeria (?), Corsica, northern Italy, Yugoslavia, Albania, and northern Greece east across Urals to the Altai and south to Kirghizstan, Caucasus, and northern Iran. Migrates south through Mediterranean region, Near East, Iraq, Iran, and Arabia to Africa; winters south to western Equatorial Africa, Cameroons, Congo, Tanganyika, and Nyasaland.

SAXICOLA MACRORHYNCHA

Saxicola macrorhyncha (Stoliczka)

Pratincola macrorhyncha Stoliczka, 1872, Journ. Asiat. Soc. Bengal, 41, p. 238 — near Rápúr, Wagur District and near Bhúj, Kutch.

Southern Afghanistan; West Pakistan and India in northern Baluchistan, Sind, West and East Punjab, south to Ambala, the Salt Range, and Rajasthan.

SAXICOLA INSIGNIS

Saxicola insignis Gray

Saxicola insignis Gray, 1846, Cat. Mamm. Birds Nepal Thibet, pp. 71, 153 — Nepal.

Breeds in Russia in the mountains of Kazakhstan, Altai, and in Tannu-Tuva and Outer Mongolia, western Khangai; migrates through western China in Tsinghai, northern Sikang, and Tibet; winters in Nepal, in India from East Punjab to Uttar Pradesh, Bihar, and West Bengal, in Sikkim, and in Bhutan.

SAXICOLA DACOTIAE

Saxicola dacotiae dacotiae (Meade-Waldo)

Pratincola dacotiae Meade-Waldo, 1889, Ibis, p. 504, pl. 15 — Fuerteventura, Mauritanice Dacos.

Fuerteventura Island, Canary Islands.

Saxicola dacotiae murielae Bannerman

Saxicola dacotiae murielae Bannerman, 1913, Bull. Brit. Orn. Club, 33, p. 37 — Allegranza.

Montaña Clara and Allegranza Islands, Canary Islands.

SAXICOLA TORQUATA

Saxicola torquata hibernans (Hartert)

Pratincola torquata hibernans Hartert, 1910, Journ. f. Orn., 58, p. 173 — Tring, England.

Saxicola torquata theresae Meinertzhagen, 1934, Ibis, p. 56 — Grogarry, South Uist, Outer Hebrides.

Breeds in the British Isles including Ireland, Orkneys, and Shetlands (rare); western Brittany and coastal Portugal; vagrant to Faeroes and southern Spain.

Saxicola torquata rubicola (Linnaeus)

(Motacilla) Rubicola Linnaeus, 1766, Syst. Nat., ed. 12, p. 332 — Europae [= France]; restricted to Seine Inférieure by Meinertzhagen, 1940, Ibis, p. 215.

Pratincola torquata insularis Parrot, 1910, Orn. Monatsb., 18, p. 155 — Corsica.

Saxicola torquata desfontainesi Blanchet, 1925, Rev. Franç. Orn., 9, p. 277 — Oued Bezirk (Cap Bon) and Hammam-Lif, Tunisia.

Saxicola torquata graecorum Laubmann, 1927, Verh. Orn. Ges. Bayern, 17, p. 351 — Corfu.

Saxicola torquata amaliae Buturlin, 1929, Syst. Notes Birds North. Caucasus, p. 16 — near Vladikavkaz.

Saxicola torquata gabrielae Neumann and Paludan, 1937, Orn. Monatsb., 45, p. 15 — Olymp(us), south of Brussa [= Bursa], Turkey.

Saxicola torquata archimedes Clancey, 1949, Bull. Brit. Orn. Club, 69, p. 84 — near Syracuse, Sicily.

Breeds in western Europe from southern Denmark east to southern Poland and southern Russia, Crimea, and Ukraine southward (except in range of *hibernans*) to Mediterranean and islands, except Malta and Cyprus (?), northwest Africa and east in Turkey and the Near East into Transcaucasia; winters in southern range to northern Libya, Egypt, Israel, Lebanon, Arabia, Iraq, and Ethiopia.

Saxicola torquata variegata (Gmelin)

Parus Variegatus S. G. Gmelin, 1774, Reise Russl., 3, p. 105, pl. 20, fig. 3 — Shemakha.

Breeds from eastern Caucasus to northern Caspian; winters south in Transcaucasia to Near East, Israel, Egypt, Eritrea, Sudan, Somaliland, Ethiopia, southwest Arabia, and Iraq.

Saxicola torquata armenica Stegmann

Saxicola torquata armenica Stegmann, 1935, Doklady Akad. Nauk. S.S.R., n.s., 3, p. 47 — Adshafana, Kurdistan.

Saxicola torquata excubitor Koelz, 1954, Contrib. Inst. Regional Explor., no. 1, p. 13 — Dorud, Luristan, Iran.

Breeds from Armenia to Iran south to northern Iraq and Syria (?) ; winters from the southern part of range through Egypt, Sudan, northern Ethiopia, and Arabia to Yemen and Aden.

Saxicola torquata maura (Pallas)

Muscicapa maura Pallas, 1773, Reise versch. Russ. Reichs, 2, p. 428 — Karassum [, Ishim River, western Siberia].

Breeds in eastern Russia and Siberia east of the Urals north to the Ob and Yenisei and to Irkutsk south from western Siberia to the Altai, Sinkiang mountains, northwestern Mongolia, Pamirs, northern Afghanistan, southern Transcaspia, northeastern Iran; winters in southern breeding range, Iran, Iraq, Afghanistan, West Pakistan, and northern India south to Bombay.

Saxicola torquata indica (Blyth)

Pr.(atincola) indica Blyth, 1847, Journ. Asiat. Soc. Bengal, 16, p. 129 — India; restricted to Kashmir by Baker, 1921, Journ. Bombay Nat. Hist. Soc., 27, p. 709.

Breeds in the Himalayas from Gilgit and Kashmir east to Nepal, Sikkim, and Assam; winters southward in central India to Hyderabad and Mysore; Andaman Islands (?).

Saxicola torquata przewalskii (Pleske)

Pratincola maura var. *Przewalskii* Pleske, 1889, Wiss. Result. Przewalski Reise, Zool., 2, Vögel, p. 46, pl. 4, figs. 1, 2, 3 — mountains of Kansu and eastern Turkestan.

Pratincola torquata yunnanensis La Touche, 1923, Bull. Brit. Orn. Club, 43, p. 134 — Shuitang and Mengtsz, southeastern Yunnan.

Breeds in China from Tsinghai and Kansu to Shensi south in Tibet, Sikang, Yunnan, and Hopeh, Kweichow, northern Vietnam, and northern Burma (?); winters in the Himalayas in Nepal, Sikkim, and Assam and on the adjacent Indian plains east to the north Burma foothills.

Saxicola torquata stejnegeri (Parrot)

Pr.(atincola) rubicola stejnegeri Parrot, 1908, Verh. Orn. Ges. Bayern, 8 (1907), p. 124 — Iterup (Etorofu) and Jesso (Hakodate), northern Japan.
Saxicola torquata kleinschmidti Meise, 1934, Abh. Ber. Mus. Dresden, 18, p. 44 — Kwanhsien, Szechwan.
Saxicola torquata delacouri David-Beaulieu, 1944, Oiseaux Tranninh, Publ. Ecole Supérieure Sciences, Univ. Indochinoise, A., p. 106 — Xieng Khoung.

Breeds in eastern Siberia from the Yenisei, Lake Baikal, and east of Khangai to the Kolyma Delta, south in Manchuria, Korea, Sakhalin, Kuriles, Hokkaido, central and northern Honshu, and Izu Islands; migrates to southern parts of breeding range and to southeastern China, eastern Assam, Burma, Thailand, northern Vietnam, Laos, Taiwan, and Riu Kiu Islands; Malaya (uncommon).

Saxicola torquata felix Bates

Saxicola torquata felix Bates, 1936, Bull. Brit. Orn. Club, 57, p. 20 — Menacha [= Manakha], Yemen.

In the mountains from southwestern Arabia south through Yemen.

Saxicola torquata albofasciata Rüppell

Saxicola albofasciata Rüppell, 1845, Syst. Uebers. Vög. Nord-ost.-Afr., p. 39 — Simen Province, Abyssinia.

Highlands of Ethiopia; southeast Sudan (breeding ?).

Saxicola torquata jebelmarrae Lynes

Saxicola torquata jebelmarrae Lynes, 1920, Bull. Brit. Orn. Club, 41, p. 17 — Jebel Marra, Darfur.

Jebel Marra, Darfur, western Sudan.

Saxicola torquata moptana Bates

Saxicola torquata moptana Bates, 1932, Bull. Brit. Orn. Club, 53, p. 8 — Mopti, French Sudan.
Upper Niger River at Mopti, Mali Republic.

Saxicola torquata nebularum Bates

Saxicola torquata nebularum Bates, 1930, Bull. Brit. Orn. Club, 51, p. 51 — near Birwa Peak, Kono District, Sierra Leone.
Mountains of Sierra Leone and Ivory Coast.

Saxicola torquata adamauae Grote

Saxicola torquata adamauae Grote, 1922, Journ. f. Orn., 70, p. 486 — Genderu Mountains.
Mountains of northern and western Cameroons.

Saxicola torquata pallidigula (Reichenow)

Pratincola pallidigula Reichenow, 1892, Journ. f. Orn., 40, p. 194 — Kamerun.
Cameroon Mountain and Fernando Po Island (subspecies ?).

Saxicola torquata axillaris (Shelley)

Pratincola axillaris Shelley, 1884, Proc. Zool. Soc. London, p. 556 — Kilimanjaro.
Pratincola emmae Hartlaub, 1890, Journ. f. Orn., 38, p. 152 — Ruganda [, Ankole, Uganda].
Eastern Congo, Uganda, Kenya, and Tanganyika, except in the range of the following subspecies.

Saxicola torquata promiscua Hartert

Saxicola torquata promiscua Hartert, 1922, Bull. Brit. Orn. Club, 42, p. 51 — Uluguru Mountains.
Tanganyika from Inpapwa to Kilosa and Uluguru Mountains.

Saxicola torquata salax (J. and E. Verreaux)

Pratincola salax J. and E. Verreaux, 1851, Rev. Mag. Zool. [Paris], 3 (2), p. 307 — Gabon.
Grasslands of middle Cameroun, Gabon south to mouth of Congo River, adjacent northern Angola in Congo District and Cuanza Norte, Congo in lower Kasai, Bateke Pla-

teau, and Bolobo, intergrading with the next form south of
this area; vagrant to Victoria Falls and Grassland, Felix-
burg, Southern Rhodesia.

Saxicola torquata stonei Bowen

Saxicola torquata stonei Bowen, 1932, Proc. Acad. Nat.
Sci. Philadelphia, 83 (1931), p. 8 — Villa General
Machado, Angola.

Interior Angola from about the Cuanza River, southeast
Congo in Katanga and Marungu, Northern Rhodesia, Nyasaland, Mount Kungwe, western Tanganyika, northern
Mozambique, south in Southern Rhodesia, except extreme
east border, Caprivi Strip, northeast South West Africa,
northern and eastern Bechuanaland, including Ngamiland,
to western and northern Transvaal, northern and eastern
Karoo regions of the Cape Province, and western Orange
Free State; vagrant in lowlands of Basutoland; nonbreeding
visitor to southern Mozambique, Swaziland, and Pondoland.

Saxicola torquata clanceyi Latimer

Saxicola torquata clanceyi Latimer, 1961 (March), in
Clancey, Durban Mus. Novit., 6, p. 90 — Wallekraal,
western Little Namaqualand, northwestern Cape Province, South Africa.

Saxicola torquata clanceyi Courtenay-Latimer, 1961
(Oct.), Bull. Brit. Orn. Club, 81, p. 116 — Wallekraal,
western Little Namaqualand, northwestern Cape Province.

Coastal district of Little Namaqualand, northwest Cape
Province from Alexander Bay to Strandfontein and Lam-
berts Bay; vagrant to Muizenberg.

Saxicola torquata torquata (Linnaeus)

(*Motacilla*) *torquata* Linnaeus, 1766, Syst. Nat., ed. 12,
1, p. 328 — Cape of Good Hope.

Pratincola caffra Keyserling and Blasius, 1840, Die Wirbelt. Europas, p. 59 — Uitenhage, eastern Cape Province.

Pratincola robusta Tristram, 1870, Ibis, p. 497 — "Mysore" [= Natal, *vide* Meinertzhagen, 1954, Birds Arabia,
p. 260].

Pratincola torquata orientalis Sclater, 1911, Ibis, p. 409
— Umfolozi Station, Zululand.

Southwestern Cape Province, coastal districts of southern
and eastern Cape northeast to Natal and Zululand, western
Swaziland, eastern Orange Free State, and highveld of
Transvaal.

Saxicola torquata oreobates Clancey

Saxicola torquata oreobates Clancey, 1956, Durban Mus.
Novit., 4, p. 281 — 40 miles east of Maseru on new
mountain road, *ca*. 8,000 ft., Basutoland.

Highlands of Basutoland and adjacent highlands of extreme eastern Cape Province, Natal and Orange Free State;
high elevations in the eastern highlands of Southern Rhodesia, Melsetter and Inyanga; nonbreeding wanderer to
Natal, Zululand, Swaziland, eastern Transval, and southern
Mozambique.

Saxicola torquata sibilla (Linnaeus)

Motacilla Sibilla Linnaeus, 1766, Syst. Nat., ed. 12, 1, p.
337 — Madagascar.
Saxicola torquata ankaratrae Salomonsen, 1934, Novit.
Zool., 39, p. 210 — Manjakatompo, Ankaratra Moun-
tains, central Madagascar.
Saxicola torquata Tsaratananae Milon, 1951, Bull. Mus.
Hist. Nat. Paris, ser. 2, 22, p. 706 — "Mont Tsara-
tanana," northern Madagascar.

Madagascar.

Saxicola torquata voeltzkowi Grote

Saxicola torquata voeltzkowi Grote, 1926, Orn. Monatsb.,
34, p. 146 — La Convalescence, 1,800 m., Grand Comoro.

Grand Comoro Island, above 500 meters.

Saxicola torquata tectes (Gmelin)

(*Muscicapa*) *tectes* Gmelin, 1789, Syst. Nat., 1 (2), 940 —
in insula Bourbon, *ex* Brisson, 1760, Ornith., 2, p. 360.
Saxicola borbonensis W. L. Sclater, 1928, Bull. Brit. Orn.
Club, 49, p. 14, new name for *Pratincola borbonica*
Sharpe, 1879, *nec Motacilla borbonica* Gmelin, 1789,
and *M. borbonica* Bory de St. Vincent, 1804 — Bourbon.

Réunion Island.

SAXICOLA LEUCURA

Saxicola leucura (Blyth)

> *Pratincola leucura* Blyth, 1847, Journ. Asiat. Soc. Bengal, 16, p. 474 — upper Sind.

West Pakistan in Sind and West Punjab; India in the Himalayan terai from Garhwal and Kumaon, through Nepal, to Assam and Manipur; recorded from Orissa in winter.

SAXICOLA CAPRATA

Saxicola caprata rossorum (Hartert)

> *Pratincola caprata rossorum* Hartert, 1910, Journ. f. Orn., 58, p. 180 — Merv, in Transcaspia.

Breeds in Transcaspia north to Syr Daria, eastern Iran, Afghanistan, intergrading with the following form in northern Baluchistan and the Mekran, and in northern Kashmir; winters in southern Iran, Iraq (straggler), southern Baluchistan, and adjacent plains of northern West Pakistan.

Saxicola caprata bicolor Sykes

> *Saxicola bicolor* Sykes, 1832, Proc. Zool. Soc. London, p. 92 — Dukhun.
> *Saxicola caprata rupchandi* Koelz, 1939, Proc. Biol. Soc. Washington, 52, p. 65 — Londa, Bombay Presidency.

Breeds in West Pakistan and India where it intergrades with *rossorum* in Baluchistan and Kashmir; North West Frontier Province, Sind, Punjab, Himachal Pradesh south to Delhi, east to Nepal and northern Bengal; winters south in Rajasthan and Kutch to Bombay, Hyderabad and northern Mysore (uncommon).

Saxicola caprata burmanica Baker

> *Saxicola caprata burmanica* Baker, 1923, Bull. Brit. Orn. Club, 43, p. 19 — Pegu.

Breeds south of the preceding subspecies from Gujarat east to Bengal, Assam, and East Pakistan south to Hyderabad, Orissa, Andhra Pradesh, Madras, Goa, and Mysore (where it grades into the next form in western Mysore), and in Burma (except Tenasserim), western Yunnan (winter), northern Thailand, Laos, Vietnam, and Cambodia.

Saxicola caprata nilgiriensis Whistler

Saxicola caprata nilgiriensis Whistler, 1940, Bull. Brit.
Orn. Club, 60, p. 90 — Ootacamund, Madras.
Western Madras and Kerala in the Nilgiri, Palni, and
Travancore ranges.

Saxicola caprata atrata (Kelaart)

Pratincola atrata Kelaart, 1851, in Blyth, Journ. Asiat.
Soc. Bengal, 20, p. 177 — Newera Elia, Ceylon.
Ceylon hills.

Saxicola caprata caprata (Linnaeus)

Motacilla Caprata Linnaeus, 1766, Syst. Nat., ed. 12, 1,
p. 335 — Luzon.
Northern Philippines on Luzon, Mindoro, and Cebu; Borneo (one old record, subsp. ?).

Saxicola caprata randi Parkes

Saxicola caprata randi Parkes, 1960, Proc. Biol. Soc.
Washington, 73, p. 59 — Bondo, Siaton, Negros, Phil-
ippine Islands.
Negros, Cebu, Bohol, and Siguijor Islands, Philippines.

Saxicola caprata anderseni Salomonsen

Saxicola caprata anderseni Salomonsen, 1953, Vidensk.
Medd. Dansk naturhist. Foren., 115, p. 260 — Del
Monte, Bukidnon Province, central Mindanao.
Central Mindanao, Philippines.

Saxicola caprata fruticola Horsfield

Saxicola fruticola Horsfield, 1821, Trans. Linn. Soc. Lon-
don, 13, p. 157 — Java.
Java, Bali, Lombok, Sumbawa, Flores, Lomblen, and Alor
Islands, Indonesia.

Saxicola caprata pyrrhonota (Vieillot)

Oenanthe pyrrhonota Vieillot, 1818, Nouv. Dict. Hist.
Nat., nouv. éd., 21, p. 428 — New Holland [*errore* = Ti-
mor].
Savu, Timor, Kisser, and Wetar Islands, Indonesia.

Saxicola caprata francki Rensch

Saxicola caprata francki Rensch, 1931, Treubia, 13, p.
380 — Laora.
Sumba Island, Indonesia.

Saxicola caprata albonotata (Stresemann)

Pratincola caprata albonotata Stresemann, 1912, Novit.
   Zool., 19, p. 321 — Indrulaman, Celebes.
Celebes (Sulawesi), Buton, and Saleyer Islands, Indonesia.

Saxicola caprata cognata Mayr

Saxicola caprata cognata Mayr, 1944, Bull. Amer. Mus.
   Nat. Hist., 83, p. 156 — Tepa, Babar Island.
Babar Islands, Indonesia.

Saxicola caprata aethiops (Sclater)

Poecilodryas aethiops Sclater, 1880, Proc. Zool. Soc. London, p. 66, pl. 7, fig. 1 — Kabakadai, New Britain.
Northern New Guinea lowlands from Humboldt Bay and
Sentani Lake to the Huon Peninsula; New Britain Island.

Saxicola caprata belensis Rand

Saxicola caprata belensis Rand, 1940, Amer. Mus. Novit.,
   no. 1072, p. 4 — Balim River, 1,600 m., Snow Mountains, Netherland New Guinea.
Central Mountains of New Guinea from Wissel Lakes
through the Nassau, Orange, and Snow Mountains.

Saxicola caprata wahgiensis Mayr and Gilliard

Saxicola caprata wahgiensis Mayr and Gilliard, 1951,
   Amer. Mus. Novit., no. 1524, p. 8 — Mafulu, Central
   Division, Papua, New Guinea.
Central mountains of Papua from Nondugl and the Wahgi
Valley, Giluwe, east to Huon Peninsula and southeastern
mountains.

SAXICOLA JERDONI

Saxicola jerdoni (Blyth)

Oreicola jerdoni Blyth, 1867, Ibis, p. 14, new name for
   Rhodophila melanoleuca Jerdon, 1863, Birds India, 2,
   p. 128 — Purneah, *nec Ocnanthe melanoleuca = Saxicola gutturalis* Vieillot, 1818, Nouv. Dict. Hist. Nat.,
   nouv. éd., 21, p. 435.
Eastern India, East Pakistan, Burma, Yunnan (one record), northeastern Laos, and northern Vietnam.

SAXICOLA FERREA

Saxicola ferrea Gray

Saxicola ferrea Gray, 1846, Cat. Mamm. Birds Nepal, Thibet, pp. 71, 153 — Nepal.

Oreicola ferrea haringtoni Hartert, 1910, Vög. pal. Fauna, 1, p. 711 — Lien-kiang, near Fu-tschau [Foochow], China.

Breeds from the Afghan-West Pakistan border in North West Frontier Province east to Kashmir and along the Himalayas of India, Nepal, and southeast Tibet to Sikang, Yunnan, western Szechwan, northern Burma, northern Laos and Vietnam, and hills of southern China to the Yangtze; winters from southern parts of breeding range south to the adjacent plains of West Pakistan, India, East Pakistan, central Burma, northern Thailand, and southern Laos and Vietnam; Formosa (accidental).

SAXICOLA GUTTURALIS

Saxicola gutturalis gutturalis (Vieillot)

Oenanthe gutturalis Vieillot, 1818, Nouv. Dict. Hist. Nat., 21, p. 421 — New Holland [*errore* = Timor].

Timor Island.

Saxicola gutturalis luctuosa (Bonaparte)

S.(axicola) luctuosa Bonaparte, 1851, Consp. Av., 1 (1850), p. 304 — Semau.

Semau Island, west of Timor.

Genus MYRMECOCICHLA Cabanis

Myrmecocichla Cabanis, 1850, Mus. Hein., 1, p. 8. Type, by subsequent designation (Gray, 1855, Cat. Gen. Birds Brit. Mus., p. 35), *Oenanthe formicivora* Vieillot.

Pentholaea Cabanis, 1850, Mus. Hein., 1, p. 40. Type, by subsequent designation (Gray, 1855, Cat. Gen. Birds Brit. Mus., p. 36), *Saxicola frontalis* Swainson.

Sciocincla Roberts, 1922, Ann. Transvaal Mus., 8, p. 232. Type, by original designation and monotypy, *Saxicola arnotti* Tristram.

MYRMECOCICHLA THOLLONI

Myrmecocichla tholloni (Oustalet)

Saxicola Tholloni Oustalet, 1886, Naturaliste, p. 300 — Lékéti, on the Alima River, French Congo.

Oenanthe chaboti Menegaux and Berlioz, 1923, Mission Rohan-Chabot Angola Rhodesia, Hist. Nat., Oiseaux, 4, p. 139, pl. 4 — Lwasinga River, Kubango District, Angola.

Myrmecocichla lynesi Bannerman, 1927, Bull. Brit. Orn. Club, 47, p. 147 — Huambo, Benguella Province, Angola.

Interior Gabon and south in Angola highlands to Lwasingwa River, near lat. 15° S.

MYRMECOCICHLA AETHIOPS

Myrmecocichla aethiops aethiops Cabanis

Myrmecocichla aethiops Cabanis, 1850, Mus. Hein., 1, p. 8 — Senegal.

Myrmecocichla buchanani Rothschild, 1920, Bull. Brit. Orn. Club, 41, p. 33 — Takoukout, Damergou, French Sudan.

Senegal, Mali, and Niger south to northern Nigeria (Sokoto and Plateau Provinces) and northern Cameroons and northeast to Chad (Mongonu).

Myrmecocichla aethiops sudanensis Lynes

Myrmecocichla aethiops sudanensis Lynes, 1920, Bull. Brit. Orn. Club, 41, p. 18 — El Fasher, Darfur.

Darfur and Kordofan (uncommon), Sudan.

Myrmecocichla aethiops cryptoleuca Sharpe

Myrmecocichla cryptoleuca Sharpe, 1891, Ibis, p. 445 — Kikuyu.

Kenya from Mount Elgon and Suk country south to Nakuru and Kikuyu, and the highlands of northern Tanganyika.

MYRMECOCICHLA FORMICIVORA

Myrmecocichla formicivora (Vieillot)

Oenanthe formicivora Vieillot, 1818, Nouv. Dict. Hist. Nat., nouv. éd., 21, p. 421 — Pays des Cafres = Sunday's

River, eastern Cape Province, *ex* Levaillant, 1805, Oiseaux Afrique, 4, pls. 186, 187.
Myrmecocichla formicivora minor Roberts, 1932, Ann.
Transvaal Mus., 15, p. 30 — Gemsbok Pan.
Myrmecocichla formicivora orestes Clancey, 1961, Durban Mus. Novit., 6, p. 85 — Wakkerstrom, *ca.* 6,500 ft.,
southeastern Transvaal-Natal boundary.
South West Africa from Ovamboland to Great Namaqualand, south in Cape Province and east in Bechuanaland
to extreme Southern Rhodesia (Wankie, once), across the
Karroo and the highveld of Orange Free State to Transvaal, Basutoland, western Swaziland, and the high interior
of Natal.

MYRMECOCICHLA NIGRA

Myrmecocichla nigra (Vieillot)
Oenanthe nigra Vieillot, 1818, Nouv. Dict. Hist. Nat.,
nouv. éd., 21, p. 431 — west coast of Africa [= Malymbe,
i.e. Malimba, Portuguese Congo, *vide* Sclater, 1930,
Syst. Av. Aethiop., p. 466].
Myrmecocichla Levaillanti Reichenow, 1882, in Hartlaub,
Abh. Nat. Ver. Bremen, 8, p. 188 — Langomeri.
Myrmecocichla stoehri Sclater, 1941, in Roberts, Ostrich,
11, p. 116 — Lavusi (Lavunsi), Serenje, near Mpika,
Northern Rhodesia; *vide* Sclater, 1906, Jour. S. African
Orn. Un., 11, pp. 98-9.
Northern Nigeria, Cameroons highlands south to Portu-
guese Congo, Kasai, and northern Angola across the Congo
in highlands to Uganda, Sudan (Equatoria), southwestern
Kenya, and western Tanganyika, and south in Katanga to
Northern Rhodesia.

MYRMECOCICHLA ARNOTTI

Myrmecocichla arnotti leucolaema Fischer and Reichenow
Myrmecocichla leucolaema Fischer and Reichenow, 1880,
Orn. Centralbl., p. 181 — Nguru Mountains [reference
not verified].
Myrmecocichla nigra var. *collaris* Reichenow, 1882, Jour.
f. Orn., 30, p. 212; 1905, Vög. Afr., 3, p. 707 — Kakoma,
vide Neunzig, 1926, Jour. f. Orn., 74, p. 754.
Southeastern Congo, Marungu, and Katanga north to

Ruanda and east to Tanganyika and the region north of Lake Nyasa.

Myrmecocichla arnotti harterti Neunzig

Myrmecocichla arnotti harterti Neunzig, 1926, Journ. f. Orn., 74, p. 754 — Malange.

Kwango District of the Congo south in western Angola to the southern edge of the Benguella Plateau.

Myrmecocichla arnotti arnotti (Tristram)

Saxicola arnotti Tristram, 1869, Ibis, p. 206, pl. 6 — Adam Kok's New Land [= Victoria Falls, northwestern Southern Rhodesia, *vide* Mackworth-Praed and Grant, 1942, Ibis, p. 521].

Saxicola shelleyi Sharpe, 1877, in Layard, Birds South Africa, ed. 2, pp. 246; 819 — Victoria Falls.

Northern and Southern Rhodesia, Nyasaland west to Bechuanaland and adjacent areas of South West Africa in Ovamboland and Ngamiland, and southern Angola, east to northeastern Transvaal.

MYRMECOCICHLA ALBIFRONS

Myrmecocichla albifrons albifrons (Rüppell)

Saxicola albifrons Rüppell, 1837, Neue Wirbelt., Vögel, p. 78 — Takeragiro, Temben [= Tembien], Abyssinia.

Northern Ethiopia, including Eritrea.

Myrmecocichla albifrons pachyrhyncha (Neumann)

Pentholaea albifrons pachyrhyncha Neumann, 1906 (January), Orn. Monatsb., 14, p. 8 — Uba, west slope, Omo District, southwestern Abyssinia.

Pentholaea macmillani Sharpe, 1906 (July), Bull. Brit. Orn. Club, 16, p. 126 — Winke Goffa, southern Abyssinia.

Southwestern Ethiopia.

Myrmecocichla albifrons clericalis (Hartlaub)

Pentholaea clericalis Hartlaub, 1882, Orn. Centralbl., 7, p. 91 — no locality; Langomeri (lat. 3° 31′ N., long. 31° 05′) and Wandi (lat. 4° 35′ N., long. 30° 27′) cited by Hartlaub, 1882, Journ. f. Orn., 30, p. 321.

Southern Sudan west of the Nile, northern Uganda and

northeastern Congo, upper Uelle River, and Ubangi-Shari Province.

Myrmecocichla albifrons frontalis (Swainson)

Saxicola frontalis Swainson, 1837, Birds W. Africa, 2, p. 46 — West Africa.

Pentholaea albifrons reichenowi Grote, 1921, Orn. Monatsb., 29, p. 93 — Satschie [= Sagdshe], near Garua, northern Cameroons.

From Senegal (winter?) east in all of West Africa to Nigeria and northeast to Chad.

Myrmecocichla albifrons limbata (Reichenow)

Pentholaea limbata Reichenow, 1921, Journ. f. Orn., 69, p. 49 — Bosum [= Bozum], Ubangi-Shari.

Northern and eastern Cameroun, intergrading with *clericalis* in the Ubangi-Shari region. Central African Republic.

MYRMECOCICHLA MELAENA

Myrmecocichla melaena (Rüppell)

Saxicola melaena Rüppell, 1837, Neue Wirbelt., Vögel, p. 77, pl. 28 — Alegua Mountain, Agame Province, Abyssinia.

Highlands of Ethiopia from Senafe to Shoa.

GENUS **THAMNOLAEA** CABANIS

Thamnolaea Cabanis, 1850, Mus. Hein., 1, p. 8. Type, by subsequent designation (Gray, 1855, Cat. Gen. Birds Brit. Mus., p. 36), *Turdus cinnamomeiventris* Lafresnaye.

THAMNOLAEA CINNAMOMEIVENTRIS

Thamnolaea cinnamomeiventris albiscapulata (Rüppell)

Saxicola albiscapulata Rüppell, 1837, Neue Wirbelt., Vögel, p. 74, pl. 26, fig. 1 — Abyssinia.

Eastern and central Ethiopia, including Eritrea, west to Sudan on the upper Blue Nile in the vicinity of Fazughli.

Thamnolaea cinnamomeiventris cavernicola Bates

Thamnolaea cinnamomeiventris cavernicola Bates, 1933, Bull. Brit. Orn. Club, 53, p. 176 — Fiko, 30 miles east of Mopti, French Sudan.

Known only from cliffs at Fiko, 300 miles northeast of Kulikoro on the upper Niger, Mali Republic.

Thamnolaea cinnamomeiventris bambarae Bates

Thamnolaea cinnamomeiventris bambarae Bates, 1928, Bull. Brit. Orn. Club, 49, p. 32 — Kulikoro, French Sudan.

Known only from Kulikoro, near Barnako, Mali Republic.

Thamnolaea cinnamomeiventris subrufipennis Reichenow

Thamnolaea subrufipennis Reichenow, 1887, Journ. f. Orn., 35, p. 78 — road near Ussure, Magala Steppe [Usukuma], Schasche [Koudoa Irangi District, Tanganyika].

Thamnolaea cinnamomeiventris usambarae Neumann, 1914, Orn. Monatsb., 22, p. 11 — Nilalo, Usambara, Tanganyika.

Extreme southeast Sudan, southwestern Ethiopia, Uganda, eastern and northeastern Congo, Kenya, and Tanganyika, and locally south to Victoria Falls area of Northern Rhodesia downstream along the Rhodesian borders to northwest Mozambique and Nyasaland.

Thamnolaea cinnamomeiventris odica Clancey

Thamnolaea cinnamomeiventris odica Clancey, 1962, Occ. Papers Nat. Mus. S. Rhodesia, no. 26b, p. 743 — Inyanga, eastern Southern Rhodesia.

Eastern Bechuanaland Protectorate (Kanye) and adjacent western and northern Transvaal northeast through Southern Rhodesia to the plateau country and adjacent highlands of Mozambique (?); also Mount Gorongoza, Mozambique.

Thamnolaea cinnamomeiventris cinnamomeiventris (Lafresnaye)

Turdus cinnamomeiventris Lafresnaye, 1836, Rev. Zool. [Paris], pls. 55, 56 — Cape of Good Hope [= Cape Province, *fide* Sclater, 1930, Syst. Av. Aethiop., p. 463].

Eastern Cape Province, Natal, except the coastal areas, Transvaal high-veld (grading into *odica* in the northern and western Transvaal), Orange Free State, Basutoland, and western Swaziland.

Thamnolaea cinnamomeiventris autochthones Clancey

Thamnolaea cinnamomeiventris autochthones Clancey, 1952, Ann. Natal Mus., 12, p. 252 — Bank of Ingwavuma River, Lebombo Foothills, northeastern Zululand.
Coastal Natal inland to about Weinen in the middle Tugela River drainage, Zululand, eastern Swaziland, and southern Mozambique in the Lebombo Range.

THAMNOLAEA CORONATA

Thamnolaea coronata coronata Reichenow

Thamnolaea coronata Reichenow, 1902, Orn. Monatsb., 10, p. 157 — Tapong [Togoland].
Togo east through Nigeria and northern Cameroons; Jebel Marra, western Sudan.

Thamnolaea coronata kordofanensis Wettstein

Thamnolaea coronata kordofanensis Wettstein, 1916, Anz. Akad. Wien, Math.-Naturwiss., 53, p. 135 — Gebel Rihal, Nuba Mountains [reference not verified].
Jebel Rihal, Nuba highlands, Kordofan, central Sudan.

THAMNOLAEA SEMIRUFA

Thamnolaea semirufa (Rüppell)

Saxicola semirufa Rüppell, 1837, Neue Wirbelt., Vögel, p. 74, pl. 25 — Zana [= Lake Tsana, *fide* Sclater, 1930, Syst. Av. Aethiop, p. 464].
Highlands of Ethiopia from Eritrea south to Lake Tsana, the Sidamo country, and to Yavello.

GENUS **OENANTHE** VIEILLOT

Oenanthe Vieillot, 1816, Analyse, p. 43. Type, by monotypy, "Motteux" Buffon = *Turdus leucurus* Gmelin.
Campicola Swainson, 1827, Zool. Journ., 3, p. 171. Type, by original designation, "Le Traquet Imitateur" Levaillant, 1805, Oiseaux Afrique, 4, pl. 181 = *Motacilla pileata* Gmelin.
Campicoloides Roberts, 1922, Ann. Transvaal Mus., 8, p. 229. Type, by monotypy, *Saxicola bifasciata* Temminck.

cf. Meinertzhagen, 1954, Birds Arabia, pp. 235-255.

OENANTHE BIFASCIATA

Oenanthe bifasciata (Temminck)

Saxicola bifasciata Temminck, 1829, Pl. Col., livr. 79, pl.
472, fig. 2 — Caffrerie [= eastern Cape Province].

Eastern Cape Province from Grahamstown to Graaff
Reinet north to Drakensberg, northern Natal, and southern
and eastern Transvaal.

OENANTHE ISABELLINA

Oenanthe isabellina (Temminck)

Saxicola isabellina Temminck, 1829, Pl. Col., livr. 79, pl.
472, fig. 1 — Nubia.

Oenanthe isabellina kargasi Koelz, 1939, Proc. Biol. Soc.
Washington, 52, p. 66 — Kargasi Pass, northeastern
Afghanistan.

Oe(nanthe) i(sabellina) sibirica Portenko, 1954, Keys
Fauna U.S.S.R., no. 54, Birds, 3, p. 162 — Kacha River,
near Krasnoyarsk, fields of Minusinsk, central Siberia.

Russia from the lower Volga east and north from Uralsk
to Kirghizstan, southern Siberia to Omsk, Altai, and Minusinsk, south of Lake Baikal in Transbaicalia, Manchuria (?),
south in Caucasus, Thrace, Island of Kos, Rhodes (?), Turkey, Syria, Lebanon, Israel, northern Arabia, Iraq, Iran, Afghanistan, West Pakistan east to Tibet (?), China in Sinkiang, Mongolia, Kansu, and Tsinghai; winters south through
central Asia to West Pakistan and north-central India,
southwest to southern Iran, Iraq, Arabia, and Cyprus and
in Africa from eastern Libya and Egypt south to Sudan,
Ethiopia, Somali Republic, Socotra, Mali (Taberréschat),
Kenya, Tanganyika, Uganda, northeastern Congo (Ishwa
Plain), Zanzibar, Northern Rhodesia (once), and Pemba;
British Isles, Japan (vagrant).

OENANTHE BOTTAE

Oenanthe bottae bottae (Bonaparte)

Campicola bottae Bonaparte, 1854, Compt. Rend. Acad.
Sci. Paris, 38, p. 7 — no locality; Yemen designated by
Sclater, 1930, Syst. Av. Aethiop., p. 455.

Yemen highlands.

— **Oenanthe bottae frenata** (Heuglin)

Saxicola frenata Heuglin, 1869, Journ. f. Orn., 17, p. 158
— Mensa, Abyssinian highlands.

Ethiopia highlands.

Oenanthe bottae heuglini (Finsch and Hartlaub)

Saxicola Heuglini Finsch and Hartlaub, 1870, Vög. Ost.-
Afr., p. 259 — "Gondar" [*errore*; Sudan suggested by
Sclater, 1930, Syst. Av. Aethiop., p. 455].

Sudan.

Oenanthe bottae campicolina (Reichenow)

Saxicola campicolina Reichenow, 1910, Orn. Monatsb., 18,
p. 175 — Garua and Sagdsche, Adamawa, Cameroons.

Mali. Niger, Chad, northern Nigeria including North
Cameroon, east to northern Cameroon and Central African
Republic (Ubangi-Shari).

OENANTHE XANTHOPRYMNA

—**Oenanthe xanthoprymna xanthoprymna** (Hemprich and
Ehrenberg)

Saxicola xanthoprymna Hemprich and Ehrenberg, 1833,
Symb. Phys. Av., fol. dd — Nubia.

Saxicola cummingi Whitaker, 1899, Bull. Brit. Orn. Club,
10, p. 17 — Fao, Persian Gulf.

Saxicola hawkeri Ogilvie-Grant, 1908, Bull. Brit. Orn.
Club, 21, p. 94 — Berber, Sudan.

Southwest Iran; on migration in Iraq and Arabia; winters in Sinai and coastal Egypt and Sudan.

— **Oenanthe xanthoprymna chrysopygia** (De Filippi)

Dromolaea chrysopygia De Filippi, 1863, Arch. Zool. Anat.
Fisiol. Genova, 2, p. 381 — Demavend, north-central
Iran.

Eastern Turkey, Erivan, Iran (except southwest), and
Russia in southern Transcaspia; migrating into the Zagros
and Persian Gulf areas, Arabia, and Iraq.

Oenanthe xanthoprymna kingi (Hume)

Saxicola kingi Hume, 1871, Ibis, p. 29 — Jodhpur.

Breeds in Afghanistan, Pamirs (?), and Baluchistan
(Amran Kwajah); winters in West Pakistan and western
India south to Gujarat.

OENANTHE OENANTHE

Oenanthe oenanthe leucorhoa (Gmelin)

Motacilla leucorhoa Gmelin, 1789, Syst. Nat., 1 (2), p. 966 — Senegal River.
Oenanthe oenanthe schiöleri, Salomonsen, 1927, Ibis, p. 203 — Hajnarfjord, Iceland.

Breeds from Ellesmere Island, Baffin Island, and northern Greenland to Iceland, Jan Mayen, and the Faeroes south to northern Quebec and Labrador; winters in west Africa in Senegal, Gambia, and Sierra Leone. On migration in British Isles and coastal western Europe east to eastern Germany, Switzerland, and Italy and south in northwest Africa, the Balearic, Azores, Madeira, Canaries, and Cape Verde Islands, and in North America casually from southeast Canada south to New York, Louisiana, Cuba, and Bermuda.

Oenanthe oenanthe oenanthe (Linnaeus)

Motacilla oenanthe Linnaeus, 1758, Syst. Nat., ed. 10, 1, p. 186 — in Europae apricis lapidosis [= Sweden, *fide* Hartert, 1910, Vög. pal. Fauna, p. 679].
Saxicola rostrata Hemprich and Ehrenberg, 1832, Symb. Phys. Av., fol. aa. — upper Egypt [= Beni Suef, *fide* Mackworth-Praed and Grant, 1951, Ibis, p. 234].
Saxicola libanotica Hemprich and Ehrenberg, 1833, Symb. Phys. Av., fol. bb. — all Syria [= Lebanon, *fide* Mackworth-Praed and Grant, 1951, Ibis, 93, p. 235].
Saxicola oenanthe argentea Lönnberg, 1909, Ark. f. Zool., 5 (9), p. 22 — Kjachta [, Transbaicalia].
Oenanthe oenanthe integer Clancey, 1950, Auk, 67, p. 392 — boulder-strewn moorlands of North Knapdale, Argyllshire, southwest Scotland.

Breeds from the British Isles across northern Europe to Lapland, Siberia, and northern islands south to the Balearic Islands, central Spain, the Mediterranean, Turkey, Syria, Lebanon, Transcaucasia, Iraq (?), Iran, and northern Afghanistan and east to Sinkiang, Mongolia, Manchuria, across to Alaska and northwestern Mackenzie; winters in Africa in Nigeria and Cameroun, Congo, Northern and Southern Rhodesia, Nyasaland, Tanganyika, Kenya, Ethiopia, Sudan, and southern Arabia; casual in West Pakistan and south China; vagrant to Sarawak (Borneo), Philippines, Pribilof Islands, and Colorado.

Oenanthe oenanthe nivea (Weigold)

Saxicola oenanthe nivea Weigold, 1913, Orn. Monatsb., 21, p. 123 — Capileira, Sierra Nevada, southern Spain.

Southern Spain and Balearic Islands (resident ?).

Oenanthe oenanthe virago Meinertzhagen

Oenanthe oenanthe virago Meinertzhagen, 1920, Bull. Brit. Orn. Club, 41, p. 20 — Mount Ida, Crete.

Eastern and south Aegean islands, including Limnos, Mytilene, Crete, Carpathos and Rhodes; winters south to Lebanon, Israel, and Egypt.

Oenanthe oenanthe seebohmi (Dixon)

Saxicola seebohmi Dixon, 1882, Ibis, p. 563, pl. 14 — Djebel Mahmel, Algeria.

Morocco and Algeria east to Tunisia (?).

Oenanthe oenanthe phillipsi (Shelley)

Saxicola phillipsi Shelley, 1885, Ibis, p. 404, pl. 12 — mountains near Berbera.

Somali Republic.

OENANTHE DESERTI

Oenanthe deserti homochroa (Tristram)

Saxicola homochroa Tristram, 1859, Ibis, p. 59 — in Sahara Tunitana [= Souf, near Algeria-Tunisia border, *vide* Vaurie, 1959, Birds Pal. Fauna, Passeriformes, p. 347].

North Africa from the Spanish Sahara east to western Egypt, grading into the following subspecies west of the Nile; south in winter to Aïr; vagrant on Canary Islands and in Great Britain.

Oenanthe deserti deserti (Temminck)

Saxicola deserti Temminck, 1825, Pl. Col., livr. 60, pl. 359, fig. 2 — Egypt.

S(axicola) atrogularis Blyth, 1847, Journ. Asiat. Soc. Bengal, 16, p. 131 — Upper Provinces, Scinde, etc. [= Agra, *vide* Baker, 1924, Fauna Brit. India, Birds, ed. 2, 2, p. 51].

Saxicola salina Eversmann, 1850, Bull. Soc. Natur. Moscow, 23, p. 567, pl. 8, fig. 2 — Kirghiz steppes.

Breeds in Egypt east of the Nile, Israel, northwest Arabia, southern Caucasus, Iran, Afghanistan, West Pakistan (northern Baluchistan), Transcaspia, and Kirghizstan north and east through Altai to Dzungaria and Mongolia to Khangai and the Gobi; winters in West Pakistan and India and west to southern Iran, Iraq, Arabia, Lebanon, Socotra Island, Sudan, Somalia, Ethiopia, and Chad; vagrant to Scotland.

Oenanthe deserti oreophila (Oberholser)

Saxicola oreophila Oberholser, 1900, Proc. U. S. Nat. Mus., 22, p. 221 — Ladakh; new name for *Saxicola montana* Gould, 1865, Birds Asia, 4, pl. 30 — Tibet, *nec Saxicola montana* Koch, 1816.

Breeds in Kashmir, Baltistan, Ladakh, and Tibet north to Pamirs, Sinkiang, Inner Monoglia, Kansu, and Tsinghai; winters in West Pakistan in Baluchistan (Mekran), Iran, Arabia, and Socotra Island; vagrant to Kurile Islands.

OENANTHE HISPANICA

Oenanthe hispanica hispanica (Linnaeus)

Motacilla hispanica Linnaeus, 1758, Syst. Nat., ed. 10, p. 186 — Hispania [= Gibraltar, *vide* Hartert, 1910, Vög. pal. Fauna, 1, p. 685].

Breeds in Spain (except northwest), Portugal, southeastern France, Italy (except in south), Sicily (?), northwestern Yugoslavia, and North Africa from Morocco to Tripoli; winters in west Africa from Senegal and Gambia to Mali.

Oenanthe hispanica melanoleuca (Güldenstädt)

Muscicapa melanoleuca Güldenstädt, 1775, Nov. Comm. Acad. Sci. Petrop., 19, p. 468 — Tiflis, Georgia, Caucasus.

Breeds in southern Italy, Sicily (?), Yugoslavia (except northwest), Rumania, Bulgaria, Greece, Crete, Turkey, Lebanon, and Syria south to Israel, Transcaucasia, and Iran; winters in Egypt, Sudan, Ethiopia, Chad, and Niger; in migration in Iraq, Arabia, Near East, and Algeria (occasional); recorded from Cyprus, Malta, and Sardinia.

OENANTHE FINSCHII

Oenanthe finschii finschii (Heuglin)

> *Saxicola Finschii* Heuglin, 1869, Orn. Nordost. Afr., 1,
> p. 350 — Siberia [= Syria, *vide* Meinertzhagen, 1930,
> in Nicoll, Birds Egypt, 1, p. 273].

Turkey, south to Israel and northern Arabia, east to Caucasus and northwest Iran, the Caspian, and southern Iran; winters in Cyprus and eastern Egypt.

Oenanthe finschii barnesi (Oates)

> *Saxicola barnesi* Oates, 1890, Fauna British India, Birds,
> 2, p. 75 — Baluchistan and Afghanistan eastwards to
> Persia; restricted to Kandahar by Baker, 1924, Fauna
> Brit. India, Birds, ed. 2, 1, p. 47.

Northeast Iran, east in Transcaspia and Kirghizstan to the Syr-Darya, and northern Afghanistan; winters in southeast Iran, southern Afghanistan, and West Pakistan in Baluchistan.

OENANTHE PICATA

Oenanthe picata (Blyth)

> *Saxicola picata* Blyth, 1847, Journ. Asiat. Soc. Bengal, 16,
> p. 131 — Sind to Ferozepore; hereby restricted to Sind.
> *Saxicola opistholeuca* Strickland, 1849, in Jardine, Contrib. Orn., p. 60 — northern India [= Punjab, *vide* Baker, 1924, Fauna Brit. India, Birds, 1, p. 44].
> *Saxicola capistrata* Gould, 1865, Birds Asia, 4, pl. 28 —
> upper Provinces of Hindostan [= Sind, *vide* Baker,
> 1924, Fauna Brit. India, Birds, ed. 2, 1, p. 43].

Iran east through Transcaspia, Afghanistan, and Kirghizstan north to Syr-Darya, south in Pamirs to West Pakistan — Afghanistan border, Gilgit, and Kashmir; winters from southern Iran to West Pakistan and India south to northern Gujarat.

OENANTHE LUGENS

Oenanthe lugens halophila (Tristram)

> *Saxicola halophila* Tristram, 1859, Ibis, p. 59 — Algerian
> Sahara.

Northern Sahara from eastern Morocco to northern Libya and Egypt (rarely; winter).

Oenanthe lugens lugens (Lichtenstein)

Saxicola lugens Lichtenstein, 1823, Verz. Doubl., p. 33 —
Nubia [= Deram, *vide* Stresemann, 1954, Abh. Deut-
schen Akad. Wiss., Math.-Naturwiss., Berlin, 1, p. 174].
Egypt east of the Nile north to Israel and Syria, Jordan,
and northern Iraq; Iran (?); wandering west across the
Nile.

Oenanthe lugens persica (Seebohm)

Saxicola persica Seebohm, 1881, Cat. Birds Brit. Mus., 5,
p. 372 — Shiraz, Persia.
Oenanthe lugens sarudnyi Härms, 1926, Journ. f. Orn.,
74, p. 40 — Sarhad District, Persian Baluchistan.
Southern Iran, wandering to Iraq (?), Israel, Egypt, and
northern Sudan.

Oenanthe lugens lugentoides (Seebohm)

Saxicola lugentoides Seebohm, 1881, Cat. Birds Brit. Mus.,
5, p. 371 — Sennaar, northeast Africa [*errore* = Yemen,
vide Sclater, 1930, Syst. Av. Aethiop., p. 445].
Southwestern Arabia, Yemen, and Aden.

Oenanthe lugens boscaweni Bates

Oenanthe lugubris boscaweni Bates, 1937, Bull. Brit. Orn.
Club, 58, p. 32 — Tarim, Wadi Hadhramaut.
Hadhramaut, southern Arabia.

Oenanthe lugens vauriei Meinertzhagen

Oenanthe lugens vauriei Meinertzhagen, 1949, Bull. Brit.
Orn. Club, 69, p. 107 — Erigavo, British Somaliland.
Northeastern Somalia.

Oenanthe lugens lugubris (Rüppell)

Saxicola lugubris Rüppell, 1837, Neue Wirbelt., Vögel, p.
77, pl. 28, fig. 1 — Simen, Abyssinia.
Highlands of central and northern Ethiopia.

Oenanthe lugens schalowi (Fischer and Reichenow)

Saxicola Schalowi Fischer and Reichenow, 1884, Jour. f.
Orn., 32, p. 57 — Lake Naivasha.
Highlands of southern Kenya and northeastern Tangan-
yika.

OENANTHE MONACHA

Oenanthe monacha (Temminck)

Saxicola monacha Temminck, 1825, Pl. Col., livr. 60, pl. 359, fig. 1 — Nubia [= Luxor, *vide* Stresemann, 1954, Abh. Deutschen Akad. Wiss., Math.-Naturwiss., Berlin, 1, p. 174].

Egypt east of the Nile, Arabia, eastern Iran, southern Afghanistan, and West Pakistan south to Sind; wanders to northern and coastal Sudan.

OENANTHE ALBONIGER

Oenanthe alboniger (Hume)

Saxicola Alboniger Hume, 1872, Stray Feathers, 1, p. 2 — stony hills which divide Kelat from Sindh . . . and Mekran Coast.

Southern Iran, Afghanistan, and West Pakistan northeast to Gilgit and Kashmir; wanders to Iraq and Oman.

OENANTHE PLESCHANKA

Oenanthe pleschanka pleschanka (Lepechin)

Motacilla pleschanka Lepechin, 1770, Nov. Comm. Acad. Sci. Petrop., 14, p. 503, pl. 14, fig. 2 — Saratov, lower Volga.

Motacilla leucomela Pallas, 1771, Nov. Comm. Acad. Sci. Petrop., 14, p. 584, pl. 22, fig. 3 — Samara, Russia.

Saxicola vittata Hemprich and Ehrenberg, 1833, Symb. Phys. Av., fol. cc — Moileh, northern Arabia.

Saxicola somalica Sharpe, 1895, Proc. Zool. Soc. London, p. 486 — Doda, western Somaliland.

Rumania and southern Rusia from Crimea to about lat. 53° N. in the Urals east to Transcaspia, Altai, and Kirghizstan south in Caucasus, Iran, Transcaspia to Afghanistan, Pamirs, West Pakistan and India to Ladakh, Sinkiang, Outer Mongolia, northern Tsinghai, Kansu, Inner Mongolia, Shansi, and Hopen; migrates from Turkey south to Egypt, Sudan, Somalia, Kenya, northern Tanganyika, and Arabia.

Oenanthe pleschanka cypriaca (Homeyer)

Saxicola cypriaca Homeyer, 1884, Zeitschr. ges. Orn., 1, p. 397 — Cyprus.

Cyprus; migrates through Israel, Egypt, and the Red Sea coasts; winters in Sudan, Ethiopia, and Kenya.

OENANTHE LEUCOPYGA

Oenanthe leucopyga leucopyga (Brehm)

> *Vitiflora leucopyga* Brehm, 1855, Der vollständige Vogel-
> fang, p. 225 — Egypt and southern Europe [= Kurusku,
> Upper Egypt (Nubia), *vide* Vaurie, 1959, Birds Pal.
> Fauna, Passeriformes, p. 353].
>
> *Oenanthe leucopyga aegra* Hartert, 1913, Novit. Zool., 20,
> p. 55 — Gara Klima, southern Algeria.

North Africa from Morocco east to Egypt and Red Sea
coast, south in the desert to northern Mali, Niger, Chad,
Sudan, and eastern Ethiopia, including Eritrea and French
Somaliland.

Oenanthe leucopyga ernesti Meinertzhagen

> *Oenanthe leucopyga ernesti* Meinertzhagen, 1930, in Ni-
> coll, Birds Egypt, p. 280 — Wadi Feiran, Sinai Penin-
> sula.

Eastern Egypt in Sinai north to Dead Sea, Arabia, Iraq,
and southwestern Iran (one record).

OENANTHE LEUCURA

— Oenanthe leucura leucura (Gmelin)

> (*Turdus*) *leucurus* Gmelin, 1789, Syst. Nat., 1 (2), p. 820
> — Gibraltar.

Portugal and Spain east along the Mediterranean coast of
France to Italy (Ponente), Sardinia, and Sicily.

Oenanthe leucura syenitica (Heuglin)

> *Saxicola syenitica* Heuglin, 1869, Journ. f. Orn., 17, p.
> 155 — El Kab, Egypt.
>
> *Saxicola leucurus riggenbachi* Hartert, 1909, Falco, (3),
> p. 36 — Rio de Oro.

Northwest coastal Africa from Rio de Oro to Morocco,
Tunisia, and Saharan Algeria; Egypt (one record).

OENANTHE MONTICOLA

Oenanthe monticola albipileata (Bocage)

> *Dromolaea albipileata* Barboza du Bocage, 1867, Jorn.
> Sci. Math. Phys. Nat. Lisboa, 1, p. 151 — Dombe, near
> Benguella, Angola.

Benguella Province, Angola.

Oenanthe monticola nigricauda Traylor

Oenanthe monticola nigricauda Traylor, 1961, Bull. Brit. Orn. Club, 81, p. 43 — Mount Moco, Huambo, Angola.

Mount Moco, Huambo, Angola.

Oenanthe monticola monticola Vieillot

Oenanthe monticola Vieillot, 1818, Nouv. Dict. Hist. Nat., nouv. éd., 21, p. 434 — Namaqualand.
Saxicola atmorii Tristram, 1869, Ibis, p. 206 — Damaraland.

Southwest and central Cape Province, South Africa, to Little Namaqualand and Griqualand West, north in South West Africa to the Kaokoveld and Namib Desert.

Oenanthe monticola griseiceps (Blandford and Dresser)

Saxicola griseiceps Blandford and Dresser, 1874, Proc. Zool. Soc. London, p. 233, pl. 37, fig. 3 — Colesberg, northern Cape Province.

Eastern Karroo north to Natal in Drakensberg, Transvaal and southern Bechuanaland.

OENANTHE MOESTA

Oenanthe moesta moesta (Lichtenstein)

Saxicola moesta Lichtenstein, 1823, Verz. Doubl., p. 33 — Egypt [= between Bir Hamam and Gassr Eschtaebi, *vide* Stresemann, 1954, Abh. Deutschen Akad. Wiss., Naturwiss., Berlin, 1, p. 173].
Oenanthe moesta theresae Meinertzhagen, 1939, Bull. Brit. Orn. Club, 59, p. 66 — Tiznit, southwestern Morocco.

Northwest Africa from southwestern Morocco east across Algeria and Libya to coastal western Egypt.

Oenanthe moesta brooksbanki Meinertzhagen

Oenanthe moesta brooksbanki Meinertzhagen, 1923, Bull. Brit. Orn. Club, 43, p. 147 — near El Jid, lat. 33° N., long. 40° E., northern Arabian Desert.

Eastern Egypt (Sinai), Jordan, Syrian desert, and Arabia to western Iraq.

OENANTHE PILEATA

Oenanthe pileata (Gmelin)

(*Motacilla*) *pileata* Gmelin, 1789, Syst. Nat., 1 (2), p. 965 — Cape of Good Hope.

Campicola livingstonii Tristram, 1867, Proc. Zool. Soc. London, p. 888 — Murchison Falls, Zambezi River.

Oenanthe pileata neseri Macdonald, 1952, Ostrich, 23, p. 161 — Erongo Mountains, Omaruru District, South West Africa.

Southeast Congo, Angola east to Kenya, and Zanzibar; visits Rhodesias in dry season; breeds in Southern Rhodesia (August-September) ; Nyasaland (one record) ; South West Africa; and South Africa in Cape Province, Orange Free State, Bechuanaland Protectorate, Transvaal, and northern Natal.

GENUS **CHAIMARRORNIS** HODGSON

Chaimarrornis Hodgson, 1844, in Gray, Zool. Misc., p. 82. Type, by monotypy, *Phoenicura leucocephala* Vigors.

cf. Goodwin, 1957, Bull. Brit. Orn. Club, 77, p. 112.

CHAIMARRORNIS LEUCOCEPHALUS

Chaimarrornis leucocephalus (Vigors)

Phoenicura leucocephala Vigors, 1831, Proc. Zool. Soc. London, p. 35 — Himalaya; restricted to Simla-Almora district by Ticehurst and Whistler, 1924, Ibis, p. 471.

Tadzhikstan and Pamirs, northeast Afghanistan, West Pakistan and India from Safed Kohs east through Kashmir and Ladakh, Nepal, Sikkim, Bhutan, and Assam hills east in southeast Tibet, China from Yunnan and Sikang to Kansu, Tsinghai, Koko Nor, and southeast to Chekiang and Anhwei; migrates to Baluchistan, northern West Pakistan and Indian foothills, Burma, northern Thailand, Tonkin, and Yangtze River lowlands.

GENUS **SAXICOLOIDES** LESSON

Saxicoloides Lesson, 1832, in Bélanger, Voy. Ind.-Orient., Zool., (4), p. 270. Type, by monotypy, *Turdus* (*Saxicoloides*) *erythrurus* Lesson.

SAXICOLOIDES FULICATA

Saxicoloides fulicata cambaiensis (Latham)

Sylvia cambaiensis Latham, 1790, Index Orn., 2, p. 554
— Gujarat, India.
Saxicoloides fulicata munda Van Tyne and Koelz, 1936,
Occ. Papers Mus. Zool. Univ. Michigan, no. 334, p. 5 —
Bhadwar, Kangra District, Punjab, British India.
Saxicoloides fulicata lucknowensis Koelz, 1939, Proc. Biol.
Soc. Washington, 52, p. 66 — Lucknow, United Province.

West Pakistan and India in Sind, Punjab, Kutch, Saurashtra, Rajasthan, Delhi, Uttar Pradesh south to northern Gujarat, northern Madhya Pradesh, and the border of Bihar;
Nepal lowlands.

Saxicoloides fulicata erythrura (Lesson)

Turdus (Saxicoloides) erythrurus Lesson, 1832, in Bélanger, Voy. Ind.-Orient., Zool., (4), p. 270 — Bengal.
Saxicoloides fulicata stuartbakeri Koelz, 1939, Proc. Biol.
Soc. Washington, 52, p. 67 — Bodhgaya, Bihar.
Plains of Bihar and West Bengal.

Saxicoloides fulicata intermedia Whistler and Kinnear

Saxicoloides fulicata intermedia Whistler and Kinnear,
1932, Journ. Bombay Nat. Hist. Soc., 36, p. 73 — Rahuri, Ahmednagar.
Central India from eastern Bombay to Madhya Pradesh,
Hyderabad, Orissa, and Andhra south to the Krishna River.

Saxicoloides fulicata fulicata (Linnaeus)

(Motacilla) fulicata Linnaeus, 1766, Syst. Nat., ed. 12, p.
336, ex Brisson, 1760, "Le Traquet des Philippines" —
Philippines [= Pondichéry, *vide* Stresemann, 1952, Ibis,
94, pp. 515, 520].
Oenanthe ptgymatura Vieillot, 1818, Nouv. Dict. Hist.
Nat., nouv. éd., 21, p. 436, ex Levaillant, 1805, "Le Traquet à queue striée" — le Bengale [= Pondichéry, *vide*
Whistler, 1935, Journ. Bombay Nat. Hist. Soc., 38, p.
286].
Th.(amnobia) rufiventer Swainson, 1832, in Swainson
and Richardson, Fauna. Bor.-Amer., Zool., p. 489, ex
Levaillant, 1805, "Le Traquet à queue striée" — le Ben-

gale; restricted to Pondichéry by Ripley, 1952, Postilla, Yale Univ., no. 13, p. 35.

India in Bombay, Goa, Madras, Andhra (north to the Krishna River), Mysore, and Kerala.

Saxicoloides fulicata leucoptera (Lesson)

M.(icropus) leucopterus Lesson, 1840, Rev. Zool. [Paris], p. 136 — Indes Orientales [= Ceylon, *fide* Baker, 1930, Fauna Brit. India, Birds, ed. 2, 7, p. 112].
Lowlands of Ceylon.

GENUS PSEUDOCOSSYPHUS SHARPE

Pseudocossyphus Sharpe, 1883, Cat. Birds Brit. Mus., 7, p. 21. Type, by monotypy, *Cossypha sharpei* Gray.

cf. Goodwin, 1956, Bull. Brit. Orn. Club, 76, pp. 143-144 (*Pseudocossyphus* relationship).

PSEUDOCOSSYPHUS IMERINUS

Pseudocossyphus imerinus erythronotus (Lavauden)

Cossypha sharpei erythronota Lavauden, 1929, Alauda, 1, p. 232 — Ambre Mountain, Madagascar.
Ambre Mountain, Madagascar.

Pseudocossyphus imerinus sharpei (Gray)

Cossypha Sharpei Gray, 1871, Ann. Mag. Nat. Hist., ser. 4, 8, p. 429 — Madagascar [= forests east of Ambatondrazaka (Lake Alaotra), *fide* Grandidier, 1879, Hist. phys. nat. pol. Madagascar, Ois., 1, p. 370].
East-central Madagascar.

Pseudocossyphus imerinus interioris (Salomonsen)

Monticola imerina interioris Salomonsen, 1934, Novit. Zool., 39, p. 211 — Manjakatompo, Ankaratra Mountains.
Central Madagascar mountains at Manjakatompo, and southern highlands (Betsileo).

Pseudocossyphus imerinus imerinus (Hartlaub)

C.(ossypha) imerina Hartlaub, 1860, Journ. f. Orn., 8, p. 97 — St. Augustine Bay, southeastern Madagascar.
Southeastern Madagascar.

Genus **MONTICOLA** Boie

Monticola Boie, 1822, Isis, col 552. Type, by subsequent designation (G. R. Gray, 1847, Gen. Birds, 1, p. 220), *Turdus saxatilis* Linnaeus.

Petrophila Swainson, 1837, Class. Birds, 2, p. 232. Type, by monotypy, *Petrocincla cinclorhyncha* Vigors.

Petronis Roberts, 1922, Ann. Transvaal Mus., 8, p. 228. Type, by monotypy, *M.(onticola) rupestris* (Vieillot).

Colonocincla Roberts, 1922, Ann. Transvaal Mus., 8, p. 228. Type, by original designation, *M.(onticola) brevipes* "Strickland and Sclater" = (Waterhouse).

Notiocichla Roberts, 1922, Ann. Transvaal Mus., 8, p. 228. Type, by monotypy, *M.(onticola) explorator* (Vieillot).

MONTICOLA RUPESTRIS

Monticola rupestris (Vieillot)

Turdus rupestris Vieillot, 1818, Nouv. Dict. Hist. Nat., nouv. éd., 20, p. 281 — near Cape Town [= Table Mountain, *fide* Sclater, 1930, Syst. Av. Aethiop., p. 449].

South Africa in Cape Province, eastern Orange Free State, Basutoland, and Natal; Transvaal, local north to Zoutpansberg.

MONTICOLA EXPLORATOR

Monticola explorator explorator (Vieillot)

Turdus explorator Vieillot, 1818, Nouv. Dict. Hist. Nat., nouv. éd., p. 260 — mountains of the Cape of Good Hope.

South Africa in southern and southwestern Cape Province, and eastern and northern Transvaal; lowlands of Natal.

Monticola explorator tenebriformis Clancey

Monticola explorator tenebriformis Clancey, 1952, Durban Mus. Novit., 4, p. 13 — near Ingwavuma, Lebombo Mountains, northeastern Zululand.

Basutoland; winters in Natal, Zululand, Swaziland, and southern Mozambique.

MONTICOLA BREVIPES

Monticola brevipes brevipes (Waterhouse)

Petrocincla brevipes Waterhouse, 1938, in Alexander, Exped. Int. Africa, 2, p. 263 — 'Tans Mountain, Damaraland, near Walvis Bay.

Monticola pretoriae Gunning and Roberts, 1911, Ann. Transvaal Mus., 3, p. 118 — near Pretoria.

Monticola brevipes kaokoensis Macdonald, 1957, Contr. Orn. W. South Africa. Brit. Mus. (Nat. Hist.), p. 122 — Kamanjab, 4,000 ft., lat. 19° 34′ S., long. 14° 48′ E.

Angola (Huxe, one record), South West Africa, northern Cape Province, and Little Namaqualand.

Monticola brevipes leucocapilla (Lafresnaye)

Petrocinela (sic) *leucocapilla* Lafresnaye, 1852, Rev. Mag. Zool. [Paris], p. 470 — no locality [= "Afr. Mer. Betzonanas"; *vide* Bangs, 1930, Bull. Mus. Comp. Zool., 70, p. 333].

Southern Bechuanaland, western Transvaal, Swaziland (Stegi, one record), and Orange Free State.

MONTICOLA RUFOCINEREUS

Monticola rufocinereus rufocinereus (Rüppell)

Saxicola rufocinerea Rüppell, 1837, Neue Wirbelt., Vögel, p. 76, pl. 27 — Simen Province, northern Abyssinia.

Petrophila rufocinerea tenuis Friedmann, 1930, Occ. Papers Boston Soc. Nat. Hist., 5, p. 325 — Uraguess, Kenya Colony.

Ethiopia, including Eritrea, Somalia, mountains of southeast Sudan (possibly not resident), eastern Uganda, western Kenya, and northeastern Tanganyika (Longido).

Monticola rufocinereus sclateri Hartert

Monticola rufocinerea sclateri Hartert, 1917, Novit. Zool., 24, p. 459 — Wasil, 4,000 ft., Yemen.

Western Arabia from Asir Tihana south in the mountains of Yemen to northern Aden.

MONTICOLA ANGOLENSIS

Monticola angolensis Sousa

Monticola angolensis Sousa, 1888, Jorn. Sci. Math. Phys. Nat. Lisboa, 12, pp. 225, 233 — Caconda, Benguella, Angola.

Monticola angolensis niassae Reichenow, 1905, Vög. Afr., 3, p. 699 — Nyasaland.

Southern and eastern Congo, Ruanda, Tanganyika, south in Angola in north and Benguella highlands, Southern Rhodesia in eastern districts (Matabeleland, rare), eastern and northwestern Northern Rhodesia, Nyasaland, and Mozambique in Quelimane District.

MONTICOLA SAXATILIS

Monticola saxatilis (Linnaeus)

[*Turdus*] *saxatilis* Linnaeus 1766, Syst. Nat., ed. 12, p. 294 — Switzerland.

Monticola saxatilis turkestanicus Zarudny, 1918, Izvest. Turkest. Otd. Russk. Geogr. Obsht., 14, p. 140 — Tian Shan [reference not verified].

Central and southern Europe north to southern Poland, south to Iberian Peninsula and northern Africa in Atlas of Morocco and Algeria, east to Greece, Turkey, Crimea, Caucasus, Lebanon, Iran, Afghanistan, West Pakistan (northern Baluchistan), Transcaspia, Tadzhikistan, Altai, Mongolia, Transbaicalia, and western China in Shansi, Kansu, and Tsinghai north to Inner Mongolia. Migrates west and south, skirting northern India and West Pakistan, in Kashmir, Ladakh, northern Punjab, and Sind to Iran, Iraq, Arabia, Egypt, and the Saharan area to winter in Africa north of the equatorial forests to northern Nigeria, upper Congo, Sudan, Ethiopia, Kenya, and Tanganyika highlands.

MONTICOLA CINCLORHYNCHUS

Monticola cinclorhynchus gularis (Swinhoe)

Oroecetes gularis Swinhoe, 1863, Ibis, p. 93, pl. 3 — Peking.

Amur Basin in southeastern Transbaicalia east to Argun River, south through Manchuria to northern Korea and northern Hopeh in China; migrates through eastern China;

winters in southeastern China, northern Vietnam, northern
and eastern Thailand, and eastern and central Burma;
straggler to Malaya.

Monticola cinclorhynchus cinclorhynchus (Vigors)

Petrocincla cinclorhyncha Vigors, 1832, Proc. Zool. Soc.
London, p. 172 — Himalayan mountains [= Simla, *fide*
Baker, 1921, Journ. Bombay Nat. Hist. Soc., 27, p.
719].

Eastern Afghanistan, West Pakistan, and India from
Safed Kohs and Kashmir east to Nepal, Sikkim, Bhutan, and
northern Assam; winters in peninsular India to Deccan and
Kerala, in East Pakistan, and in the Arakan in Burma.

MONTICOLA RUFIVENTRIS

Monticola rufiventris (Jardine and Selby)

Turdus erythrogaster Vigors, 1832, Proc. Zool. Soc. London, p. 171 — Himalayan Mountains; *nec Turdus erythrogaster* Boddaert, 1783.
Petrocincla rufiventris Jardine and Selby, 1833, Ill. Orn.,
3, pl. 129 — Himalayan district; restricted to Simla by
Ripley, 1961, Synopsis Birds India Pakistan, p. 523.
Monticola (= *Petrophila*) *semicastanea* Collin and Hartert, 1927, Novit. Zool., 34, p. 51, *nom. nov.* for *M.
erythrogaster* (Vigors).
Monticola rufiventris sinensis A. C. Meinertzhagen, 1927,
Bull. Brit. Orn. Club, 47, p. 148 — Kuatun, northwestern Fukien.

West Pakistan and India from the western Himalayas to
Nepal, Sikkim, Bhutan, southeast Tibet, Assam, Burma,
China in Yunnan, Sikang, and Szechwan southeast to Fu-
kien; winters to northern Thailand, Laos, and Vietnam.

MONTICOLA SOLITARIUS

Monticola solitarius solitarius (Linnaeus)

[*Turdus*] *solitarius* Linnaeus, 1758, Syst. Nat., ed. 10, p.
170 — Oriente [= Italy, ex Willoughby, *vide* Hartert,
1910, Vög. pal. Fauna, p. 674].
Monticola solitarius scorteccii Moltoni, 1934, Atti Soc.
Italiana Sci. Nat. Milano, 73, p. 366 — Gat, southwestern Libya.

Monticola solitarius behnkei Niethammer, 1943, Anz. Akad. Wiss. Wien, Math.-Naturwiss., 80 (3), p. 9 — Samaria, Crete.

Southern Switzerland, south-central France, and Italy south and east to Spain, Portugal, Yugoslavia, Albania, Greece, Mediterranean islands, and north Africa from Morocco to Tunis, Libya east to Turkey, Lebanon, and Syria; Caucasus; Israel (?); winters in its breeding range and also south to Egypt, southern Arabia, Sudan, and in the Sahara as far as Chad, Senegal, and Central African Republic (Bamako).

Monticola solitarius longirostris (Blyth)

P.(*etrocincla*) *longirostris* Blyth, 1847, Journ. Asiat. Soc. Bengal, 16, p. 150 — from Sind to Ferozepore.

Monticola cyanus transcaspicus Hartert, 1909, Bull. Brit. Orn. Club, 23, p. 43 — Sirax, Aschabad (near Tedjen).

Northern Iraq mountains, Iran, Transcaspia, and Afghanistan; intergrading with *pandoo* in eastern Afghanistan and on Afghanistan-West Pakistan border; winters in Arabia, Egypt, Sudan, Ethiopia, and Somalia east to Iraq, West Pakistan (Kohat, Sind), and India (Saurashtra ?).

Monticola solitarius pandoo (Sykes)

Petrocincla Pandoo Sykes, 1832, Proc. Zool. Soc. London, p. 87 — Ghats, Deccan.

P.[*etrocincla*] *affinis* Blyth, 1843, Journ. Asiat. Soc. Bengal, 12, p. 177 — Tenasserim and Darjeeling.[1]

Pamirs north to Tian Shan and mountains of the Tadzhikistan-Chinese border south to Chitral and along Himalayas of India and Tibet to Nepal, Sikkim, Bhutan, Assam, and China in Sikang, Yunnan, Szechwan, southeast China, and Shensi; migrates through West Pakistan and India to peninsular India, Ceylon, Andaman and Nicobar Islands, Burma, Thailand, Laos, Vietnam, and Hainan, south to Malaya, Sumatra, and Borneo.

Monticola solitarius philippensis (Müller)

Turdus Philippensis Müller, 1776, Natursyst., Suppl., p. 145 — Philippines.

---

[1] A population of birds with variable shades of color and washed with chestnut on the lower parts, intermediate between *longirostris* and *pandoo*, was named *affinis* by Blyth.

Petrophila solitaria magna La Touche, 1920, Bull. Brit. Orn. Club, 40, p. 97 — Shaweishan Island.

Monticola (= *Petrophila*) *solitaria latouchei* Kuroda, 1922, Ibis, p. 92 — Sasu-mura, Tsushima Island.

Monticola saxatilis centralasiae Serebrovskij, 1928, Compt. Rend. Acad. Sci. Leningrad, (1927), p. 325 — northwest Mongolia.

Monticola philippensis taivanensis Momiyama, 1930, Bull. Soc. Biogeogr., 1, p. 177 — Sinhi-syô, Tyôsiu-gun, Takao District, South Formosa.

Northeastern Russia in Ussuriland, Korea, China from Manchuria south to Shantung, Japan, and Formosa; offshore Japanese islands from Bonins to Riu Kiu chain; winters south from Yangtze valley to Hainan, northeast Thailand, Laos, Vietnam, Malaya, Borneo and adjacent islands, Philippines, Celebes (Sulawesi), Talaud, Moluccas, Southeast Islands, and Palau Archipelago.

Monticola solitarius madoci Chasen

Monticola solitarius madoci Chasen, 1940, Bull. Brit. Orn. Club, 60, p. 97 — Batu Caves near Kuala Lumpur, Selangor, Malay States.

Malaya.

GENUS **MYIOPHONEUS** TEMMINCK

Myiophoneus[1] Temminck, 1822, Pl. Col., livr. 29, pl. 170. Type, by original designation, *Turdus flavirostris* Horsfield.

Arrenga Lesson, 1831, Traité Orn., p. 388. Type, by monotypy, *Turdus cyaneus* Horsfield = *Pitta glaucina* Temminck.

cf. Delacour, 1942, Auk, 59, pp. 246-264 (review of genus).

MYIOPHONEUS BLIGHI

Myiophoneus blighi (Holdsworth)

Arrenga blighi Holdsworth, 1872, Proc. Zool. Soc. London, p. 444, pl. 19 — "Nuwara Eliya"; corrected to banks

---

[1] Temminck (*ibid.*), in his generic description, spells the new genus *Myiophoneus*, and on the following page accompanying plate 170 spells it "*Myophonus.*"

of the Lemastota oya, 4,200 ft., Haputale District (Uva), by Legge, 1880, Birds Ceylon, p. 464.
Hills of Ceylon.

MYIOPHONEUS MELANURUS

Myiophoneus melanurus (Salvadori)
Arrenga melanurus Salvadori, 1879, Ann. Mus. Civ. Genova, 14, p. 227 — Mount Singalan, West Sumatra.
Mountains of Sumatra, above 4,000 feet.

MYIOPHONEUS GLAUCINUS

Myiophoneus glaucinus castaneus Ramsay
Myiophoneus castaneus Ramsay, 1880, Proc. Zool. Soc. London, p. 16 — Mount Sago, neighborhood of Padang, West Sumatra.
Sumatra, in the foothills.

Myiophoneus glaucinus glaucinus (Temminck)
Pitta glaucina Temminck, 1823, Pl. Col., livr. 33, pl. 194 — Java.
Java and Bali, in the mountains.

Myiophoneus glaucinus borneensis Slater
Myiophoneus borneensis Slater, 1885, Ibis, p. 124 — Bungal Hills, near Sarawak.
Foothills and mountains of Borneo.

MYIOPHONEUS ROBINSONI

Myiophoneus robinsoni Ogilvie-Grant
Myiophoneus robinsoni Ogilvie-Grant, 1905, Bull. Brit. Orn. Club, 15, p. 69 — Gunung Menkuanghebah, Selangore.
Malaya mountains.

MYIOPHONEUS HORSFIELDII

Myiophoneus horsfieldii Vigors
Myophonus Horsfieldii Vigors, 1831, Proc. Zool. Soc. London, p. 35 — Himalayan mountains; restricted to Malabar by Baker, 1923, Hand-list Indian Birds, p. 93.

Hills of Peninsular India from Mount Abu south to Kerala; east in the central hills from Pachmarhi and Surguja to Orissa.

MYIOPHONEUS INSULARIS

Myiophoneus insularis Gould

Myiophoneus insularis Gould, 1862, Proc. Zool. Soc. London, p. 280 — Formosa.

Mountains of Formosa (Taiwan).

MYIOPHONEUS CAERULEUS

Myiophoneus caeruleus temminckii Gray

Myiophoneus Temminckii G. R. Gray, in Temminck, 1822, Pl. Col., livr. 29, text preceding pl. 170 — India in the Himalaya mountains; restricted to Simla-Almora district, by Ticehurst and Whistler, 1924, Ibis, p. 471.

Myiophoneus tibetanus Madarász, 1886, Ibis, p. 145 — (central) Tibet.

Myiophoneus temmincki turcestanicus Zarudny, 1909, Orn. Monatsb., 17, p. 168 — Tashkent.

Myophonus caeruleus rileyi Deignan, 1938, Proc. Biol. Soc. Washington, 51, p. 25 — Doi Ang Ka, northern Siam.

Myophonus caeruleus euterpe Koelz, 1954, Contrib. Inst. Regional Exploration, no. 1, p. 12 — Mawphlang, Khasi Hills.

Southern Kirgizstan and Tadzhikistan from Bukhara and Fergana east through the Tian Shan, south in Afghanistan from the Iran border through northern and south-central hills to Baluchistan ; West Pakistan and India from Safed Kohs east to Kashmir, Ladakh along the Himalayas to Nepal, Sikkim, Bhutan, southeast Tibet, and Assam hills as far as the Dibang and Lohit river valleys, where intergrades with *eugenei;* Sikang, northern Burma to Laukkang, south along hills to Arakan and northern Shan States; winters in adjacent foot hills and plains in the northern part of the range, northern Punjab, and, in the east, in northern Thailand.

Myiophoneus caeruleus eugenei Hume

Myiophoneus Eugenei Hume, 1873, Stray Feathers, 1, p. 475 — Thayetmyo and the western Pegu Hills.

Myiophoneus klossii Robinson,[1] 1915, Ibis, p. 750 — Koh
Melisi, West Island, southeastern Siam.

Myiophoneus stonei de Schauensee, 1929, Proc. Acad.
Nat. Sci. Philadelphia, 87, p. 469 — Chieng Mai, north
Siam.

Mishmi Hills of northeast Assam, Burma (except the
range of *temminckii*) south to Tenasserim, northern Thailand, Laos, and northern Vietnam, China in Yunnan, southern Sikang, and southern Szechwan, where intergrades with
caeruleus in the Kangting and Mount Omei areas.

Myiophoneus caeruleus caeruleus (Scopoli)

Gracula (*caerulea*) Scopoli, 1786, Del. Flor. Fauna Insubr., fasc. 2, p. 88; ex Sonnerat "merle bleu de la
Chine," 1782, Voy. Ind. Orient., p. 188, pl. 108 — China;
restricted to Canton by Stresemann, 1924, Abh. Ber.
Mus. Dresden, 16, no. 2, p. 28.

Myophonus caeruleus immansuetus Bangs and Penard,
1925, Occ. Papers Boston Soc. Nat. Hist., 5, p. 147 —
Ichang, Province of Hupeh, China.

China from Kansu and western Szechwan south to southern Yunnan and Kwangtung, east to northern Hupeh; winters in the Yangtze valley and south to Laos and northern
Vietnam (Tonkin).

Myiophoneus caeruleus crassirostris Robinson

Myiophoneus crassirostris Robinson, 1910, Bull. Brit. Orn.
Club, 25, p. 99 — Trang, northern Malay Peninsula.

Myophonus temminckii changensis Riley, 1928, Proc. Biol.
Soc. Washington, 41, p. 207 — Koh Chang, Siam.

Peninsular and extreme southeast Thailand and Islands
in the Gulf of Siam, south to the Langkawi Islands, Perlis,
and Kedah, northern Malaya.

Myiophoneus caeruleus dicrorhynchus Salvadori

Myophonus dicrorhynchus Salvadori, 1879, Ann. Mus. Civ.
Genova, 14, p. 227 — Ajermantcior, Padangpanjang,
West Sumatra.

Malaya from Patani and the Selangor hills south to Gunong Tahan in Pahang; western Sumatra hills.

---

[1] Based on an aberrant, partially albinistic specimen; see Delacour,
1942, *tom. cit.*, p. 255.

Myiophoneus caeruleus flavirostris (Horsfield)

Turdus flavirostris Horsfield, 1821, Trans. Linn. Soc. London, 13, p. 149 — Java.

Java hills.

GENUS GEOMALIA STRESEMANN

Geomalia Stresemann, 1931, Orn. Monatsb., 39, p. 10. Type, by monotypy, *Geomalia heinrichi* Stresemann.

GEOMALIA HEINRICHI

Geomalia heinrichi heinrichi Stresemann

Geomalia heinrichi Stresemann, 1931, Orn. Monatsb., 39, p. 11 — Mount Latimodjong, 2,800 m., Celebes.

Known only from Mount Latimodjong, south-central Celebes, from 5,900 to 8,500 feet altitude.

Geomalia heinrichi matinangensis Stresemann

Geomalia heinrichi matinangensis Stresemann, 1931, Orn. Monatsb., 39, p. 82 — Ile-Ile, 1,700 m., Mount Matinang.

Known from the Tentolomatinan range south of Paleleh, northern Celebes and southeastern Celebes on Mount Tanke Salokko, above 5,000 feet.

GENUS ZOOTHERA VIGORS

Zoothera Vigors, 1832, Proc. Zool. Soc. London, p. 172. Type, by monotypy, *Zoothera monticola* Vigors.

Geokichla Müller, 1835, Tijdschr. Natuur. Gesch. Phys., 2, pl. 3, p. 348. Type, by original designation, *Turdus citrinus* Latham.

Geocichla "Kuhl, 1820" = Gould, 1836, Proc. Zool. Soc. London, pt. 4, p. 7. Type, by monotypy, *Geocichla rubecula* Gould.

Oreocincla Gould, 1838, Synops. Birds Australia, pl. 55, app. p. 3. Type, by monotypy, *Turdus varius* Horsfield = *Oreocincla horsfieldi* Bonaparte.

Ixoreus Bonaparte, 1854, Compt. Rend. Acad. Sci. Paris, 38, no. 1, p. 3 (note). Type, by original designation, *Turdus naevius* Gmelin.

Cichlopasser Bonaparte, 1854, *ibid.*, p. 6. Type, by monotypy, *Turdus terrestris* Kittlitz.

Hesperocichla Baird, 1864, Rev. Amer. Birds, 1, p. 12. Type, by monotypy, *Turdus naevius* Gmelin.

Ridgwayia Stejneger, 1883, Proc. U. S. Nat. Mus., 5, p. 460. Type, by original designation, *Turdus pinicola* Sclater.

Aegithocichla Sharpe, 1903, Hand-list, 4, p. 134. Type, by monotypy, *Turdus terrestris* kittlitz.

Oreocichla "Gould" = Sharpe, 1903, Hand-list, 4, p. 136; emendation of *Oreocincla*.

Pseudoturdus Roberts, 1922, Ann. Transvaal Mus., 8, p. 227. Type, by monotypy, *Turdus guttatus* Vigors.

cf. Mackworth-Praed and Grant, 1937, Ibis, pp. 874-876 (*piaggiae; gurneyi*).

Vaurie, 1955, Amer. Mus. Novit., no. 1706, pp. 1-8 (*mollissima; dixoni*).

ZOOTHERA SCHISTACEA

Zoothera schistacea (Meyer)

Geocichla schistacea Meyer, 1884, Zeitschr. ges. Orn., 1, p. 211, pl. 8 — Tanimbar.

Tanimbar Island, Indonesia.

ZOOTHERA DUMASI

Zoothera dumasi dumasi (Rothschild)

Geocichla dumasi Rothschild, 1898, Bull. Brit. Orn. Club, 8, p. 30 — Mount Mada, 3,000 ft., Buru.

Mountains of Buru Island, Moluccas.

Zoothera dumasi joiceyi (Rothschild and Hartert)

Turdus joiceyi Rothschild and Hartert, 1921, Bull. Brit. Orn. Club, 41, p. 74 — high mountains of Ceram.

Ceram Island, Moluccas.

ZOOTHERA INTERPRES

Zoothera interpres interpres (Temminck)

Turdus interpres "Kuhl" = Temminck, 1826, Pl. Col., livr., 78, pl. 458 — Java and Sumatra.

Geokichla interpres minima Hachisuka, 1934, Tori, 8, p. 221 — Basilan.

Peninsular Thailand (Trang) and Malaya (vagrant, four specimens); Sumatra, Borneo, Java, Bali (?), Lombok, Sumbawa, and Flores Islands; Sulu Archipelago and Basilan Island, Philippines.

Zoothera interpres leucolaema (Salvadori)
> Geocichla leucolaema Salvadori, 1892, Ann. Mus. Civ.
> Genova, 32, p. 135 — Bua-Bua, Enggano Island.
> Enggano Island, West Sumatra Islands.

ZOOTHERA ERYTHRONOTA

Zoothera erythronota dohertyi (Hartert)
> Geocichla dohertyi Hartert, 1896, Novit. Zool., 3, p. 555,
> pl. 11, fig. 3 — Lombok (type) and Sumbawa.
> Lombok, Sumbawa, Sumba, Flores, and Timor Islands,
Lesser Sunda Islands.

Zoothera erythronota erythronota (Sclater)
> Geocichla erythronota Sclater, 1859, Ibis, p. 113 — Ma-
> cassar.
> Geocichla frontalis Madarász, 1899, Termés. Füzet., 22,
> p. 111, pl. 8 — Celebes.
> Celebes (not known from the eastern peninsula).

Zoothera erythronota mendeni (Neumann)
> Turdus (Geokichla) mendeni Neumann, 1939, Bull. Brit.
> Orn. Club, 59, p. 47 — Peling.
> Peling Island, east of Celebes.

ZOOTHERA WARDII

Zoothera wardii (Blyth)
> T.(urdus) Wardii Blyth, 1842, Journ. Asiat. Soc. Bengal,
> 11, p. 882 — Mysore.
> Breeds in the Indian Himalayas from Kulu to Simla, east
through Nepal, Sikkim, and Bhutan to Assam; winters in the
Ceylon hills; on passage in the Himalayan foothills, eastern
Ghats, Mysore, and the southern hill ranges of the Indian
Peninsula.

ZOOTHERA CINEREA

Zoothera cinerea (Bourns and Worcester)
> Geocichla cinerea Bourns and Worcester, 1894, Occ. Pa-
> pers Minnesota Acad. Nat. Sci., 1, p. 23 — Mindoro.
> Mindoro and northern Luzon, Philippine Islands.

ZOOTHERA PERONII

Zoothera peronii peronii (Vieillot)

Turdus peronii Vieillot, 1818, Nouv. Dict. Hist. Nat., nouv. éd., 20, p. 276 — New Holland [*errore* = Timor; restricted to Kupang by Mayr, 1944, Bull. Amer. Mus. Nat. Hist., 83, p. 155].
Western Timor.

Zoothera peronii audacis (Hartert)

Geocichla audacis Hartert, 1899, Bull. Brit. Orn. Club, 8, p. 43 — Dammar Island, in the south of the Banda Sea.
Damar Island, Indonesia and Portuguese Timor.

ZOOTHERA CITRINA

Zoothera citrina citrina (Latham)

Turdus citrinus Latham, 1790, Index Orn., 1, p. 350 — India; restricted to Cachar by Baker, 1921, Journ. Bombay Nat. Hist. Soc., 27, p. 718.
West Pakistan and India along the Himalayas to Nepal, Sikkim, Bhutan, Assam, and northern Burma; winters in the foothills south to Rajasthan, Madras, and Ceylon.

Zoothera citrina cyanotus (Jardine and Selby)

Turdus cyanotus Jardine and Selby, 1828, Illus. Orn., 1, pl. 46 — Bangalore, India.
Turdus citrinus amadoni Biswas, 1951, Journ. Bombay Nat. Hist. Soc., 49, p. 661 — Chanda, Chanda District, Central Provinces.
Peninsular India from Gujarat east to Andhra and south; more common in the western hill ranges.

Zoothera citrina innotata (Blyth) 3

Geocichla innotota (sic) Blyth, 1846, Journ. Asiat. Soc. Bengal, 15, p. 370 — Nicobar Islands and Malaya; restricted to Malay Peninsula by Blyth, 1847, *ibid.*, 16, p. 146.
Southern Burma, China in southern Yunnan, Thailand, Laos, southern Vietnam, and Cambodia; winters south into peninsular Burma, Thailand, and Malaya.[1]

---

[1] Chasen's (1935, Bull. Raffles Mus., no. 11, p. 242) and Gibson-Hill's (1949, Bull. Raffles Mus., no. 20, p. 195-196) records of *Z. c. citrina* as an uncommon winter resident in Sumatra and Malaya presumably refer to the subspecies *innotata*.

Zoothera citrina melli (Stresemann)

Turdus citrinus melli Streseman, 1923, Journ. f. Orn., 71,
p. 365 — Dragon's Head, Kwangtung Province, China.
China, breeding in West Fukien (?) ; winters to Lung
T'ou Shan, northern Kwangtung.

Zoothera citrina courtoisi (Hartert)

Turdus citrinus courtoisi Hartert, 1919, Bull. Brit. Orn.
Club, 40, p. 52 — Leoufang, Anhwei Province, eastern
eastern China.
Anhwei, China.

Zoothera citrina aurimacula (Hartert)

Turdus citrinus aurimacula Hartert, 1910, Novit. Zool.,
17, p. 236 — Hainan.
Hainan Island and central Annam, South Vietnam.

Zoothera citrina andamanensis (Walden)

Geocichla andamanensis Walden, 1874, Ann. Mag. Nat.
Hist., 14, p. 156 — Andamans.
Andaman Islands.

Zoothera citrina albogularis (Blyth)

G.(cocichla) albogularis Blyth, 1847, Journ. Asiat. Soc.
Bengal, 16, p. 146 — Nicobar Islands.
Nicobar Islands.

Zoothera citrina gibsonhilli (Deignan)

Geokichla citrina gibson-hilli Deignan, 1950, Zoologica,
New York, 35, p. 127 — Sungei Balik (*ca.* lat. 10° 31'
N., long. 98° 33' E.), Mergui District, Tenasserim Di-
vision, Burma.
Central part of the Malay Peninsula from southern Tenas-
serim, Burma to Trang Province, Thailand.

Zoothera citrina aurata (Sharpe)

Geocichla aurata Sharpe, 1888, Ibis, p. 478 — Mount
Kinabalu.
British North Borneo on Mount Kinabalu and Mount
Trus Madi.

Zoothera citrina rubecula (Gould)

> *Geocichla rubecula* Gould, 1836, Proc. Zool. Soc. London,
> p. 7 — Java; restricted to West Java by Bartels, 1938,
> Orn. Monatsb., 46, p. 115.
> West Java.

Zoothera citrina orientis (Bartels)

> *Geocichla citrina orientis* Bartels, 1938, Orn. Monatsb.,
> 46, p. 115 — Gunong Raoeng, Idjen Mountains, East
> Java.
> East Java and Bali (subspecies ?).

ZOOTHERA EVERETTI

Zoothera everetti (Sharpe)

> *Geocichla everetti* Sharpe, 1892, Ibis, p. 323 — Mount
> Dulit, northwest Borneo.
> High mountains of Sarawak and British North Borneo
> (Mounts Kinabalu and Trus Madi).

ZOOTHERA SIBIRICA

Zoothera sibirica sibirica (Pallas)

> *Turdus sibiricus* Pallas, 1776, Reise versch. Prov. Russ.
> Reichs, 3, p. 694 — Siberia [= Konda River, Transbai-
> calia, *vide* Pallas, *ibid.*, p. 186].
> Siberia in the Yenisei and Lena River areas north to lat.
> 69° N., west to Mariinsk, east to Transbaicalia, northeast
> Mongolia, northwest Manchuria, Amur River region, Us-
> suriland, and Sakhalin (uncertain ?); migrates through
> Manchuria, Korea, and north and east China; winters in
> Vietnam, Laos, Thailand, central Burma, India in the Mani-
> pur Hills (rare) and on the Andaman Islands, Malaya,
> Sumatra, Nias Island (once), Borneo (very rare), and Java.

Zoothera sibirica davisoni (Hume)

> *Turdulus Davisoni* Hume, 1877, Stray Feathers, 5, p. 63
> — Mooleyit, Tenasserim.
> Sakhalin Island, northern Japan from Hokkaido to Hon-
> shu; migrates through Shikoku and Kyushu and the south
> China coast; winters in north Vietnam (Tonkin), Burma,
> Tenasserim, Malaya, and Sumatra.

ZOOTHERA NAEVIA

Zoothera naevia naevia (Gmelin)

Turdus naevius Gmelin, 1789, Syst. Nat., 1 (2), p. 817,
based on the Spotted Thrush of Latham, 1783, Gen.
Syn., 2 (1), p. 27 — in sinu Americae Natcae [= Noot-
ka Sound, Vancouver Island, British Columbia].

Southeastern Alaska (Yakutat Bay) south along the
coastal areas through the Cascade Range in British Columbia and islands through Washington and Oregon to northwestern California; winters from extreme southern Alaska
south to southwestern California as far as San Diego and
Santa Cruz Island; casual inland in California to Eagle
Lake.

Zoothera naevia meruloides (Swainson)

Orpheus meruloides Swainson, 1832, in Swainson and
Richardson, Fauna Bor.-Amer., 2 (1831), p. 187, pl. 38
— Fort Franklin, lat. 65¼° [= Mackenzie; restricted to
Great Bear Lake, Mackenzie, by Amer. Orn. Union,
1931, Check-list North Amer. Birds, ed. 4, p. 257].

Northern Alaska from northern Yukon east to northern
and western Mackenzie south to the base of the Alaska Peninsula, Kodiak Island, Prince William Sound, central and
southeast British Columbia, southwest Alberta, eastern
Washington, northeast Oregon, northern Idaho, and northwestern Montana; winters from southern British Columbia
and northern Idaho south in California to Death Valley,
Santa Cruz and San Clemente Islands, and northeastern
Baja California; casual in Montana; accidental in northern
Alaska (Point Barrow), Guadeloupe Island, and a number
of western states east to Minnesota and Kansas, and on the
east coast in Quebec, Massachusetts, New York, and New
Jersey.

ZOOTHERA PINICOLA

Zoothera pinicola (Sclater)

Turdus pinicola Sclater, 1859, Proc. Zool. Soc. London,
p. 334 — southern Mexico, pine-forests of the tableland
above Jalapa, Veracruz.

High mountains of Mexico from southern Chihuahua and
Coahuila through Sinaloa, Durango, Nayarit, Jalisco, Mi-

choacán, Guerrero, Oaxaca, Durango, Distrito Federal, Hidalgo, and Puebla to west-central Veracruz.

ZOOTHERA PIAGGIAE

Zoothera piaggiae piaggiae (Bouvier)

Turdus piaggiae Bouvier, 1877, Bull. Soc. Zool. France, 2, p. 456 — Uganda (M'Tésa's country) [= Lake Tana, northern Abyssinia, *vide* Chapin, 1953, Bull. Amer. Mus. Nat. Hist., 75A, p. 579].

Geocichla gurneyi tanganjicae Sassi, 1914, Anz. Akad. Wiss. Wien, Math.-Naturwiss., 51, p. 311 — forest northwest of Tanganyika, 2,000 m.

Ethiopia, Sudan (Borna Hills), northern and western Kenya, Uganda and eastern Congo mountains west of Lake Tanganyika almost to Albertville.

Zoothera piaggiae hadii (Macdonald)

Geokichla piaggiae hadii Macdonald, 1940, Bull. Brit. Orn. Club, 60, p. 98 — Emogadung, Dongotona Mountains, southeastern Sudan.

Imatong and Dongotona Mountains, southeastern Sudan.

Zoothera piaggiae kilimensis (Neumann)

Geocichla gurneyi kilimensis Neumann, 1900, Journ. f. Orn., 48, pp. 188; 310 — Kifinika, Kilimanjaro.

Geocichla piaggiae keniensis Mearns, 1913, Smiths. Misc. Coll., 61 (10), p. 3 — Mount Kenya, Kenya Colony.

Central and southern Kenya from Mount Kenya and the Aberdares south to Mount Kilimanjaro, Tanganyika.

Zoothera piaggiae rowei (Grant and Mackworth-Praed)

Geokichla piaggiae rowei Grant and Mackworth-Praed, 1937, Bull. Brit. Orn. Club, 57, p. 101 — Loliondo Forest, northern Arusha District, northern Tanganyika.

Loliondo and Magaidu Forests, Arusha District, northern Tanganyika.

Zoothera piaggiae williamsi (Macdonald)

Geokichla piaggiae williamsi Macdonald, 1948, Bull. Brit. Orn. Club, 69, p. 16 — Mount Muhavura, southwestern Kigezi, Uganda.

Muhavura Mountain, southwestern Uganda.

ZOOTHERA OBERLAENDERI

Zoothera oberlaenderi (Sassi)

Geocichla gurneyi oberlaenderi Sassi, 1914, Anz. Akad. Wiss. Wien, Math.-Naturwiss., 51, p. 310 — between Beni and Mawambi, northeastern Congo.

Northeastern Congo from the Semliki Valley to the northern edge of the Congo forest near Arebi; Uganda, Toro District (Bwamba Forest).

ZOOTHERA GURNEYI

Zoothera gurneyi chuka (van Someren)

Geocichla gurneyi chuka van Someren, 1930, Journ. East Africa Uganda Nat. Hist. Soc., no. 37, p. 195 — Chuka, Mount Kenya, central Kenya.

Mount Kenya, Kenya.

Zoothera gurneyi chyulu (van Someren)

Geokichla gurneyi chyulu van Someren, 1939, Journ. East Africa Uganda Nat. Hist. Soc., 14, p. 77 — Chyulu Range, southeastern Kenya.

Chyulu Mountains, southeastern Kenya.

Zoothera gurneyi otomitra (Reichenow)

Geocichla gurneyi otomitra Reichenow, 1904, Orn. Monatsb., 12, p. 95 — Bulongwa and Tandalla, Kondeland; restricted to Bulongwa, northeast of Lake Nyasa, by Sclater, 1930, Syst. Av. Aethiop., p. 444.

Geocichla gurneyi raineyi Mearns, 1913, Smiths. Misc. Coll., 61 (10), p. 4 — Mount Mbololo, east of Mount Kilimanjaro, southeastern Kenya.

Geocichla gurneyi usambarae Neumann, 1920, Journ. f. Orn., 68, p. 82 — Mlalo, near Wilhelmstal, Usambara.

Taita area of southeast Kenya south through Tanganyika mountains, and northern Nyasaland south to Mzumara and Mussissi; Northern Rhodesia (Nyika, one specimen) ; Mount Moco, Angola.

Zoothera gurneyi gurneyi (Hartlaub)

Turdus gurneyi Hartlaub, 1864, Ibis, p. 350, pl. 9 — near Pietermaritzburg, Natal.

South Africa in Natal and eastern Cape Province, Pondoland.

Zoothera gurneyi disruptans (Clancey)

Turdus gurneyi disruptans Clancey, 1955, Bull. Brit. Orn. Club, 75, p. 74 — Vumba Highlands, 5,500 ft., near Umtali, eastern Southern Rhodesia.

Central and southern Nyasaland, Mozambique, eastern Southern Rhodesia from Melsetter to Inyanga, and eastern and northern Transvaal.

ZOOTHERA CAMERONENSIS

Zoothera cameronensis (Sharpe) [1]

Geocichla cameronensis Sharpe, 1905 (July), Ibis, p. 472 — Efulen, Cameroon.

Known only from Efulen, Cameroons.

ZOOTHERA PRINCEI

Zoothera princei princei (Sharpe)

Chamaetylas princei Sharpe, 1873, Proc. Zool. Soc. London, p. 625 — Denkera, in the interior of Fantee.

Forests of Sierra Leone (?), upper Guinea, Liberia, Ivory Coast, and Ghana.

Zoothera princei batesi (Sharpe)

Geocichla batesi Sharpe, 1905 (Nov.), Bull. Brit. Orn. Club, 16, p. 36 — Efulen, Cameroons.

Geocichla princei graueri Sassi, 1914, Anz. Akad. Wiss. Wien, Math.-Naturwiss., 51, p. 309 — Moera forest west, of Semliki, Congo.

Coastal southern Cameroons, Cameroun Republic, south nearly to Río Muni, northeastern Congo, Avakubi, Babeyru, and Uganda (Bugoma forest; one specimen).

ZOOTHERA CROSSLEYI

Zoothera crossleyi crossleyi (Sharpe)

Turdus crossleyi Sharpe, 1871, Proc. Zool. Soc. London, p. 607, pl. 47 — Cameroon Mountain.

Cameroon Mountain and Mount Kupe, southern Cameroons, Cameroun Republic.

---

[1] Appears to be a distinct species, *fide* B. P. Hall (pers. comm.).

Zoothera crossleyi pilettei (Schouteden)

G. (*eocichla*) *Gurneyi Pilettei* Schouteden, 1918, Rev. Zool. Africaine, 5, p. 294 — Lesse, Semliki Valley.

Northeastern Congo, Semliki Valley, near Ruwenzori and near Beni; Mount Wago, west of Lake Albert (identity ?).

ZOOTHERA GUTTATA

Zoothera guttata guttata (Vigors)

Turdus guttatus Vigors, 1831, Proc. Zool. Soc. London, p. 92 — Algoa Bay, Africa [= Durban, Natal, *apud* Smith, 1839, Ill. Zool. South Africa, Aves, 2 (8), pl. 39].

Turdus fischeri natalicus Grote, 1938, Falco, 34, p. 14, *nom. nov.* for *Turdus guttatus* Vigors, 1831, *nec Muscicapa guttata* Pallas, 1814, = *Catharus guttatus* (Pallas).

Turdus fischeri belcheri Benson, 1950, Ostrich, 21 (2), p. 58 — Soche hill, near Blantyre, southern Nyasaland.

Southern Nyasaland, Natal, Pondoland, and Cape Province near East London (sight records).

Zoothera guttata fischeri (Hellmayr)

Turdus guttatus fischeri Hellmayr, 1901, Orn. Monatsb., 9, p. 54 — Pangani River, Tanganyika.

Coastal area of Kenya and Tanganyika from Kipini to the Pangani River.

ZOOTHERA SPILOPTERA

Zoothera spiloptera (Blyth)

O. (*reocincla*) *spiloptera* Blyth, 1847, Journ. Asiat. Soc. Bengal, 16, p. 142 — Ceylon.

Ceylon.

ZOOTHERA ANDROMEDAE

Zoothera andromedae (Temminck)

Myiothera andromedae Temminck, 1826, Pl. Col., livr. 66, pl. 392 — Java and Sumatra.

Geocichla mindanensis Mearns, 1907, Philippine Journ. Sci., 2, sec. A, p. 359 — Mount Malindang, 6,500 ft., northwestern Mindanao, Philippine Islands.

Sumatra, Enggano, western Java, Lombok, Timor, Wetar, and Roma Islands, Indonesia; Mindoro and Mindanao, Philippine Islands.

ZOOTHERA MOLLISSIMA

Zoothera mollissima whiteheadi (Baker)

Oreocincla whiteheadi Baker, 1913, Bull. Brit. Orn. Club, 31, p. 79 — Khagan Valley, Afridi Country, North West Frontier Province, India.

Oreocincla mollissima simlaensis Baker, 1924, Fauna Brit. India, Birds, 2, p. 164 — Simla, Punjab.

North West Frontier Province of West Pakistan east along the Himalayas to India in Almora, Garhwal, and west-central Nepal, where intergrades with *mollissima*.

Zoothera mollissima mollissima (Blyth)

T.(urdus) mollissimus Blyth, 1842, Journ. Asiat. Soc. Bengal, 11, p. 188 — Darjeeling.

Nepal (east of *whiteheadi*), Darjeeling, Sikkim, Bhutan, southeast Tibet, Assam, and northern Burma (?); winters in southern Assam, northeast Burma, China in northwest Yunnan, and northern Vietnam (Tonkin).

Zoothera mollissima griseiceps (Delacour)

Oreocincla griseiceps Delacour, 1930, Ibis, p. 581 — Chapa.

Borders of western Szechwan, southern and central Sikang, Yunnan, and northern Vietnam (Tonkin).

ZOOTHERA DIXONI

Zoothera dixoni (Seebohm)

Geocichla dixoni Seebohm, 1881, Cat. Birds Brit. Mus., 5, p. 161 — Himalayas, (specimens from) Nepal (and) Darjeeling.

Indian Himalayas from Himachal Pradesh east through Nepal, Darjeeling, Sikkim, Bhutan, and Assam, southeast Tibet and Sikang, to the borders of western Szechwan, northwestern Yunnan, and northern Burma (Adung Valley); winters in southern Assam, northern and central Burma, northern Thailand, and northern Vietnam (Tonkin).

ZOOTHERA DAUMA

Zoothera dauma aurea (Holandre)

Turdus varius Pallas, 1811, Zoogr. Rosso-Asiat., 1, p. 449 — Krasnoyarsk; *nec Turdus varius* Vieillot, 1803.

Turdus aureus Holandre, 1825, Ann. Moselle, p. 60 —
Metz, eastern France [reference not verified].

Conifer areas of Siberia from the Yenisei east to Tunguska and north to 62°-64° N. lat., south to Krasnoyarsk and Lake Baikal, south and west irregularly across western Siberia and the Urals to eastern Russia (Perm), along the Tian Shan, northeast Mongolia (?), northwest Manchuria, and Korea; migrates through western Siberia, Sinkiang, Mongolia, Manchuria, Korea, Tsushima, and Quelpart Islands, eastern China, and the Riu Kiu Islands; winters from the Yangtze valley south in China west to Yunnan, north and northeastern Burma (irregularly), northern Thailand (rare), Laos, Vietnam, Philippines (Luzon, Mindoro, rare) ; vagrant to western Europe, western Mediterranean and British Isles.

Zoothera dauma dauma (Latham)

Turdus Dauma Latham, 1790, Index Orn., 1, p. 362 —
India; restricted to Kashmir by Baker, 1921, Journ.
Bombay Nat. Hist. Soc., 27, p. 720.

Oreocincla dauma socia Thayer and Bangs, 1912, Mem.
Mus. Comp. Zool., 40, p. 174 — Tatsienlu, western
Szechwan [= Kangting, eastern Sikang].

Turdus aureus angustirostris Gyldenstolpe, 1916, Orn.
Monatsb., 24, p. 28 — Khun Tan, northern Siam.

Himalayas of West Pakistan and India from Kashmir and Murree east through Nepal, Darjeeling, Sikkim, Bhutan, Assam in the southern hills, Burma, and southeast Tibet; north in China in eastern Sikang, north Yunnan and western Szechwan; Thailand, Vietnam, Laos, and Formosa; winters south into the plains of India, East Pakistan, and southern Burma; vagrant to Malaya.

Zoothera dauma neilgherriensis (Blyth)

O.(reocincla) neilgherriensis Blyth, 1847, Journ. Asiat.
Soc. Bengal, 16, p. 141 — Nilgiris.

Southern peninsular Indian hills in Mysore, Madras, and Kerala.

Zoothera dauma imbricata Layard

Zoothera imbricata Layard, 1854, Ann. Mag. Nat. Hist.,
ser. 2, 13, p. 212 — Ceylon.
Ceylon hills.

Zoothera dauma toratugumi (Momiyama)

Turdus aureus toratugumi Momiyama, 1940, Dobuts. Zas-
shi, 52, p. 462 — Tosa, Shikaku [reference not verified].
Turdus aureus miharagokko Momiyama, 1940, Dobuts.
Zasshi, 52, p. 462 — Hachijo Island [reference not veri-
fied].
O.(reocincla) d.(auma) exorientis Portenko, 1954, Keys
Fauna U.S.S.R., no. 54, Birds, 3, p. 221 — Suputinka
River, Ussuriland.

Breeds in Manchuria (where integrades with *aureus*),
Ussuriland, Korea (transient ?), and Japan from Hokkaido
to Honshu; winters in Honshu, Shikoku (breeds ?), Kyushu Island, and Miyake Island (one record), and south to
Formosa (Pescadores Islands).

Zoothera dauma major (Ogawa)

Geocichla major Ogawa, 1905, Annot. Zool. Japan, 5, p.
178 — Amami-O-Shima.
Turdus dauma amami Hartert, 1922, Vög. pal, Fauna, 3, p.
2155, *nom. nov. Turdus dauma major* (Ogawa) *nec
Turdus major* Brehm, 1831.
Amami-O-Shima Island, northern Riu Kiu Islands, Japan.

Zoothera dauma hancii (Swinhoe)

Oreocincla hancii Swinhoe, 1863, Ibis, p. 275 — North
Formosa.
Oreocincla horsfieldi affinis Richmond, 1902, Proc. Biol.
Soc. Washington, 15, p. 158 — Khow Nok Ram, 3,000
ft., Trong, lower Siam.
Hills of Tenasserim (?) and peninsular Thailand
(Trang), South Vietnam (Langbian Peaks, Dalat), and
Formosa; southern Riu Kiu Islands (Iriomotejima, va-
grant?).

Zoothera dauma horsfieldi (Bonaparte)

Oreocincla horsfieldi Bonaparte, 1857, Rev. Mag. Zool.
[Paris], p. 205 — Java.
Sumatra, Java, and Lombok; Bali (?).

Zoothera dauma machiki (Forbes)

Geocichla machiki Forbes, 1883, Proc. Zool. Soc. London,
p. 589, pl. 52 — Timor-Laut, Tanimbar Islands.
Tanimbar Islands.

Zoothera dauma papuensis (Seebohm)

Geocichla papuensis Seebohm, 1881, Cat. Birds Brit. Mus., 5, p. 158, pl. 9 — southeastern New Guinea.

Nassau range, mountains of the Huon Peninsula, and mountains of southeast New Guinea.

Zoothera dauma eichhorni (Rothschild and Hartert)

Turdus dauma eichhorni Rothschild and Hartert, 1924, Bull. Brit. Orn. Club, 44, p. 52 — St. Matthias Island.

Saint Matthias (Mussau) Island, Bismarck Archipelago.

Zoothera dauma choiseuli (Hartert)

Turdus dauma choiseuli Hartert, 1924, Novit. Zool., 31, p. 273 — Choiseul Island.

Choiseul Island, Solomon Islands.

Zoothera dauma cuneata (De Vis)

Geocichla cuneata De Vis, 1890, Proc. Royal Soc. Queensland, 6 (1889), p. 242—Herberton, north Queensland.

Mountains of Herberton district, north Queensland, Australia.

Zoothera dauma heinei (Cabanis)

Oreocincla Heinei Cabanis, 1851, Mus. Hein., 1, p. 6 — Japan [*errore* = north Australia, Queesland, *idem*, 1872, Journ. f. Orn., 20, p. 237].

Southern Queensland.

Zoothera dauma lunulata (Latham)

Turdus lunulata Latham, 1801, Index Orn., suppl., p. xlii — Nova Hollandia; restricted to Sydney by Mathews, 1920, Birds Australia, suppl. 1, check-list, p. 147.

Geocichla lunulata halmaturina Campbell, 1906, Emu, 5, p. 142 — Kangaroo Island.

Turdus lunulatus dendyi Mathews, 1912, Novit. Zool., 18, p. 340 — (Sassafras,) Victoria.

Turdus australasianus Mathews, 1920, Birds Australia, suppl. 1, check-list, p. 147, ex Cotton, 1848, Tasmanian Journ. Nat. Sci., 3, p. 363, *nomen nudum*.

New South Wales, Victoria, and South Australia; Kangaroo Island.

Zoothera dauma macrorhyncha (Gould)
Oreocincla macrorhyncha Gould, 1838, Synops. Birds
  Australia, pt. 4, app., p. 3 — New Zealand, or Van Die-
  men's Land (i.e. Tasmania).
Tasmania.

ZOOTHERA TALASEAE

Zoothera talaseae (Rothschild and Hartert)
Turdus talaseae Rothschild and Hartert, 1926, Bull. Brit.
  Orn. Club, 46, p. 53 — Talasea, New Britain.
New Britain Island.

ZOOTHERA MARGARETAE

Zoothera margaretae turipavae Cain and Galbraith
Zoothera margaretae turipavae Cain and Galbraith, 1955,
  Bull. Brit. Orn. Club, 75, p. 92 — Turipava, 4,100 ft.,
  mountains of Guadalcanal.
Guadalcanal, Solomon Islands.

Zoothera margaretae margaretae (Mayr)
Turdus margaretae Mayr, 1935, Amer. Mus. Novit., no.
  820, p. 4 — San Cristobal, 1,900 ft., Solomon Islands.
San Cristobal, Solomon Islands.

ZOOTHERA MONTICOLA

Zoothera monticola monticola Vigors
Zoothera monticola Vigors, 1832, Proc. Zool. Soc. London,
  p. 172 — Himalayan Mountains [= Simla-Almora area,
  vide Ticehurst and Whistler, 1924, Ibis, p. 472].
Zoothera monticola tenebricola Koelz, 1954, Contrib. Inst.
  Regional Explor., no. 1, p. 13 — Sangau, Lushai Hills.
Indian Himalayas from Kulu, Nepal, Darjeeling, Sikkim,
Bhutan, and Assam to northeastern Burma in Myitkina
District and the northern Chin Hills.

Zoothera monticola atrata Delacour and Greenway
Zoothera monticola atrata Delacour and Greenway, 1939,
  Bull. Brit. Orn. Club, 59, p. 131 — Chapa, 5,000 ft.,
  Tonkin.
North Vietnam (Tonkin).

ZOOTHERA MARGINATA

Zoothera marginata Blyth

Z.(oothera) marginata Blyth, 1847, Journ. Asiat. Soc.
Bengal, 16, p. 141 — Arakan.
Zoothera marginata parva Delacour and Jabouille, 1930,
Oiseau Rev. Franç. Orn., 11, p. 397 — Long-Phinh,
Pakha (Tonkin).
Zoothera marginata tenebrosa Koelz, 1952, Journ. Zool.
Soc. India, 4, p. 41 — Nichuguard, Naga Hills.

Hills of Nepal, Darjeeling, Sikkim, Bhutan, Assam,
Burma, Thailand (except the Peninsula), Laos, and North
and South Vietnam.

ZOOTHERA TERRESTRIS

Zoothera terrestris (Kittlitz)

Turdus terrestris Kittlitz, 1831, Mém. Acad. Imp. Sci.
St. Pétersbourg, 1, p. 245, pl. 17 — Bonin Island.
Bonin Island; extinct.

GENUS **AMALOCICHLA** DE VIS

Amalocichla De Vis, 1892, Ann. Rept. British New
Guinea, 1890-1891, app. cc., p. 95. Type, by monotypy,
Amalocichla sclateriana De Vis [reference not veri-
fied].
Pseudopitta Reichenow, 1915, Journ. f. Orn., 63, p. 129.
Type, by monotypy, Eupetes incertus Salvadori.

AMALOCICHLA SCLATERIANA

Amalocichla sclateriana occidentalis Rand

Amalocichla sclateriana occidentalis Rand, 1940, Amer.
Mus. Novit., no. 1074, p. 1 — Lake Habbema, 9 km.
northeast, 2,800 m., Snow Mountains, Netherlands New
Guinea.
Oranje Mountains, West New Guinea.

Amalocichla sclateriana sclateriana De Vis

Amalocichla sclateriana De Vis, 1892, Ann. Rept. British
New Guinea, 1890-1891, p. 95 — Mount Owen Stanley,
southeastern New Guinea [reference not verified].
Mountains of southeastern New Guinea.

AMALOCICHLA INCERTA

Amalocichla incerta incerta (Salvadori)

Eupetes incertus Salvadori, 1875, Ann. Mus. Civ. Genova, 7, p. 967 — Mount Arfak.

Arfak Mountains, West New Guinea.

Amalocichla incerta olivascentior Hartert

Amalocichla incerta olivascentior Hartert, 1930, Novit. Zool., 36, p. 85 — Wondiwoi Mountain.

Wandammen and Weyland Mountains, Wissel Lakes district east to Nassau and Oranje Mountains, West New Guinea.

Amalocichla incerta brevicauda (De Vis)

Drymoedus brevicauda De Vis, 1894, Ann. Rept. British New Guinea, 1893-1894, p. 103 — Mount Maneao, southeast New Guinea [reference not verified].

Eastern New Guinea in the Schraderberg, Bismarck, Saruwaged (Huon Peninsula), Herzog, and southeastern mountains.

GENUS **CATAPONERA** HARTERT

Cataponera Hartert, 1896, Novit. Zool., 3, p. 70. Type, by monotypy, *Cataponera turdoides* Hartert.

CATAPONERA TURDOIDES

Cataponera turdoides abditiva Riley

Cataponera abditiva Riley, 1918, Proc. Biol. Soc. Washington, 31, p. 158 — Rano Rano, Celebes.

North-central Celebes.

Cataponera turdoides tenebrosa Stresemann

Cataponera turdoides tenebrosa Stresemann, 1938, Orn. Monatsb., 46, p. 46 — Latimodjong Mountains.

Latimodjong Mountains, southern Celebes.

Cataponera turdoides turdoides Hartert

Cataponera turdoides Hartert, 1896, Novit. Zool., 3, p. 70 — Bonthain Peak.

Mountains of southwestern Celebes.

Cataponera turdoides heinrichi Stresemann

Cataponera turdoides heinrichi Stresemann, 1938, Orn.
Monatsb., 46, p. 46 — Tanke Salokko, 2,000 m., Meng-
koka Mountains.
Mengkoka Mountains, southeast Celebes.

Genus NESOCICHLA Gould

Nesocichla Gould, 1855, Proc. Zool. Soc. London, p. 165.
Type, by montypy, *Nesocichla eremita* Gould.

NESOCICHLA EREMITA

Nesocichla eremita eremita Gould

Nesocichla eremita Gould, 1855, Proc. Zool. Soc. London.
p. 165 — Tristan da Cunha Island.
Tristan da Cunha Island, South Atlantic.

Nesocichla eremita gordoni Stenhouse

Nesocichla eremita gordoni Stenhouse, 1924, Scottish
Naturalist, p. 95 — Inaccessible Island.
Inaccessible Island, South Atlantic.

Nesocichla eremita procax Elliott

Nesocichla eremita procax Elliott, 1954, Bull. Brit. Orn.
Club, 74, p. 22 — Nightingale Island.
Nightingale Island, South Atlantic.

Genus CICHLHERMINIA Bonaparte

Cichlherminia Bonaparte, 1854, Compt. Rend. Acad. Sci.
Paris, 38, p. 2. Type, by subsequent designation (Gray,
1855, Cat. Gen. Subgen. Birds, p. 43), *Turdus lherminieri* Lafresnaye.

CICHLHERMINIA LHERMINIERI

Cichlherminia lherminieri lawrencii Cory

Cichlherminia lawrencii Cory, 1891, Auk, 8, p. 44 —
Montserrat, West Indies.
Montserrat, West Indies.

Cichlherminia lherminieri lherminieri (Lafresnaye)
　　Turdus L'Herminieri Lafresnaye, 1844, Rev. Zool.
　　[Paris], 7, p. 167 — Guadeloupe.
　　Guadeloupe, West Indies.

Cichlherminia lherminieri dominicensis (Lawrence)
　　Margarops dominicensis Lawrence, 1880, Forest and
　　Stream, 14, no. 9, p. 165 — Dominica.
　　Dominica, West Indies.

Cichlherminia lherminieri sanctaeluciae (Sclater)
　　Margarops sanctae-luciae Sclater, 1880, Ibis, p. 73 —
　　St. Lucia.
　　St. Lucia, West Indies.

GENUS **PHAEORNIS** SCLATER

Phaeornis Sclater, 1859, Ibis, p. 327. Type, by monotypy,
　　Taenioptera obscura = *Muscicapa obscura* Gmelin.

PHAEORNIS OBSCURUS

Phaeornis obscurus myadestinus Stejneger
　　Phaeornis myadestina Stejneger, 1887, Proc. U.S. Nat.,
　　Mus., 10, p. 90 — Kauai Island, Hawaiian Archipelago.
　　Kauai Island.

Phaeornis obscurus oahensis Wilson and Evans
　　Phaeornis oahensis Wilson and Evans, 1899, Aves Ha-
　　waiienses, introd., p. xiii — Oahu.
　　Oahu Island; extinct.

Phaeornis obscurus rutha Bryan
　　Phaeornis rutha Bryan, 1908, Occ. Papers Bishop Mus.
　　[Honolulu], 4, pp. 43; 81 — Kilohana Mountain, Pu-
　　nalu Mountain, Halawa, Molokai.
　　Molokai Island.

Phaeornis obscurus lanaiensis Wilson
　　Phaeornis lanaiensis Wilson, 1891, Ann. Mag. Nat. Hist.,
　　ser. 6, 7, p. 460 — Lanai.
　　Lanai Island.

Phaeornis obscurus obscurus (Gmelin)

(Muscicapa) obscura Gmelin, 1789, Syst. Nat., 1 (2), p.
945 — Sandwich Islands.
Hawaii Island.

PHAEORNIS PALMERI

Phaeornis palmeri Rothschild

Phaeornis palmeri Rothschild, 1893, Avifauna Laysan,
p. 67 — Halemanu, Kauai.
Kauai Island.

GENUS CATHARUS BONAPARTE

Catharus Bonaparte, 1851, Consp. Av., 1 (1850), p. 278.
Type, by monotypy, *Turdus immaculatus* Bonaparte =
Turdus aurantiirostris Hartlaub.

cf. Zimmer, 1944, Auk, 61, pp. 404-408 (*aurantiirostris*
and *griseiceps*).
Bond, G. M., 1963, Proc. U. S. Nat. Mus., 114, pp. 373-
387 (*ustulatus*).

CATHARUS GRACILIROSTRIS

Catharus gracilirostris gracilirostris Salvin

Catharus gracilirostris Salvin, 1865, Proc. Zool. Soc. London, (1864), p. 580 — Costa Rica (Volcán de Cartago).
Mountains of Costa Rica.

Catharus gracilirostris accentor Bangs

Catharus gracilirostris accentor Bangs, 1902, Proc. New
England Zool. Club, 3, p. 50 — Volcán de Chiriquí,
Chiriquí, Panama.
Mountains of western Panama.

CATHARUS AURANTIIROSTRIS

Catharus aurantiirostris melpomene (Cabanis)

T.(urdus) Melpomene Cabanis, 1851, Mus. Hein., 1, p.
5 — Xalapa [= Jalapa, Veracruz, Mexico].
Oaxaca, Chiapas, northeast Puebla, and west-central
Veracruz, Mexico.

Catharus aurantiirostris clarus Jouy

Catharus melpomene clarus Jouy, 1894, Proc. U. S. Nat. Mus., 16, p. 773 — Barranca Ibarra, Jalisco, Mexico.

Southern Sinaloa, Nayarit, Jalisco, Michoacán, Guerrero, central Chihuahua, Durango, Guanajuato, México, San Luis Potosí, Hidalgo, western Puebla, and southwestern Tamaulipas, Mexico.

Catharus aurantiirostris aenopennis Moore

Catharus aurantiirostris aenopennis Moore, 1938, Proc. Biol. Soc. Washington, 50, p. 96 — floor of Arroyo Hondo, about twenty miles north of junction of Ríos Chinipas and Fuerte, southwestern Chihuahua, Mexico.

Northern Sinaloa and southwestern Chihuahua; winters to lowlands of Sinaloa, Mexico.

Catharus aurantiirostris bangsi Dickey and van Rossem

Catharus melpomene bangsi Dickey and van Rossem, 1925, Proc. Biol. Soc. Washington, 38, p. 135 — Volcán de San Salvador, El Salvador.

El Salvador, Honduras and Guatemala.

Catharus aurantiirostris costaricensis Hellmayr

Catharus melpomene costaricensis Hellmayr, 1902, Journ. f. Orn., 50, p. 45 — Costa Rica; restricted to San José by Hellmayr, 1934, Field Mus. Nat. Hist. Pub., Zool. Ser., 13, pt. 7, p. 471.

Catharus melpomene bathoica Bangs and Griscom, 1932, Proc. New England Zool. Club, 13, p. 51 — Ojo Ancho, 500 ft., tip of Nicoya Peninsula, Costa Rica.

Nicaragua and northwest Costa Rica.

Catharus aurantiirostris russatus Griscom

Catharus griseiceps russatus Griscom, 1924, Amer. Mus. Novit., no. 141, p. 6 — Boruca, Costa Rica.

Mountains of southwestern Costa Rica and western Panama (Chiriquí).

Catharus aurantiirostris griseiceps Salvin

Catharus griseiceps Salvin, 1866, Proc. Zool. Soc. London, p. 68 — Veraguas.

Mountains of western Panama in eastern Chiriquí and Veraguas.

Catharus aurantiirostris phaeopleurus Sclater and Salvin

Catharus phaeopleurus Sclater and Salvin, 1875, Proc. Zool. Soc. London, p. 541 — Medellin, Antioquia, Colombia.

Colombia in the Cauca and upper Patía valleys south to the upper Guáitara Valley, Nariño.

Catharus aurantiirostris aurantiirostris (Hartlaub)

Turdus aurantiirostris Hartlaub, 1850, Rev. Zool. [Paris], ser. 2, 1, p. 158 — Venezuela; restricted to Caracas by Hartlaub, 1851, Contrib. Orn., p. 80, pl. 72.

Eastern Colombia in the Santa Marta range and Magdalena valley; coastal Venezuela in Distrito Federal, Miranda, Aragua, Carabobo, Lara, Falcón, and the Andes of Táchira.

Catharus aurantiirostris birchalli Seebohm

Catharus birchalli Seebohm, 1881, Cat. Birds Brit. Mus., 5, p. 289 — Orinoco valley; restricted to mountains inland of Cumaná by Hellmayr, 1919, Verh. Orn. Ges. Bayern, 14, p. 126.

Northeast Venezuela, in Sucre, and on Trinidad (Aripo).

Catharus aurantiirostris barbaritoi Aveledo and Gines

Catharus aurantiirostris barbaritoi Aveledo and Gines, 1952, Novedades Cient., Contrib. Ocas. Mus. Hist. Nat. La Salle, Caracas, ser. Zool., no. 6, p. 11 — Kunana, 1,140 m., Río Negro, Sierra de Perijá, Zulia, Venezuela.

Venezuela in Sierra de Perijá and valley of upper Río Negro.

Catharus aurantiirostris inornatus Zimmer

Catharus aurantiirostris inornatus Zimmer, 1944, Auk, 61, p. 404 — San Gil, Santander, Colombia.

Colombia on western slope of eastern Andes in Sogamoso Valley, Santander.

Catharus aurantiirostris insignis Zimmer

Catharus aurantiirostris insignis Zimmer, 1944, Auk, 61, p. 406 — near San Agustín, Huila, Colombia.

Colombia in the upper Magdalena valley; "Bogotá" region.

CATHARUS FUSCATER

Catharus fuscater hellmayri Berlepsch

Catharus fuscater hellmayri Berlepsch, 1902, Orn. Monatsb., 10, p. 69 — Chiriquí.

Costa Rica and western Panama in Chiriquí and Veraguas.

Catharus fuscater mirabilis Nelson

Catharus fuscater mirabilis Nelson, 1912, Smiths. Misc.
Coll., 60 (3), p. 24 — Mount Pirre, near head of Río
Limon, eastern Panama.

Eastern Panama in the area of Mount Pirre.

Catharus fuscater sanctaemartae Ridgway

Catharus fuscater sanctae-martae Ridgway, 1904, Smiths.
Misc. Coll., 47, p. 112 — Elhibano [= El Libano], Santa
Marta, Colombia.

Northern Colombia in the Santa Marta Mountains.

Catharus fuscater fuscater (Lafresnaye)

M.(yioturdus) fuscater Lafresnaye, 1845, Rev. Zool.
[Paris], 8, p. 341 — Bogotá.

Eastern Panama on Mount Tacarcuna; eastern Andes of
Colombia (La Palmita, Pueblo Nuevo, Cachiri, Peña Blan-
ca) ; Venezuela in Trujillo, Barinas, Mérida, and Táchira;
Ecuador.

Catharus fuscater opertaneus Wetmore

Catharus fuscater opertaneus Wetmore, 1955, Noved.
Colomb., Contr. Cient. Mus. Hist. Nat., Univ. Cauca,
Colombia, no. 2, p. 46 — Hacienda Potreros, 6,500 ft.,
Río Herradura, 15 mi. southwest of Frontino, Antio-
quia, Colombia.

Western Andes of Colombia on the Río Herredura, Frontino, Antioquia.

Catharus fuscater caniceps Chapman

Catharus fuscater caniceps Chapman, 1924, Amer. Mus.
Novit., no. 138, p. 14 — Palambla, 5,000-6,500 ft., Piura, Peru.

Northern and central Peru, west to Río Saña valley.

Catharus fuscater mentalis Sclater and Salvin

Catharus mentalis Sclater and Salvin, 1876, Proc. Zool. Soc. London, p. 352 — "Suape" (Suapi), near "Tilotilo," Yungas, Bolivia.

Southeastern Peru (Dept. Puno) and northern Bolivia (La Paz).

CATHARUS OCCIDENTALIS

Catharus occidentalis olivascens Nelson

Catharus olivascens Nelson, 1899, Proc. Biol. Soc. Washington, 13, p. 31 — Sierra Madre, Chihuahua (65 miles east of Batopilas), Mexico.

Northern Mexico in northern Sinaloa, western Chihuahua, and northwestern Durango.

Catharus occidentalis fulvescens Nelson

Catharus occidentalis fulvescens Nelson, 1897, Auk, 14, p. 75 — Amecameca, México.

Mountains of Mexico in Jalisco, southern Sinaloa, Michoacán, Guerrero, southern Durango, Guanajuato, southern Tamaulipas, and western Puebla, and Mexico.

Catharus occidentalis occidentalis Sclater

Catharus occidentalis Sclater, 1859, Proc. Zool. Soc. London, p. 323 — Totontepec, Oaxaca.

Mountains of southeastern Mexico in Oaxaca, eastern San Luis Potosí, Puebla, and western Veracruz.

Catharus occidentalis alticola Salvin and Godman

Catharus alticola Salvin and Godman, 1879, Biol. Centr.-Amer., Aves, 1, p. 3 — Volcán de Fuego, Guatemala.

Southern Mexico (Chiapas), Guatemala, El Salvador, and Honduras.

Catharus occidentalis frantzii Cabanis

Catharus Frantzii Cabanis, 1861, Journ. f. Orn., 8 (1860), p. 323 — Volcán del Irazú.

Mountains of Costa Rica and western Panama (Chiriquí).

CATHARUS MEXICANUS

Catharus mexicanus mexicanus (Bonaparte)

Mal.(acocychla) mexicana Bonaparte, 1856, Compt. Rend.

Acad. Sci. Paris, 43, p. 998 — Jalapa, Veracruz, Mexico.

Catharus mexicanus smithi Nelson, 1909, Proc. Biol. Soc. Washington, 22, p. 49 — Carricitos, Sierra Madre of the East, fifty miles northwest of Victoria, Tamaulipas, Mexico.

Mexico in Tamaulipas, Hidalgo, México, Veracruz, and western Chiapas.

Catharus mexicanus cantator Griscom

Catharus mexicanus cantator Griscom, 1930, Amer. Mus. Novit., no. 438, p. 4 — Finca Sepacuite, 3,500 ft., about 50 miles east of Coban, Vera Paz, Guatemala.

Highlands of Mexico (Chiapas), eastern Guatemala, and and Honduras (San Pedro mountains).

Catharus mexicanus fumosus Ridgway

Catharus fumosus Ridgway, 1888, Proc. U. S. Nat. Mus., 10, p. 505 — Costa Rica.

Nicaragua, Costa Rica, and western Panama in Chiriquí and Veraguas.

CATHARUS DRYAS

Catharus dryas ovandensis Brodkorb

Catharus dryas ovandensis Brodkorb, 1938, Occ. Papers Mus. Zool. Univ. Michigan, no. 369, p. 4 — Mount Ovando, 1,775 m., Chiapas.

Highlands of Chiapas, Mexico.

Catharus dryas dryas (Gould)

Malacocichla dryas Gould, 1855, Proc. Zool. Soc. London, (1854), p. 285, pl. 75 — Guatemala.

Guatemala, on the Pacific slope of the Cordillera and Sierra de las Minas, and Honduras (Volcán de Puca); western Ecuador.

Catharus dryas maculatus (Sclater)

Malacocichla maculata Sclater, 1858, Proc. Zool. Soc. London, p. 64 — Río Napo, Ecuador [= eastern Ecuador].

Eastern Colombia, at the head of Magdalena Valley and on the eastern slope of Andes and Macarena Mountains, eastern Ecuador, Peru, and Bolivia.

Catharus dryas ecuadoreanus Carriker

Catharus dryas ecuadoreanus Carriker, 1935, Proc. Acad. Nat. Sci. Philadelphia, 87, p. 355 — Alamor, Ecuador. Andes of western Ecuador.

CATHARUS FUSCESCENS

Catharus fuscescens fuscescens (Stephens)

Turdus Fuscescens Stephens, 1817, in Shaw, Gen. Zool., 10 (1), p. 182 — Pennsylvania.

Eastern Canada in Ontario, southern Quebec, New Brunswick, and Nova Scotia; United States from northeastern Ohio through mountains of Pennsylvania, West Virginia, Maryland, Virginia, Kentucky, Tennessee, and North Carolina to northern Georgia, and New England states south to Pennsylvania, New Jersey, and the District of Columbia; migrates through eastern United States west to Nebraska, Oklahoma, and eastern Texas, and through eastern Mexico (Quintana Roo; Yucatán), the Bahamas, Cuba, Honduras, Costa Rica, and Panama; winters in Colombia, Venezuela, and British Guiana to south-central Brazil.

Catharus fuscescens fuliginosa (Howe)

Hylocichla fuscescens fuliginosa Howe, 1900, Auk, 17, p. 271 — Codroy, Newfoundland.

Breeds in southwestern Newfoundland and south-central Quebec; winters presumably in South America; migrates along the eastern United States coast in Massachusetts, Maryland, and Virginia.

Catharus fuscescens salicicola (Ridgway)

Hylocichla fuscescens salicicola Ridgway, 1882, Proc. U.S. Nat. Mus., 4, p. 374 — Colorado; restricted to Fort Garland, Colorado by Amer. Orn. Union, 1910, Check-list North Amer. Birds, ed. 3, p. 360.

Canada from southern British Columbia, southern central Alberta, Saskatchewan, and Manitoba to southern Ontario and south in United States from Nevada, Colorado, northeast Arizona, Wyoming, and South Dakota to Minnesota, southern Wisconsin, Illinois, Indiana, Michigan, and southwestern Ohio; winters from Colombia and Venezuela to Brazil (Mato Grosso).

Catharus fuscescens subpallidus (Burleigh and Duvall)

Hylocichla fuscescens subpallida Burleigh and Duvall, 1959, Proc. Biol. Soc. Washington, 72, p. 33 — Moscow, Latah County, Idaho.

Northern Washington, northeastern Oregon, northern and central Idaho, and western Montana; migrates through Colorado; winter range unknown.

CATHARUS MINIMUS

Catharus minimus minimus (Lafresnaye)

Turdus minimus Lafresnaye, 1848, Rev. Zool. [Paris], 11, p. 5 — Bogotá, Colombia.

Turdus aliciae Baird, 1858, in Baird, Cassin, and Lawrence, Rept. Expl. Surv. R.R. Pacific, 9, p. 217 — Illinois and upper Missouri; restricted to West Northfield, Illinois, by Ridgway, 1907, Bull. U.S. Nat. Mus., 50, pt. 4, p. 61.

Northeast Siberia, Alaska; Canada in Yukon, Mackenzie, British Columbia, Manitoba, Ontario, eastern Quebec, central Labrador, and Newfoundland; migrates through Mississippi Valley and eastern United States, rarely in Montana, Wyoming, Oklahoma, Texas, Mexico (Campeche and Yucatán), Guatemala, Cuba (rare), and Martinique (once); winters in Nicaragua, Colombia, Venezuela, and British Guiana to northern Peru and northwestern Brazil.

Catharus minimus bicknelli (Ridgway)

Hylocichla aliciae bicknelli Ridgway, 1882, Proc. U. S. Nat. Mus., 4, p. 377 — near the summit of Slide Mountain, Ulster County, New York.

Eastern United States in New York and New England; Canada north to southern Nova Scotia, Gaspé Peninsula, and Gulf of St. Lawrence area to Cape St. Charles, southern Labrador; migrates through eastern states, casually west to Illinois, Indiana, Ohio, Tennessee, and Louisiana, and also in the Bahamas; winters in Hispaniola.

CATHARUS USTULATUS

Catharus ustulatus almae (Oberholser)

Hylocichla ustulata almae Oberholser, 1898, Auk, 15, p. 304 — East Humboldt Mts., opposite Franklin Lake, Nevada.

Hylocichla ustulata incana Godfrey, 1952, Canadian Field-Nat., **65** (1951), p. 173 — Lapie River, Canol Road, mile 132, Yukon Territory.

Southern and eastern Alaska (except southeast coast), Yukon, western Mackenzie, west-central Alberta south through the mountains to Colorado and northwest Utah; migrates in Mississippi Valley to winter in southeast and Gulf Coast United States; Costa Rica (Bonilla, one record).

Catharus ustulatus ustulatus (Nuttall)

Turdus ustulatus Nuttall, 1840, Man. Orn. U.S. and Canada, ed. 2, pp. vi, 400, 830 — forests of the Oregon [= Fort Vancouver, Washington].

Southeastern coastal Alaska from Juneau south, coastal British Columbia, west of the Cascade Mountains in Washington and Oregon; migrates through California; winters in western Mexico.

Catharus ustulatus oedicus (Oberholser)

Hylocichla ustulata oedica Oberholser, 1899, Auk, **16**, p. 23 — Santa Barbara, California.

Breeds in California (except southeast) and southwest Oregon (Klamath Mountains) north along the east slopes of the Cascade range to northern Washington; migrates south to winter in Arizona and in Mexico in Baja California, Tamaulipas, Puebla, San Luis Potosí, Guanajuato, Durango, Chihuahua, Sonora, Sinaloa, Nayarit, Guerrero, Oaxaca, and Chiapas.

Catharus ustulatus swainsoni (Tschudi)

Turdus Swainsoni Tschudi, 1845, Faun. Peru., Aves, p. 28; new name for *Merula wilsonii* Swainson, *nec Turdus wilsonii* Bonaparte — Carlton House, Saskatchewan.

Hylocichla ustulata clarescens Burleigh and Peters, 1948, Proc. Biol. Soc. Washington, **61**, p. 118 — Glenwood, Newfoundland.

Breeds in Canada from eastern Alberta (Athabaska River) east through Saskatchewan, Manitoba, Ontario, Quebec, and southern Labrador to Newfoundland and Nova Scotia, south in the United States from northern Minnesota, Wisconsin, and northern Michigan to northern New England, south in the Appalachians of New York, Pennsylvania and West Virginia; migrates south of its breeding range through British Columbia, Idaho, and Illinois to Alabama, Florida, Oklahoma, and Texas; winters rarely in Mexico (Veracruz, Tabasco, Islas Tres Marias, Yucatán, and Chiapas) and in El Salvador, Guatemala, Costa Rica, Jamaica, Swan Island, Cuba, and the Bahamas, and more commonly in Panama, Colombia, Venezuela, British Guiana, Ecuador, Peru, Bolivia, Brazil (upper Rio Negro), and Argentina (Tucumán).

CATHARUS GUTTATUS

Catharus guttatus guttatus (Pallas)

Muscicapa guttata Pallas, 1811, Zoogr. Rosso-Asiat., 1, p. 465 — Kodiak Island, Alaska.

Turdus Pallasii Cabanis, 1847, Arch. f. Naturg., 13, p. 205 — new name for *Musicapa guttata* Pallas, nec *Turdus guttatus* Vigors, 1831.

Hylocichla guttata euboria Oberholser, 1956, Proc. Biol. Soc. Washington, 69, p. 69 — Lewes River, Yukon River, Yukon, Canada.

Alaskan Peninsula and adjacent islands east to southwest Yukon and south to central British Columbia; winters on Vancouver Island and in Washington, Oregon, California, Colorado, Nevada, Texas, Baja California, Tamaulipas, Nuevo León, Sonora, Sinaloa, Hidalgo, San Luis Potosí, Coahuila, Guanajuato, Durango, Chihuahua, Michoacán, and Guadalupe Island (subspecies ?).

Catharus guttatus nanus (Audubon)

Turdus Nanus Audubon, 1839, Orn. Biog., 5, p. 201 — Columbia River [= Fort Vancouver, Washington, *vide* Amer. Orn. Union, 1931, Check-list. Birds North Amer., ed. 4, p. 258; fig. as *"Turdus minor* Gm.", 1838, Birds Amer. (folio), 4, pl. 419, fig. 1].

Hylocichla guttata vaccinia Cumming, 1933, Murrelet, 14, p. 79 — Seymour Mountain, Vancouver District, British Columbia.

Catharus guttatus munroi Phillips, 1962, Anal. Inst. Biol. México, 32 (1961), p. 351 — Nulki Lake, British Columbia.

Coastal southeastern Alaska and western British Columbia; winters from southern British Columbia south to south-

ern Baja California and Sonora (one record), and east as
far as eastern California, Nevada, Arizona, and New Mexico.

Catharus guttatus slevini (Grinnell)

Hylocichla aonalaschkae slevini Grinnell, 1901, Auk, 18,
p. 258 — vicinity of Point Sur, Monterey County,
California.

Hylocichla guttata oromela Oberholser, 1932, Sci. Pub.
Cleveland Mus. Nat. Hist., 4, no. 1, p. 8 — north base
of Brook Peak, Warner Mountain, 15 miles northeast
of Lakeview, Oregon.

Catharus guttatus jewetti Phillips, 1962, Anal. Inst. Biol.
México, 32 (1961), p. 356 — Hurricane Ridge and El-
wha River, Olympic Mountains, Clallam Co., Washington.

Washington, Oregon, and northern California south to
Monterey County; winters in Baja California, Sonora, and
Sinaloa.

Catharus guttatus sequoiensis (Belding)

Turdus sequoiensis Belding, 1889, Proc. California Acad.
Sci., ser. 2, 2, p. 18 — Big Trees, Calaveras County,
California.

Sierra Nevada Range of California and western Nevada
south to mountains of southern California; winters in Arizona, Texas, Tamaulipas and Sonora; accidental in North
Dakota, Kansas, Oklahoma, Texas, and Louisiana.

Catharus guttatus polionotus (Grinnell)

Hylocichla guttata polionota Grinnell, 1918, Condon, 20,
p. 89 — Wyman Creek, 8,000 ft., White Mountains,
Inyo County, California.

Eastern Washington, eastern Oregon, Nevada, southwestern Utah and east-central California; migrates through
Arizona, Oklahoma, and Texas; winters in southeastern
Arizona, in Sonora, Sinaloa, Jalisco, Michoacán, Tamaulipas, Nuevo León, Chihuahua, Guerrero, Oaxaca, Durango,
Guanajuato, México, Hidalgo, Puebla, Veracruz, and Chiapas.

Catharus guttatus auduboni (Baird)

Turdus auduboni Baird, 1864, Rev. Amer. Birds, 1, p.
16 — "Fort Bridger ?", Wyoming.

Hylocichla guttata dwighti Bishop, 1933, Proc. Biol. Soc. Washington, 46, p. 201 — Lion Creek, Priest Lake, Idaho.

Rocky Mountains from southeastern British Columbia and western Montana to Idaho, Wyoming, eastern Nevada, Arizona, New Mexico, and western Texas; winters in Tamaulipas, Nuevo León, Sonora, Baja California, Chihuahua, Durango, Guanajuato, México, Morelos, Hidalgo, Puebla, Veracruz, Chiapas, and Guatemala.

Catharus guttatus faxoni (Bangs and Penard)

Hylocichla guttata faxoni Bangs and Penard, 1921, Auk, 38, p. 443 — Shelburne, New Hampshire.

Central Yukon, Mackenzie, northeastern British Columbia, Alberta, Saskatchewan, and Manitoba through Ontario and Quebec to southern Labrador and Nova Scotia, south from Minnesota, Wisconsin, and northern Michigan to Ohio, Pennsylvania, New York, and the New England states and in the Appalachians to West Virginia and Maryland; winters from the southern part of the breeding range in United States south to Texas, Louisiana, Alabama, Georgia, and Florida; Bermuda (casual).

Catharus guttatus crymophilus (Burleigh and Peters)

Hylocichla guttata crymophila Burleigh and Peters, 1948, Proc. Biol. Soc. Washington, 61, p. 117 — Badger, Newfoundland.

Newfoundland; winter range uncertain, recorded from Virginia and Georgia.

GENUS **HYLOCICHLA** BAIRD

Hylocichla Baird, 1864, Rev. Amer. Birds, 1, p. 12. Type, by original designation, *Turdus mustelinus* Gmelin.

cf. Dilger, 1957, Syst. Zool., 5, pp. 174-182 (*Hylocichla* behavior, relationships).

HYLOCICHLA MUSTELINA

Hylocichla mustelina (Gmelin)

T.(urdus) mustelinus Gmelin, 1789, Syst. Nat., 1 (2), p. 817, ex Latham, "Tawny Thrush" — in Noveboraco [= New York, *vide* Amer. Orn. Union, 1931, Check-list North Amer. Birds, ed. 4, p. 257].

Eastern United States from southeastern South Dakota south to Oklahoma and southeastern Texas, eastward and south to northern Florida and north to southern Ontario and Quebec; casual in New Brunswick and North Dakota; migrates south through eastern Mexico in San Luis Potosí, Oaxaca, Veracruz, Tabasco, Campeche, Yucatán, Quintana Roo, and Chiapas; Cuba and Bahamas (casual) ; winters from southern Texas through eastern Mexico, Guatemala, El Salvador, Honduras, Costa Rica, and Panama.

GENUS **PLATYCICHLA** BAIRD

Platycichla Baird, 1864, Rev. Amer. Birds, 1, p. 32. Type, by original designation, *Platycichla brevipes* Baird = *Turdus flavipes* Vieillot.

PLATYCICHLA FLAVIPES

Platycichla flavipes venezuelensis (Sharpe)

Merula venezuelensis Sharpe, 1902, in Seebohm, Monog. Turdidae, 2, p. 83 — Venezuela [= vicinity of Caracas; *vide* Hellmayr, 1934, *ibid.*, p. 427].

Colombia in Santa Marta and the east slope of the eastern Andes in Norte de Santander; northern and western Venezuela to the Gran Sabana and Alto Paragua.

Platycichla flavipes melanopleura (Sharpe)

Merula melanopleura Sharpe, 1902, *in* Seebohm, Monog. Turdidae, 2, p. 87, pl. 103, fig. 2 — Trinidad.

Northeast Venezuela in Anzoátegui, Sucre, Monagas, and Margarita Island; Trinidad.

Platycichla flavipes xanthoscelus (Jardine)

Turdus xanthoscelus Jardine, 1847, Ann. Mag. Nat. Hist., ser. 1, 20, p. 329 — Tobago.

Tobago Island.

Platycichla flavipes polionota (Sharpe)

Merula polionota Sharpe, 1902, in Seebohm, Monog. Turdidae, 2, p. 85, pl. 103, fig. 1 — Roraima, British Guiana.

Southern Venezuela (Bolívar) and British Guiana.

Platycichla flavipes flavipes (Vieillot)

Turdus flavipes Vieillot, 1818, Nouv. Dict. Hist. Nat., nouv. éd., 20, p. 277 — Brazil [= Rio de Janeiro; *vide* Hellmayr, 1934, *ibid.*, p. 425].

Southeastern Brazil from southern Bahia to Rio Grande do Sul, Argentina in Misiones, and northeastern Paraguay.

PLATYCICHLA LEUCOPS

Platycichla leucops (Taczanowski)   4

Turdus leucops Taczanowski, 1877, Proc. Zool. Soc. London, p. 331 — Ropaybamba, Peru.

Colombian Andes, except Santa Marta range; Venezuela in western Lara and isolated mountains in Miranda, Bolívar, and southern Amazonas; British Guiana, northern Brazil (Rio Padauiri), Ecuador, Peru south to Marcapata, and Bolivia in La Paz and Santa Cruz.

GENUS TURDUS LINNAEUS

Turdus Linnaeus, 1758, Syst. Nat., ed. 10, 1, p. 168. Type, by subsequent designation (Gray, 1840, List Gen. Birds, p. 27), *Turdus viscivorus* Linnaeus.

Merula Leach, 1816, Syst. Cat. Spec. Mamm. Birds Brit. Mus., p. 20. Type, by monotypy, *Merula nigra* Leach = *Turdus merula* Linnaeus.

Arceuthornis Kaup, 1829, Skizz. Entw. Europ. Thierw., p. 93. Type, by subsequent designation (Gray, 1842, List Gen. Birds, app., p. 8), *Turdus pilaris* Linnaeus,

Ixocossyphus Kaup, 1829, *ibid.*, p. 145. Type, by original designation, *Turdus viscivorus* Linnaeus.

Planesticus Bonaparte, 1854, Compt. Rend. Acad. Sci. Paris, 38, p. 3. Type, by subsequent designation (Baird, 1864, Rev. Amer. Birds, 1, p. 12), *Turdus lereboulleti* Bonaparte = *Turdus jamaicensis* Gmelin.

Semimerula Sclater, 1859, Proc. Zool. Soc. London, p. 332. Type, by original designation, *Turdus gigas* Fraser.

Mimocichla Sclater, 1859, Proc. Zool. Soc. London, p. 336. Type, by subsequent designation (Baird, 1864, Rev.

Amer. Birds, 1, p. 35), *Turdus rubripes* Temminck.

Psophocichla Cabanis, 1860, Journ. f. Orn., 8, p. 182.
Type, by original designation, *Turdus strepitans* Smith
= *Merula litsipsirupa* Smith.

Peliocichla Cabanis, 1882, Journ. f. Orn., 30, p. 318. Type,
by original designation, *Turdus pelios* Bonaparte.

Streptocichla Cabanis, 1882, Journ. f. Orn., 30, p. 321.
Type, by original designation, *Turdus strepitans* Smith
= *Merula litsipsirupa* Smith.

Cossyphopsis Stejneger, 1883, Proc. U. S. Nat. Mus., 5,
p. 478. Type, by monotypy, *Turdus reevei* Lawrence.

Haplocichla Ridgway, 1905, Proc. Biol. Soc. Washington,
18, p. 212. Type, by monotypy, *Turdus aurantius*
Gmelin.

Afrocichla Roberts, 1922, *ibid.*, p. 228. Type, by original
designation, *Turdus olivaceus* Linnaeus.

cf. Rensch, 1923, Journ. f. Orn., 71, pp. 95-104 (*olivaceus* and *libonyanus*).

Dorst, 1950, Oiseau Rev. Franç. Orn., 20, pp. 212-248
(review of *Turdus*).

TURDUS BEWSHERI

Turdus bewsheri comorensis Milne-Edwards and Oustalet

Turdus comorensis Milne-Edwards and Oustalet, 1885,
Compt. Rend. Acad. Sci. Paris, 101, p. 221 — Grand
Comoro.

Grand Comoro Island, Comoro Islands.

Turdus bewsheri moheliensis Benson

Turdus bewsheri moheliensis Benson, 1960, Ibis, 103b, p.
76 — Nioumachoua, Moheli.

Moheli Island, Comoro Islands.

Turdus bewsheri bewsheri Newton

Turdus bewsheri Newton, 1877, Proc. Zool. Soc. London,
p. 299, pl. 34 — Anjouan Island.

Anjouan Island, Comoro Islands.

TURDUS OLIVACEOFUSCUS

Turdus olivaceofuscus olivaceofuscus Hartlaub

Turdus olivaceofuscus Hartlaub, 1852, Abh. Geb. Naturw. Hamburg, 2 (2), p. 49, pl. 3 — São Tomé.

São Tomé Island, Gulf of Guinea.

Turdus olivaceofuscus xanthorhynchus Salvadori

Turdus xanthorhynchus Salvadori, 1901, Boll. Mus. Zool. Anat. Comp. Torino, **16**, no. 414, p. 2 — Principe.

Principe Island, Gulf of Guinea.

TURDUS OLIVACEUS

Turdus olivaceus chiguancoides Seebohm

Turdus chiguancoides Seebohm, 1881, Cat. Birds Brit. Mus., 5, p. 231 — plains of the Gambia River.

P.(eliocichla) cryptopyrrha Cabanis, 1882, Journ. f. Orn., 30, p. 320 — Casamanze, Gambia River.

West Africa in Senegal, Portuguese Guinea, Guinea, Sierra Leone, Liberia, and Ivory Coast; merging into *saturatus* in western Ghana.

Turdus olivaceus saturatus (Cabanis)

P.(eliocichla) saturata Cabanis, 1882, Journ. f. Orn., 30, p. 320 — Camerun and Chinchoxo [= Duala, *vide* Rensch, 1923, Journ. f. Orn., 71, p. 97].

Turdus tessmanni Reichenow, 1921, Journ. f. Orn., 69, p. 49 — Bosum (Kamerun).

Western Ghana, where intergrades with *chiguancoides*, east through Nigeria, Cameroons (except upper slopes of Mount Cameroon and Adamawa lowlands), Rio Muni (?), Gabon, and Congo from the bend of the Ubangi south to the Lower Congo River and east in the middle Congo area to northern Kasai (?).

Turdus olivaceus adamauae Grote

Turdus libonyanus adamauae Grote, 1922, Journ. f. Orn., 70, p. 404 — Badda (Adamawa, North Kamerun).

Northern Cameroon, north of Yola, Nigeria.

Turdus olivaceus nigrilorum Reichenow

Turdus nigrilorum Reichenow, 1892, Journ. f. Orn., **40**, p. 194 — Buea, Cameroon Mountain.

Cameroon Mountain, above 2,500 feet.

Turdus olivaceus poensis Alexander

Turdus poensis Alexander, 1903, Bull. Brit. Orn. Club, **13**, p. 37 — Bakaki.

Fernando Po Island, Gulf of Guinea.

Turdus olivaceus bocagei (Cabanis)

P.(eliocichla) Bocagei Cabanis, 1882, Journ. f. Orn., **30**, p. 320 — Angola.

Congo, in southern Kasai west to Kwango, and south in northern Angola and the Benguela highlands.

Turdus olivaceus centralis Reichenow

Turdus pelios centralis Reichenow, 1905, Vög. Afr., **3**, p. 690 — Wadelai on Bahr el Jebel, *vide* Chapin, 1953, Bull. Amer. Mus. Nat. Hist., **75A**, p. 587.

Turdus icterorhynchus "Pr. Würt" = Emin, 1894, Proc. Zool. Soc. London, p. 601 — Ipoto.

Turdus albipectus Reichenow, 1908, Orn. Monatsb., **16**, p. 191 — Mboga, near lower Semliki River.

Upper Congo in the Uele east in Uganda to Mount Elgon, south to Kivu Volcanos and as far as Lake Kivu (?); extreme southern Sudan, southwestern Ethiopia, and northwestern Kenya.

Turdus olivaceus pelios Bonaparte

Turdus pelios Bonaparte, 1851, Consp. Av., **1** (1850), p. 273 — *ex Asia centrali* [= Fazoglu (Fazughli, Sudan), *vide* Rensch, 1923, Journ. f. Orn., **71**, p. 99].

Sudan south of lat. 14° N., except southern Equatoria; Ethiopia in the central and northwestern areas, and Eritrea.

Turdus olivaceus graueri Neumann

Turdus graueri Ncumann, 1908, Bull. Brit. Orn. Club, **21**, p. 56 — Nsasa, southeastern Ruanda.

Turdus pelios ubendeensis Moreau, 1944, Bull. Brit. Orn. Club, **64**, p. 65 — Upper Nyamansi River, Ushamba area of Ubende, western Tanganyika.

Extreme eastern Congo and Ruanda from Bukoba and Lake Burigi south to the northern end of Lake Tanganyika and the Ubende district of Tanganyika.

Turdus olivaceus stormsi Hartlaub

> *Turdus stormsi* Hartl.(aub), 1886, Bull. Mus. Hist. Nat. Belg. 4, p. 143, pl. 3 — Mpala (on west shore of Lake Tanganyika).

Southeast Congo in Katanga, northern Mwinilunga District of Northern Rhodesia, adjacent Angola, and south and west shores of Lake Tanganyika.

Turdus olivaceus williami White

> *Turdus olivaceus williami* White, 1949, Bull. Brit. Orn. Club, 69, p. 57 — Kansoku Forest, Mwinilunga, Northern Rhodesia.

Northern Rhodesia, except northern part of Mwinilunga District.

Turdus olivaceus swynnertoni Bannerman

> *Turdus swynnertoni* Bannerman, 1913, Bull. Brit. Orn. Club, 31, p. 56 — Chirinda Forest.

Eastern border of southern Rhodesia.

Turdus olivaceus transvaalensis (Roberts)

> *Afrocichla olivacea transvaalensis* Roberts, 1936, Ostrich, 7, p. 109 — Woodbush Forest Reserve, Transvaal.

Northern Transvaal.

Turdus olivaceus smithi Bonaparte

> *T.(urdus) smithi* Bonaparte, 1851, Consp. Av., 1 (1850), p. 274 — "ex *Afr. mer.*": new name for *Merula obscura* Smith, *nec T. obscurus* Gmelin.
> *Turdus Cabanisi* Cabanis (ex Bonaparte MS), 1851, Mus. Hein., 1, p. 3 — Kafferland.

South Africa from Little Namaqualand, central and northern Cape Province, Orange Free State to western and southern Transvaal, and Ngamiland.

Turdus olivaceus olivaceus Linnaeus

> *Turdus olivaceus* Linnaeus, 1766, Syst. Nat., ed. 12, 1, p. 292 — Cape of Good Hope.

Southwestern Cape Province east to Grahamstown, South Africa.

Turdus olivaceus pondoensis Reichenow

Turdus pondoensis Reichenow, 1917, Journ. f. Orn., 65, p. 391 — Pondoland in southeast Africa.

Cape Province from Transkei north through Natal and Swaziland.

TURDUS ABYSSINICUS

Turdus abyssinicus abyssinicus Gmelin

Turdus abyssinicus Gmelin, 1789, Syst. Nat., 1 (2), p. 824 — Abyssinia.

Merula elgonensis Sharpe, 1891, Ibis, p. 445 — Mount Elgon.

Planesticus olivaceus polius Mearns, 1913, Smiths. Misc. Coll., 61 (10), p. 2 — Mount Lololokui, 6,000 ft., north of the Northern Guaso Nyiro River.

Turdus olivaceus chyuluensis van Someren, 1939, Journ. East Africa Uganda Nat. Hist. Soc., 14, p. 75 — Chyulu Camp 3, 5,600 ft., Chyulu Range, Kenya.

Ethiopia, including Eritrea, south in the western Kenya highlands, and northeastern Tanganyika at Loliondo and Olonoti.

Turdus abyssinicus baraka (Sharpe)

Merula baraka Sharpe, 1903, Bull. Brit. Orn. Club, 14, p. 19 — Ruwenzori.

Merula johnstoni Sharpe, 1906, Ibis, p. 543 — Entebbe.

Turdus sylvestris Reichenow, 1908, Orn. Monatsb., 16, p. 191 — Bugoye Forest.

Eastern Congo in Ruwenzori hills, west and northwest of Lake Albert; western and southern Uganda to Entebbe.

Turdus abyssinicus deckeni Cabanis

Turdus Deckeni Cabanis, 1868, Journ. f. Orn., 16, p. 412 — East Africa [= Kilimanjaro, *vide* Reichenow, 1894, Vög. Deutsch-Ost-Afr., p. 233].

Northeast Tanganyika in the Kilimanjaro area.

Turdus abyssinicus oldeani Sclater and Moreau

Turdus olivaceus oldeani Sclater and Moreau, 1935, Bull. Brit. Orn. Club, 56, p. 13 — Oldeani Forest, 6,500 ft., Mbulu District, Tanganyika.

Northeast Tanganyika around Ngorongoro and Mounts Meru, Gerui, and Ufiome.

Turdus abyssinicus bambusicola Neumann

Turdus olivaceus bambusicola Neumann, 1908, Bull. Brit.
Orn. Club, 21, p. 56 — Bamboo Forest, 2,300-2,400 m.,
western Kivu Volcanos.

Highlands of the Kivu District, eastern Congo; Ruanda;
and mountains northwest of Lake Tanganyika.

Turdus abyssinicus roehli Reichenow

Turdus roehli Reichenow, 1905, Orn. Monatsb., 13, p.
182 — Mlalo, near Lushota, Usambara Mountains.

Northeast Tanganyika from north Pare to Usambara
Mountains.

Turdus abyssinicus nyikae Reichenow

Turdus nyikae Reichenow, 1904, Orn. Monatsb., 12, p.
95 — Nyika Plateau, northern Nyasaland.
Turdus milanjensis uluguru Hartert, 1923, Bull. Brit.
Orn. Club, 44, p. 6 — Bagito, Uluguru Mountains, Tanganyika.

Highlands of central Tanganyika south to northern Nyasaland; extreme northeast Northern Rhodesia in the
Mafinga Mountains, and Nyika Plateau, Nyasaland.

Turdus abyssinicus milanjensis Shelley

Turdus milanjensis Shelley, 1893, Ibis, p. 12 — Mlanje
Plateau, 6,000 ft., southern Nyasaland.

Southern Nyasaland from Dedza south to Mlanje and
Cholo; Mozambique.

TURDUS HELLERI

Turdus helleri (Mearns)

Planesticus helleri Mearns, 1913, Smiths. Misc. Coll., 61,
no. 10, p. 1 — Mount Mbololo, east of Mount Kilimanjaro, southeastern Kenya.

Taita hills area of southeastern Kenya.

TURDUS LIBONYANUS

Turdus libonyanus verrauxi Bocage

Turdus Verrauxi Barboza du Bocage, 1869, Jorn. Sci.
Math. Phys. Nat. Lisboa, 3, p. 341 — Caconda.

Peliocichla Schuetti Cabanis, 1882, Journ. f. Orn., 30, p.
    319 — Malange, Angola.
Kasai, southern Congo, south through Angola, to northern Damaraland, South West Africa.

Turdus libonyanus chobiensis (Roberts)

Peliocichla libonyanus chobiensis Roberts, 1932, Ann.
    Transvaal Mus., 15, p. 29 — Kabulabula, Chobe River,
    Bechuanaland.
Bechuanaland in Ngamiland and adjacent South West
Africa, Northern Rhodesia in Barotseland and Balovale
District, and Southern Rhodesia in northwestern Matabele-
land.

Turdus libonyanus libonyanus (Smith)

Merula libonyanus Smith, 1836, Rep. Exped. Centr.
    Africa, p. 45 — near Kurrichane, western Transvaal.
Eastern Bechuanaland and Transvaal; winters into northern Natal.

Turdus libonyanus peripheris Clancey

Turdus libonyanus peripheris Clancey, 1952, Bonn. Zool.
    Beitr., 3, p. 17 — Pietermaritzburg, Natal, South
    Africa.
Central Natal.

Turdus libonyanus tropicalis Peters

Turdus tropicalis Peters, 1881, Journ. f. Orn., 29, p. 50
    — Inhambane, Mozambique.
Merula cinerascens Reichenow, 1898, Orn. Monatsb., 6,
    p. 82 — Tabora and Kakoma, interior German East
    Africa.
Turdus tephrinus Oberholser, 1921, Proc. Biol. Soc. Wash-
    ington, 34, p. 102, *nom. nov.* for *Merula cinerascens*
    Reichenow, *nec Turdus cinerascens* Latham.
Turdus libonyanus costae Rensch, 1923, Journ. f. Orn.,
    71, p. 99 — Magogoni on the lower Ruvu, Tanganyika.
Turdus libonyanus niassae Rensch, 1923, Journ. f. Orn.,
    71, p. 100 — Somba, Nyasaland.
Mozambique; Southern Rhodesia, except Matabeleland;
Nyasaland, Tanganyika, and southeast Congo.

TURDUS TEPHRONOTUS

Turdus tephronotus Cabanis

Turdus tephronotus Cabanis, 1878, Journ. f. Orn., 26, p. 205, 218, pl. 3, fig. 2 — Tiva River and Ndi, Taita.
Turdus tephronotus australoabyssinicus Benson, 1942, Bull. Brit. Orn. Club, 63, p. 13 — near Yavello, southern Abyssinia.

Central Ethiopia and Somalia south to Kenya, from the Endoto Mountains and Taita east to the coast, south to Mount Kilimanjaro and the Dodoma District, eastern Tanganyika.

TURDUS MENACHENSIS

Turdus menachensis Ogilvie-Grant

Turdus menachensis Ogilvie-Grant, 1913, Bull. Brit. Orn. Club, 31, p. 86 — Menacha, Yemen.

Yemen and Asir Tihama, southern Saudi Arabia.

TURDUS LUDOVICIAE

Turdus ludoviciae (Phillips)

Merula ludoviciae Phillips, 1895, Bull. Brit. Orn. Club, 4, p. 36 — Somaliland (Darass).

Northern plateau of Somalia, in the Golis hills and Warsangeli escarpment.

TURDUS LITSIPSIRUPA

Turdus litsipsirupa simensis (Rüppell)

Merula (Turdus) simensis Rüppell, 1840, Neue Wirbelt., Vögel, p. 81, pl. 29, fig. 1 — Angetkat, northern Abyssinia.

Highlands of Ethiopia east to Eritrea, south to Gardulla and the Kullo District.

Turdus litsipsirupa litsipsirupa (Smith)

Merula litsipsirupa A. Smith, 1836, Rep. Exped. Centr. Africa, p. 45 — between the Orange River and the Tropic.

South Africa in Cape Province from Colesburg and the northeast Cape, Orange Free State, Natal, and Zululand to eastern Bechuanaland, Transvaal, Southern Rhodesia,

Northern Rhodesia in Balovali District and Barotseland, Swaziland, and southern Mozambique.

Turdus litsipsirupa pauciguttatus Clancey

> *Turdus litsipsirupa pauciguttatus* Clancey, 1956, Durban Mus. Novit., 4, p. 290 — Okahandja, Damaraland, South West Africa.

South West Africa in northern Great Namaqualand, Damaraland, the Kaokoveld, Ovamboland, and northwestern Bechuanaland, as well as southern Angola (?).

Turdus litsipsirupa stierlingi (Reichenow)

> *Geocichla litsipsirupa stierlingi* Reichenow, 1900, Orn. Monatsb., 8, p. 5 — Iringa.
> *Turdus simensis kösteri* Neumann, 1929, Orn. Monatsb., 37, p. 177 — Chipepe, Bailundo, Angola.

Congo in Katanga north to Kabalo on the Lualaba and the Lomani District; Marungu; northern Angola; Northern Rhodesia, except the range of *litsipsirupa;* Nyasaland west of the Rift; and western and south-central Tanganyika.

TURDUS DISSIMILIS

Turdus dissimilis dissimilis Blyth

> *T.(urdus) dissimilis* Blyth, 1847, Journ. Asiat. Soc. Bengal, 16, p. 144 — lower Bengal.
> *Turdus protomomelas* Cabanis, 1867, Journ. f. Orn., 15, p. 286 — Himalaya.

East Pakistan, India (Assam), Burma, Yunnan, extreme northern Thailand, northeastern Laos and Tonkin, North Vietnam.

Turdus dissimilis hortulorum Sclater

> *Turdus hortulorum* Sclater, 1863, Ibis, p. 196 — Camoëns Garden, Macao.

Eastern Siberia in Yakutia south to the Amur, Manchuria, Ussuriland, and probably northern Korea; migrates in Manchuria, Korea, Quelpart Island, and China; winters in southeast China, North Vietnam in Tonkin, and South Vietnam in North Annam; straggler to Japan on Honshu and to Formosa.

TURDUS UNICOLOR

Turdus unicolor Tickell

T.(urdus) unicolor T.(ickell), 1833, Journ. Asiat. Soc. Bengal, 2, p. 577 — Bansigar in Borabhúm.

Turdus unicolor subbicolor Koelz, 1954, Contrib. Inst. Regional Exploration, no. 1, p. 12 — Mawphlang, Khasi Hills.

Breeds in West Pakistan and India from Chitral and Kashmir east to the Nepal valley, and possibly Sikkim; winters in the plains from the northwest to Assam and south to the Eastern Ghats of Andhra.

TURDUS CARDIS

Turdus cardis Temminck

Turdus cardis Temminck, 1831, Pl. Col., livr. 87, pl. 518 — Japan.

Turdus cardis lateus Thayer and Bangs, 1909, Bull. Mus. Comp. Zool., 52, p. 140 — Ichang, Hupeh, China.

Turdus cardis merulinus Stresemann, 1929, Orn. Monatsb., 37, p. 140 — Yao-shan, Kwangsi.

Breeds in Japan on Hokkaido, Honshu, and Shikoku; Yangtze valley in China, in Kweichow, Hupeh, and Anhwei; straggler to Sakhalin, Korea, and Quelpart Island; migrates in southern Japan and Riu Kiu Islands; winters in southeastern China in Yunnan, Hainan Island, North Vietnam in Tonkin, South Vietnam in Annam, and Laos.

TURDUS ALBOCINCTUS

Turdus albocinctus Royle

Turdus Albicollis Royle, 1839, Illus. Bot. Himalayan Mountains, 2 (1835), pl. 8, fig. 3 — Himalayas, nec Turdus albicollis Vieillot, 1818.

T.(urdus) albocinctus Royle, 1840, Illus. Bot. Himalayan Mountains, 1 (1839), pp. 77-78, applied to plate of T. albicollis (nec T. albicollis Vieillot) — "Hills" [= Himalayas; restricted to Dehra Dun by Ripley, 1961, Synopsis Birds India Pakistan, p. 531].

Indian Himalayas and Nepal east to southeastern Tibet and southwestern Sikang; straggling in winter to the Assam foothills and northern Burma (Hukawng Valley).

TURDUS TORQUATUS

Turdus torquatus torquatus Linnaeus

Turdus torquatus Linnaeus, 1758, Syst. Nat., ed. 10, p.
170 — *Europa*; restricted to Sweden by Hartert, 1910,
Vög. pal. Fauna, 1, p. 663.

Europe from Scandinavia across to northwestern Russia,
south through Baltic islands and British Isles and Ireland
to northern France; migrates in western and central Europe
and the Faeroe Islands; winters in the Mediterranean region
from Portugal and northwestern Africa east to Greece.

Turdus torquatus alpestris (Brehm)

Merula alpestris Br.(ehm), 1831, Naturg. Vög. Deutsch.,
p. 337 — Alpine Tyrol.

High elevations of central and eastern Europe from
southern Germany, southern Poland, Czechoslovakia, and
western Ukraine south to eastern and central France, Spain,
Portugal, Italy, Yugoslavia, Bulgaria, and Cyprus (one
record); may breed in Greece (Mt. Alibotush); winters at
lower elevations and south to Mediterranean islands, western Turkey, Algeria, and Tunisia; straggler to England,
Lebanon, and Sinai Peninsula.

Turdus torquatus amicorum Hartert

M.(erula) torquata orientalis Seebohm, 1888, Ibis, p.
311 — Caucasus and Persia; restricted to Kislovodsk
by Hartert, 1910, Vög. pal. Fauna, 1, p. 664, *nec Turdus
orientalis* Gmelin, 1789.

Turdus torquatus amicorum Hartert, 1923, Vög. pal.
Fauna, Nachtrag, p. 57, *nom. nov.* for *Merula torquata
orientalis* Seebohm, preoccupied.

Turdus torquatus caucasicus Collin and Hartert, 1927,
Novit. Zool., 34, p. 52, *nom. nov.* for *Merula torquatus
orientalis* Seebohm, preoccupied.

Turkey in eastern Anatolia, Caucasus, and northern Iran
to Kopet Dagh in Transcaspia; winters to southern Iran;
straggler to Sinai Peninsula.

TURDUS BOULBOUL

Turdus boulboul (Latham)

Lanius boulboul Latham, 1790, Index Orn., 1, p. 80 —

India; restricted to Darjeeling by Baker, 1924, Fauna
Brit. India, Birds, ed. 2, 2, p. 130.

Turdus boulboul yaoschanensis Yen, 1932, Bull. Mus. Hist.
Nat. Paris, 4, p. 380 — Yaoschan, 700-2,000 m., Kwangsi.

Himalayas of West Pakistan, India, and Nepal to Yunnan, Laos, North Vietnam in Tonkin, and China in Kwangsi; winters in northern Burma and Thailand (rarely).

TURDUS MERULA

Turdus merula merula Linnaeus

Turdus Merula Linnaeus, 1758, Syst. Nat., ed. 10, p. 170
— *in Europae sylvis;* restricted to Sweden by Hartert,
1910, Vög. pal. Fauna, 1, p. 665.

Turdus rüdigeri Kleinschmidt, 1919, Falco, 14, no. 2, p.
15 — Chambley, eastern France.

Turdus merula ticehursti Clancey, 1938, Ibis, p. 750 —
Darnley, east Renfrewshire, Scotland.

Turdus merula mallorcae Jordans, 1950, Syll. Biol., Leipzig, p. 172 — Arta, Mallorca.

Western Europe from Scandinavia, Finland, and Russia
to the Urals, south to British Isles, Portugal, Spain, Balearic and western Mediterranean islands, France, Italy,
and Yugoslavia; grading into *mauritanicus* in southern
Spain and the Balearics, and into *aterrimus* in southeastern
Europe; straggler to Spitsbergen, Jan Mayen, Bear Island,
Faeroes, Iceland, Greenland, Malta, and Egypt; introduced
in New Zealand and associated islands.

Turdus merula azorensis Hartert

Turdus merula azorensis Hartert, 1905, Novit. Zool., 12,
p. 116 — south of Santa Cruz, Graciosa, Azores.

Azore Islands.

Turdus merula cabrerae Hartert

Turdus merula cabrerae Hartert, 1901, Novit. Zool., 8, p.
313 — Tenerife.

Turdus merula agnetae Volsøe, 1949, Dansk Orn. For.
Tidskr., 43, p. 82 — Cubo de Galga, La Palma, Canary
Islands.

Madeira and western Canary Islands.

Turdus merula mauritanicus Hartert

Turdus merula mauritanicus Hartert, 1902, Novit. Zool.,
9, p. 323 — Mhoiwla, near Mazagan, Morocco.
Merula algira Madarász, 1903, Ann. Mus. Natl. Hungarici,
1, p. 559 — Bône, Algeria.
Northwestern Africa from Morocco to Tunisia.

Turdus merula aterrimus (Madarász)

Merula aterrima Madarász, 1903, Orn. Monatsb., 11, p.
186 — Vladikavkaz, northern Caucasus.
Southeastern Europe, intergrading with *merula* in northern Yugoslavia, southern Ukraine, and Crimea; south in Rumania, Bulgaria, Greece, Turkey, and Aegean Islands (except *insularum*), east to Caucasus, Transcaucasia, and
northern Iran; winters south in Crete, Rhodes, and Cyprus.

Turdus merula insularum Niethammer

Turdus merula insularum Niethammer, 1943, Anz. Akad.
Wis. Wien, Math.-Naturwiss., 80, no. 3, p. 8 — Sama-
ria, Crete.
Islands of Crete, Rhodes, Mytilene, Samos, Icaria, and
Andros (?).

Turdus merula syriacus Hemprich and Ehrenberg

Turdus Merula var. *syriaca* Hemprich and Ehrenberg,
1833, Symb. Phys. Avium, fol. bb — Syria [= Lebanon, *vide* Kumerloeve, 1962, Iraq Nat. Hist. Mus. Publ.,
no. 20, p. 30].
Southern Turkey in Cilician Taurus (intergrades with
aterrimus), Lebanon, Syria, and Israel east in northern
Iraq to western and southern Iran; winter straggler to
northern Egypt (Damietta).

Turdus merula intermedius (Richmond)

Merula merula intermedia Richmond, 1896, Proc. U.S.
Nat. Mus., 18, p. 585 — Aksu, Eastern Turkestan.
Turdus merula brodkorbi Koelz, 1939, Proc. Biol. Soc.
Washington, 52, p. 67 — Farakar, Afghanistan.
Tadzhikistan and Kirghizstan, northeastern Afghanistan,
and Pamirs east to Sinkiang; winters in southern Afghan-
istan and southern Iraq, perhaps reaching West Pakistan
(Quetta and Peshawar, sight records?).

Turdus merula maximus (Seebohm)

Merula maxima Seebohm, 1881, Cat. Birds Brit. Mus., 5, p. 405 — Kashmir [= Gulmarg, *vide* Meinertzhagen, 1926, Bull. Brit. Orn. Club, 46, p. 99].

Turdus merula buddae R. and A. Meinertzhagen, 1926, Bull. Brit. Orn. Club, 46, p. 98 – Gyangtse, southern Tibet.

West Pakistan, from Safed Kohs (?), and India from Kashmir east to Garhwal (?), alpine Sikkim, Bhutan, and southeastern Tibet.

Turdus merula sowerbyi Deignan

Turdus merula sowerbyi Deignan, 1951, Proc. Biol. Soc. Washington, 64, p. 135 — Loshan (Kiating), Szechwan Province, China.

Szechwan Province, China.

Turdus merula mandarinus Bonaparte

T. (urdus) mandarinus Bonaparte, 1851, Consp. Av., 1 (1850), p. 275 — Asia or China [= Amoy to Shanghai; *vide* Swinhoe, 1860, Ibis, p. 56].

Turdus wulsini Riley, 1925, Proc. Biol. Soc. Washington, 38, p. 115 — Hingi (Hwangtsaopa).

Kweichow, China.

Turdus merula nigropileus (Lafresnaye)

Merula nigropileus Lafresnaye, 1840, Rev. Zool. [Paris], p. 65 — Nilgiris; restricted to Kalhatti, northern Nilgiri Plateau, by Ripley, 1950, Journ. Bombay Nat. Hist. Soc., 49, p. 50.

Turdus simillimus mahrattensis Whistler and Kinnear, 1932, Journ. Bombay Nat. Hist. Soc., 36, p. 76 — Mahableshwar.

India in the Western Ghats from Gujarat south to Malabar, Mysore, and northern edge of Nilgiri Plateau east in Andhra to Nallamalai Hills; wanders south in winter to Kerala.

Turdus merula spencei Whistler and Kinnear

Turdus simillimus spencei Whistler and Kinnear, 1932, Journ. Bombay Nat. Hist. Soc., 36, p. 77 — Jeypore Agency.

Eastern India in the Eastern Ghats from Madhya Pradesh south to the Seshachalam Hills in Andhra.

Turdus merula simillimus Jerdon

Turdus simillimus Jerdon, 1839, Madras Journ. Lit. Sci., 10, p. 253 — Nilgiris; restricted to Avalanche, higher southern Nilgiri Plateau, by Ripley, 1950, Journ. Bombay Nat. Hist. Soc., 49, p. 51.

Western Madras and Mysore in central and southern Nilgiri Plateau, Brahmagiris south to Palni Hills, where intergrades with *bourdilloni*.

Turdus merula bourdilloni (Seebohm)

Merula bourdilloni Seebohm, 1881, Cat. Birds Brit. Mus., 5, p. 251 — Travancore [= Colathoorpolay Patnas, Travancore, *vide* Whistler and Kinnear, 1932, Journ. Bombay Nat. Hist. Soc., 36, p. 76].

Kerala in the Palni and Nelliampathi hills, where intergrades with *simillimus*, south to the tip of the hill ranges.

Turdus merula kinnisii (Kelaart)

Merula Kinnisii Kelaart, 1851, in Blyth, Journ. Asiat. Soc. Bengal, 20, p. 177 — Newera Elia.

Hills of Ceylon.

TURDUS POLIOCEPHALUS

Turdus poliocephalus erythropleurus Sharpe

Turdus erythropleurus Sharpe, 1887, Proc. Zool. Soc. London, p. 515 — Christmas Island.

Christmas Island, Indian Ocean, and Cocos-Keeling Islands (introduced).

Turdus poliocephalus loeseri de Schauensee

Turdus javanicus löseri de Schauensee, 1939 (June 12), Notulae Naturae, no. 18, p. 1 — Blangbeké, 6,940 ft., Acheen, north Sumatra.

Turdus javanicus hoogerwerfi Chasen, 1939 (July), Treubia, 17, p. 137 — Mount Leuser, 3,000 m., Atjeh, north Sumatra.

Mountains of north Sumatra.

Turdus poliocephalus indrapurae Robinson and Kloss

Turdus indrapurae Robinson and Kloss, 1916, Journ. Straits Branch Roy. Asiat. Soc., no. 73, p. 277 — Korinchi Peak, 10,000 ft., west Sumatra.
Mountains of southwestern central Sumatra.

Turdus poliocephalus biesenbachi Stresemann

Turdus javanicus biesenbachi Stresemann, 1930, Orn. Monatsb., 38, p. 149 — Mount Papandajan, 2,600 m.
Mount Papandajan, western Java.

Turdus poliocephalus fumidus Müller

Turdus (Merula) fumidus S. Müller, 1843, Verh. Nat. Gesch. Land Volk, p. 201 — Mount Gedeh.
Mount Gedeh and Mount Pangerango, western Java.

Turdus poliocephalus stresemanni Bartels

Turdus javanicus stresemanni Bartels, 1938, Orn. Monatsb., 46, p. 113 — Mount Lawoe, ca. 2,000 m.
Mount Lawoe, central Java.

Turdus poliocephalus javanicus Horsfield

Turdus Javanicus Horsfield, 1821, Trans. Linn. Soc. London, 13, p. 148 — central Java; restricted to Mount Tjerimai by Stresemann, 1930, Orn. Monatsb., 38, p. 149.
Central Java.

Turdus poliocephalus whiteheadi (Seebohm)

Merula whiteheadi Seebohm, 1893, Bull. Brit. Orn. Club, 1, p. 25 — near Tozari, 7,000 ft., eastern Java.
Mountains of eastern Java.

Turdus poliocephalus seebohmi (Sharpe)

Merula seebohmi Sharpe, 1888, Ibis, p. 386 — Kinabalu, northern Borneo.
Mounts Kinabalu and Trus Madi, northern Borneo.

Turdus poliocephalus niveiceps (Hellmayr)

Turdus albiceps Swinhoe, 1864, Ibis, p. 363 — Formosa.
Planesticus niveiceps Hellmayr, 1919, Verh. Orn. Ges. Bayern, 14, p. 133, *nom. nov.* for *Turdus albiceps* Swin-

hoe, *nec Turdus albiceps* Pucheran = *Cossypha albicapilla* (Vieillot).
Mountains of Formosa; Botel Tobago (?).

Turdus poliocephalus thomassoni (Seebohm)
Merula thomassoni Seebohm, 1894, Bull. Brit. Orn. Club,
3, p. 51 — mountains of northern Luzon.
Mountains of northern Luzon, Philippines.

Turdus poliocephalus mayonensis (Mearns)
Merula mayonensis Mearns, 1907, Philippine Journ. Sci.,
2, sec. A, p. 358 — Mount Mayon, 4,000 ft., Albay Province, Luzon.
Mountains of southern Luzon, Philippines.

Turdus poliocephalus mindorensis Ogilvie-Grant
Turdus mindorensis Ogilvie-Grant, 1896, Ibis, p. 465 —
Mindoro.
Mountains of Mindoro, Philippines.

Turdus poliocephalus nigrorum Ogilvie-Grant
Turdus nigrorum Ogilvie-Grant, 1896, Ibis, p. 544 — vol-
cano of Canloon, Negros.
Mountains of Negros, Philippines.

Turdus poliocephalus malindangensis (Mearns)
Merula malindangensis Mearns, 1907, Philippine Journ.
Sci., 2, sec. A, p. 357 — Mount Malindang, Lebo Peak,
5,750 ft., northwestern Mindanao.
Mount Malindang, northwest Mindanao, Philippines.

Turdus poliocephalus katanglad Salomonsen
Turdus poliocephalus katanglad Salomonsen, 1953, Vi-
densk. Medd. Dansk naturhist. Foren., 115, p. 227 —
Mount Katanglad, 1,450 m., Bukidnon Province, central
Mindanao.
Mount Katanglad, central Mindanao, Philippines.

Turdus poliocephalus kelleri (Mearns)
Merula kelleri Mearns, 1905, Proc. Biol. Soc. Washington,
18, p. 6 — Mount Apo, 6,000 ft., southern Mindanao.
Mount Apo, southeastern Mindanao, Philippines.

Turdus poliocephalus hygroscopus Stresemann

Turdus celebensis hygroscopus Stresemann, 1931, Orn.
Monatsb., 39, p. 44 — Latimodjong Mountains, 2,800 m.
Latimodjong Mountains, southern Celebes.

Turdus poliocephalus celebensis (Büttikofer)

Merula celebensis Büttikofer, 1893, Notes Leyden Mus.,
15, p. 109 — district of Macassar.
Bonthain Peak and Wawa Kareng, southwestern Celebes.

Turdus poliocephalus schlegelii Sclater

Turdus schlegelii Sclater, 1861, Ibis, p. 280 — Timor [=
western Timor, *vide* Mayr, 1944, Bull. Amer. Mus. Nat.
Hist., 83, p. 155].
Mount Mutis, western Timor.

Turdus poliocephalus sterlingi Mayr

Turdus poliocephalus sterlingi Mayr, 1944, Bull. Amer.
Mus. Nat. Hist., 83, pp. 135, 155 — Mount Ramelan,
2,600 m., Portugese Timor.
Mount Ramelan, eastern Timor.

Turdus poliocephalus deningeri Stresemann

Turdus deningeri Stresemann, 1912, Bull. Brit. Orn. Club,
31, p. 4 — Gunung Pinaia, 7,500 ft., Ceram.
Mountains of Ceram.

Turdus poliocephalus versteegi Junge

Turdus poliocephalus versteegi Junge, 1939, Nova Guinea,
(n.s.), 3, p. 8 — Kajan Mountains (Oranje Range).
Oranje Mountains, West New Guinea.

Turdus poliocephalus carbonarius Mayr and Gilliard

Turdus poliocephalus carbonarius Mayr and Gilliard,
1951, Amer. Mus. Novit., no. 1524, p. 7 — Mount Wil-
helm, Bismarck Mountains, central highlands, Man-
dated Territory of New Guinea.
Turdus poliocephalus erebus Mayr and Gilliard, 1952,
Amer. Mus. Novit., no. 1577, p. 7, *nom. nov.* for *Turdus
poliocephalus carbonarius* Mayr and Gilliard, *nec Turdus carbonarius* Lichtenstein = *Platycichla flavipes*
Vieillot.
Bismarck Mountains, Territory of New Guinea.

Turdus poliocephalus keysseri Mayr

Turdus poliocephalus keysseri Mayr, 1931, Mitt. Zool. Mus. Berlin, **17**, p. 692 — Mongi-Busu, Saruwaged Mountains.

Saruwaged Mountains, Huon Peninsula, Territory of New Guinea.

Turdus poliocephalus papuensis (De Vis)

Merula papuensis De Vis, 1890, Ann. Rept. British New Guinea, 1888-89, p. 60 — Mount Victoria, southeastern New Guinea [reference not verified].

Mountains of southeastern New Guinea.

Turdus poliocephalus canescens (De Vis)

Merula canescens De Vis, 1894, Ann. Rept. British New Guinea, 1893-94, p. 105 — Goodenough Island [reference not verified].

Goodenough Island, D'Entrecasteaux Archipelago.

Turdus poliocephalus heinrothi Rothschild and Hartert

Turdus melanarius heinrothi Rothschild and Hartert, 1924, Bull. Brit. Orn. Club, **44**, p. 53 — St. Matthias Island.

St. Matthias (Mussau) Island, Bismarck Archipelago.

Turdus poliocephalus bougainvillei Mayr

Turdus poliocephalus bougainvillei Mayr, 1941, Amer. Mus. Novit., no. 1152, p. 6 — Bougainville Island.

Bougainville Island, Solomon Islands.

Turdus poliocephalus kulambangrae Mayr

Turdus poliocephalus kulambangrae Mayr, 1941, Amer. Mus. Novit., no. 1152, p. 6 — Kulambangra Island.

Kulambangra Island, Solomon Islands.

Turdus poliocephalus sladeni Cain and Galbraith

Turdus poliocephalus sladeni Cain and Galbraith, 1955, Bull. Brit. Orn. Club, **75**, p. 92 — near Turipava, 4,100 ft., mountains of Guadalcanal.

Guadalcanal Island, Solomon Islands.

Turdus poliocephalus rennellianus Mayr

Turdus poliocephalus rennellianus Mayr, 1931, Amer. Mus. Novit., no. 486, p. 21 — Rennell Island.

Rennell Island, Solomon Islands.

Turdus poliocephalus vanikorensis Quoy and Gaimard

Turdus vanikorensis Quoy and Gaimard, 1830, Voy. Astrolabe, Zool., 1, p. 188, pl. 7, fig. 2 — Vanikoro.

Vanikoro and Utupua Islands, Santa Cruz Islands and Espiritu Santo and Malo, New Hebrides.

Turdus poliocephalus placens Mayr

Turdus poliocephalus placens Mayr, 1941, Amer. Mus. Novit., no. 1152, p. 5 — Vanua Lava Island, Banks Islands.

Ureparapara (Bligh) and Vanua Lava Islands, Banks Islands.

Turdus poliocephalus whitneyi Mayr

Turdus poliocephalus whitneyi Mayr, 1941, Amer. Mus. Novit., no. 1152, p. 5 — Gaua Island, Banks Islands.

Santa Maria (Gaua) Island, Banks Islands.

Turdus poliocephalus malekulae Mayr

Turdus poliocephalus malekulae Mayr, 1941, Amer. Mus. Novit., no. 1152, p. 5 — Malekula Island, New Hebrides.

Pentecost, Malekula, and Ambrim Islands, New Hebrides.

Turdus poliocephalus becki Mayr

Turdus poliocephalus becki Mayr, 1941, Amer. Mus. Novit., no. 1152, p. 4 — Mai Island, New Hebrides.

Paama, Lopevi, Epi, and Emae (Mai) Islands, New Hebrides.

Turdus poliocephalus efatensis Mayr

Turdus poliocephalus efatensis Mayr, 1941, Amer. Mus. Novit., no. 1152, p. 4 — Efate Island, New Hebrides.

Efate (Vate) and Nguna Islands, New Hebrides.

Turdus poliocephalus albifrons (Ramsay)

Merula albifrons Ramsay, 1879, Proc. Linn. Soc. New South Wales, 3, p. 336 — mountainous parts of Eromanga, New Hebrides.

Eromanga (Erromango) Island, New Hebrides.

Turdus poliocephalus pritzbueri Layard

Turdus Pritzbueri Layard, 1878, Ann. Mag. Nat. Hist., ser. 5, 1, p. 374 — Lifu, New Caledonia.

Tana (Tanna) Island, New Hebrides and Lifu Island, Loyalty Islands.

Turdus poliocephalus mareensis Layard and Tristram
    Turdus mareensis Layard and Tristram, 1879, Ibis, p. 472 — Maré.
    Merula mareensis larochensis Sarasin, 1913, Vögel Neu-Caladonien Loyalty-Inseln, Nova Caledonia, Zool., 1 (1), p. 42 — Maré.
Maré Island, Loyalty Islands.

Turdus poliocephalus xanthopus Forster
    Turdus xanthopus J. R. Forster, 1844, Descr. Anim., p. 266 — New Caledonia.
New Caledonia.

Turdus poliocephalus poliocephalus Latham
    Turdus poliocephalus Latham, 1801, Index Orn., suppl., p. 44 — Norfolk Island.
Norfolk Island.

Turdus poliocephalus vinitinctus (Gould)
    Merula vinitincta Gould, 1855, Proc. Zool. Soc. London, p. 165 — Lord Howe Island.
Lord Howe Island.

Turdus poliocephalus layardi (Seebohm)
    Merula layardi Seebohm, 1890, Proc. Zool. Soc. London, p. 667 — Viti Levu.
Viti Levu, Ovalau, Yasawa, and Koro Islands, Fiji Islands.

Turdus poliocephalus ruficeps (Ramsay)
    Merula ruficeps Ramsay, 1876 (Feb.), Proc. Linn. Soc. New South Wales, 1, p. 43 — Fiji Islands [= Kandavu, *vide* Seebohm, 1881, Cat. Birds Brit. Mus., 5, p. 256].
    Merula bicolor Layard, 1876 (April), Ibis, p. 153 — Fiji.
Kandavu Island, Fiji Islands.

Turdus poliocephalus vitiensis Layard
    Turdus vitiensis Layard, 1876, Ann. Mag. Nat. Hist., ser. 4, 17, p. 305 — Bua, Vanua Levu, Fiji Islands.
    Merula vanuensis Seebohm, 1890, Proc. Zool. Soc. Lon-

don, pp. **666-667**, *nom. nov.* for *Merula vitiensis* Layard.
Vanua Levu Island, Fiji Islands.

Turdus poliocephalus hades Mayr
Turdus poliocephalus hades Mayr, 1941, Amer. Mus. Novit., no. 1152, p. 4 — Ngau Island, Fiji.
Ngau Island, Fiji Islands.

Turdus poliocephalus tempesti Layard
Turdus tempesti Layard, 1876, Proc. Zool. Soc. London, p. 420 — south end of Taveuni, at Selia Levu, Vuna Point.
Taveuni Island, Fiji Islands.

Turdus poliocephalus samoensis Tristram
Turdus samoensis Tristram, 1879, Ibis, p. 188 — Samoa Islands.
Savaii and Upolu Islands, Samoan Islands.

TURDUS CHRYSOLAUS

Turdus chrysolaus orii Yamashina
Turdus chrysolaus orii Yamashina, 1929, Tori, 6, no. 28, p. 155 — Paramushir Island, northern Kurile Islands.
Northern and central Kurile Islands; winters south to main Japanese Islands and Riu Kiu Islands (Ishigaki).

Turdus chrysolaus chrysolaus Temminck
Turdus chrysolaus Temminck, 1831, Pl. Col., livr. 87, pl. 537 — Japan.
Turdus jouyi Stejneger, 1887, Proc. U. S. Nat. Mus., 10, p. 4 — Hondo, Japan.
Sakhalin, Hokkaido, and northern Honshu; migrates southward in Fukien, southeastern China; winters in western Honshu, Shikoku, Kyushu, Riu Kiu Islands, southeastern China, Formosa, Hainan, and northern Philippines (Calayan, northern Luzon); uncommon spring and autumn migrant in South Korea.

TURDUS CELAENOPS

Turdus celaenops Stejneger
Turdus celaenops Stejneger, 1887, Science, 10, p. 108; also

1887, Proc. U. S. Nat. Mus., 10, p. 484 — Miyakeshima, "The Seven Islands," Idzu, Japan.

Merula celaenops yakushimensis Ogawa, 1905, Annot. Zool. Japon., 5, p. 180 — Yakushima.

Merula celaenops kurodai Momiyama, 1923, Dobuts. Zasshi, 35, p. 404 — Hachijo, Izu Islands.

Seven Islands of Izu and Yakushima Island (status ?), Japan; vagrant to Honshu.

TURDUS RUBROCANUS

Turdus rubrocanus rubrocanus Hodgson

Merula castanea Gould, 1835, Proc. Zool. Soc. London, p. 185 — Himalayas, *nec Turdus castaneus* Müller, 1776.

Turdus rubrocanus "Hodgs.," 1846, in Gray, Cat. Mamm. Birds Nepal Thibet, p. 81 — Nepal.

Turdus gouldi cinereiceps Collin and Hartert, 1927, Novit. Zool., 34, p. 52, *nom. nov.* for *Turdus castaneus* Gould, preoccupied.

West Pakistan and India from the Safed Kohs, on the Afghanistan-North West Frontier Province boundary, east to the Kashmir hills and east to Garhwal and Nepal, Sikkim (?), and Bhutan (?); straggler to Assam, and Chin Hills of Burma.

Turdus rubrocanus gouldi (Verreaux)

Merula Gouldi Verreaux, 1871, Nouv. Arch. Mus. Hist. Nat. [Paris], 6, p. 34 — Setchuan occidental [= Paohing, eastern Sikang].

Southeastern Tibet, Sikang, and Yunnan, east in eastern Tsinghai, Szechwan, Kansu, and Tsinking mountains, and Shensi; straggler to Nepal, Assam Valley, and northern Burma (Hukawng, Northern Shan States).

TURDUS KESSLERI

Turdus kessleri (Przewalski)

Merula Kessleri Przewalski, 1876, Mongol. i Strana Tangut., 2, p. 62, pl. 10 — Kansu [reference not verified].

China in Szechwan, eastern Tsinghai, and Kansu west to eastern Sikang; winters to southwestern Sikang, and southeastern Tibet; straggler to Sikkim.

TURDUS FEAE

Turdus feae (Salvadori)

Merula Feae Salvadori, 1887, Ann. Mus. Civ. Genova, 5, p. 514 — Mulayit Mountain, Tenasserim.

Turdus subpallidus Hume, 1888, Stray Feathers, 11, p. 132 — eastern hills, Manipur.

Northern China in Hopeh, north of Peiping; winters south in Burma irregularly and in India in southeastern Assam rarely.

TURDUS PALLIDUS

Turdus pallidus Gmelin

Turdus pallidus Gmelin, 1789, Syst. Nat., 1 (2), p. 815 — "Sibiria, *ultra lacum* Baikal."

Northeastern Siberia from the Amur and Ussuri River areas north; Manchuria?; migrates through Manchuria, Shaweishan, Korea (commonly in spring, rarely in fall), and adjacent islands, Sakhalin, Hokkaido (uncommon), Honshu, and southern Japanese islands; winters from south-central Honshu, south in Japan to Riu Kius, Formosa, and southeastern China from the Yangtse Valley south to Kwantung; recorded from Yunnan.

TURDUS OBSCURUS

Turdus obscurus Gmelin

Turdus obscurus Gmelin, 1789, Syst. Nat., 1 (2), p. 816 — "Sibiriae *silvis, ultra lacum* Baical."

Merula subobscura Salvadori, 1889, Ann. Mus. Civ. Genova, 7, p. 413 — Tahò, Karen Hills.

Turdus obscurus buturlini Domaniewski, 1918, Compt. Rend. Soc. Sci. Varsovie, 11, fasc. 4, pp. 444, 459 — Ascold Island [reference not verified].

Northern Siberia from the Yenisei and northeastern Altai (irregular breeding), north and east to Kamchatka south to south of Lake Baikal, northern Mongolia, and the Sayan range; Sakhalin, Kurile Islands, and Honshu (rare breeder; Mount Fuji, 1930) ; migrates through Mongolia, Manchuria, Korea and adjacent islands, Japanese Islands including Riu Kius and China, south to winter in southeastern China and Yunnan, Formosa, Philippines, Palau Islands, North and South Vietnam, Laos, Cambodia, Thailand, Malaya, Suma-

tra, Billiton Island, and Java; British Borneo south to
Satang Island; India in Assam, straggling west to West
Bengal, Sikkim, and Nepal; south Andaman Island; strag-
gler to Alaska (Amchitka Island), and to western Europe.

TURDUS RUFICOLLIS

Turdus ruficollis atrogularis Jarocki

Turdus atrogularis Jarocki, 1819, Spis. Plaków Król.
Warszawa, p. 14 — Poland [reference not verified].

Eastern Russia across the Urals north to lat. 60° N. and
Siberia to the Yenisei basin and the lower Tunguska south
to northern Altai; hybridizing with *ruficollis* in middle Siberia from the Altai to the lower Tunguska east toward
Lake Baikal; migrates through Mongolia, Sinkiang, Kirghizstan, and Tadzhikistan, northern Iran, and Afghanistan; winters in Iraq (rare), southern Iran and Afghanistan, West Pakistan and India along the Himalayas to Nepal,
Sikkim, Bhutan, Assam, and south to East Pakistan; Indian
plains (rarely) ; Burma and southwestern China in Yunnan; straggler to western Europe including British Isles,
and in Arabia and Yemen, China in Chihli, and Riu Kiu
Islands (Ishigaki).

Turdus ruficollis ruficollis Pallas

Turdus ruficollis Pallas, 1776, Reise Versch. Prov. Russ.
Reichs, 3, p. 694 — Dauria [= Transbaicalia].

Eastern Siberia from the Altai and Sayan ranges east to
Baikal, north to lat. 60° N. on the lower Tunguska, northern
Manchuria; migrates through Mongolia, northern Manchuria, China west to Sinkiang and south to Sikang, and
Tibet; winters from the eastern Indian Himalayas, uncom-
monly in Sikikm, Bhutan, and Assam, and more abundantly
from northern Burma to Yunnan.

TURDUS NAUMANNI

Turdus naumanni eunomus Temminck

Turdus eunomus Temminck 1831, Pl. Col., livr. 87, pl.
514 — Japan.
Turdus eunomus ni Momiyama, 1927, Annot. Orn. Orient,
1, p. 141 — Akira-mura, Higasi-Katusika-gun, Prov.
Simosa (Shimosa) in Hondo.

Turdus eunomus turuchanensis Johansen, 1954, Journ. f. Orn., 95, p. 329 — Dudinka, Turushan District, lower Yenisei.

Siberia from the Taz and Yenisei Rivers south to the Tunguska, east across the northern taiga to Lena and Kolyma Rivers, Kamchatka, and the Commander Islands; migrates in the Kurile Islands, Sakhalin, Hokkaido, Korea and adjacent islands, Manchuria, and Mongolia; winters in southern Honshu and southern Japanese islands, Riu Kius, Formosa, southern China from Hopeh west to Yunnan, reaching northern Burma in small numbers, and India in Assam occasionally, Nepal, and West Pakistan (two records); straggler to western Europe including the British Isles.

Turdus naumanni naumanni Temminck

Turdus naumanni Temminck, 1820, Man. Orn., ed. 2, 1, p. 170 — Silesia and Austria. . . . Hungary.

Siberia south of breeding range of *eunomus*, from the upper Yenisei and southern Tunguska Rivers east to the Lena and the Sea of Okhotsk, Lake Baikal, and northern Manchuria and Amurland; Stanovoy west to Yablonovy ranges; Sakhalin (?); migrates through Mongolia, and Korea; winters from southern Manchuria, northern China, and Korea south to the Yangtse and Japanese Islands (rare); straggler to Fukien, Yunnan, Quelpart Island, Riu Kiu Islands, and Formosa.

TURDUS PILARIS

Turdus pilaris Linnaeus

Turdus pilaris Linnaeus, 1758, Syst. Nat., ed. 10, p. 168 — Europa; restricted to Sweden by Hartert, 1910, Vög. pal. Fauna, 1, p. 646.

Turdus pilaris zarudnyi Loudon, 1912, Orn. Monatsb., 20, p. 5 — Transcaspia.

Turdus ultrapilaris Kleinschmidt, 1919, Falco, 14, no. 2, p. 16 — Issykkul.

Turdus pilaris tertius Johansen, 1936, Animad. Syst. Mus. Zool. Univ. Tomsk, no. 5, p. 3 — Yakutsk.

Southern Greenland (colonized in 1937), northern Europe in Norway, Sweden, Finland, Poland, Germany, Switzerland (uncommon), Austria, Hungary, east in Russia, south

to Ukraine, across Siberia north to lat. 70° N., south to
Altai, east to Sayan Range and Lake Baikal; sporadic rare
breeder in Faeroes, Holland, and mountains of France;
migrates throughout northern Europe, western Russia, and
Siberia; winters from Iceland, Faeroes, and southern Scandinavia south through British Isles and Europe to the
Mediterranean, Turkey, Caucasus, and Iran; less common
on Mediterranean islands, North Africa, and Lebanon;
rarely to Egypt, Syria, India, Canary Islands, and Madeira;
rare straggler to arctic islands (Spitzbergen, Jan Mayen),
Jens Munk Island, Foxe Basin off Baffin Island.

TURDUS ILIACUS[1]

Turdus iliacus coburni Sharpe

Turdus coburni Sharpe, 1901, Bull. Brit. Orn. Club, 12,
p. 28 — Iceland.

Iceland and Faeroe Islands; winters from the Faeroes to
the British Isles, Holland, and France; occasional on Greenland, Bear Island, and Jan Mayen.

Turdus iliacus iliacus Linnaeus

Turdus iliacus Linnaeus, 1766, Syst. Nat., ed. 12, p. 292 —
Europe [= in sylvis acerosis (Sweden), *vide*, 1761,
Fauna Svecica, p. 79].

Turdus musicus gerchneri Zarudny, 1918, Izvest. Turkest. Otd. Russk. Geogr. Obsht., 14, p. 126 — Turkestan
(winter migrant) [reference not verified].

Northern Europe from Scandinavia north to lat. 70° N.,
and south in the Baltic area to Poland and northern Germany, rarely south to northeastern France; Russia south
to northern Ukraine and Orenburg, across Siberia north
to the taiga limits, and east to Lake Baikal and the Kolyma
River south to the Altai; winters from the Faeroes and
British Isles, Europe to the Mediterranean, Canaries, Madeira, North Africa, Egypt (straggler), Turkey, Lebanon,
Syria, Caucasus, Iran (uncommon), and India (?); recorded from Greenland, Jan Mayen, Bear Island, and Spitzbergen.

[1] By Opinion 551, 1959, Ops. Int. Comm. Zool. Nomencl., 20, pp. 199-
210, *Turdus musicus* Linnaeus has been placed on the Official Index
of Rejected and Invalid Names in Zoology.

TURDUS PHILOMELOS

Turdus philomelos hebridensis Clarke

> *Turdus musicus hebridensis* Clarke, 1913, Scottish Nat.,
> p. 53, pl. 1 — Barra, Outer Hebrides.
> British Isles in the Outer Hebrides, and Isle of Skye.

Turdus philomelos clarkei Hartert

> *Turdus Ericetorum* Turton, 1807, Brit. Fauna, 1, p. 35,
> based on Lewin's "Heaththrush," 1796, Birds Great
> Brit., 2, p. 68, pl. 63 — restricted to Darenth, north
> Kent by Clancey, 1943, Ibis, p. 90.[1]
> *Turdus philomelos clarkei* Hartert, 1909, Bull. Brit. Orn.
> Club, 23, p. 54 — Tring, Herts (England).
> *Turdus ericetorum catherinae* Clancey, 1938, Ibis, p. 749
> — Cathcart, east Renfrewshire, Scotland.
> British Isles, except for range of *hebridensis*, and western
> Europe in southern Holland, northwestern France, inter-
> grading into *philomelos* in a broad zone in Denmark, cen-
> tral Germany and southern Czechoslovakia, south in the
> remainder of France, northern Spain, northern Portugal
> to central Italy, southeastern Europe to northern Greece,
> Bulgaria, and Rumania; winters locally and moves west and
> southwest; northern British birds into Ireland or western
> France, European birds to Spain and Portugal, south to
> Corsica, and Ballearic Islands; straggler to Faeroe Islands,
> Algeria (one record), and Madeira.

Turdus philomelos philomelos Brehm

> *Turdus philomelos* Br.(ehm), 1831, Handb. Naturgesch.
> Vög. Deutschl., p. 382 — wanders in April and October
> through central Germany.
> *Turdus philomelus distinctus* Zarudny, 1918, Izvest. Tur-
> kest. Otd. Russk. Geogr. Obsht., 14, p. 126 — Gilan,
> northern Iran [reference not verified].
> Breeds in Scandinavia north to lat. 70° N., across Siberia
> to the Angara and lower Tunguska Rivers and the south-
> western Lake Baikal area; south in Europe to the area of
> intergradation listed under *clarkei*, Ukraine, Rumania (?),
> east in Turkey, Caucasus, and northwestern Iran to south-

---

[1] By Opinion 405, 1956, Ops. Int. Comm. Zool. Nomencl., **13**, pp.
107-118, *Turdus ericetorum* Turton has been placed on the Official
Index of Rejected and Invalid Names in Zoology.

eastern Caspian; south in Siberia to Barnaul, the Sinkiang border mountains and the Sayan range; migrates south to southern Europe, southern British Isles, Canary Islands, Morocco, Egypt, Lebanon, Israel, Iraq, Saudi Arabia (scarce), Transcaspia, and Tadzhikistan to southern Iran; straggler to Faeroes, Madeira, and Sudan.

Turdus philomelos nataliae Buturlin

> *Turdus philomelus nataliae* Buturlin, 1929, Syst. Notes Birds North. Caucasus, p. 15 — Krasnoyarsk [reference not verified].

Breeding range uncertain; may inhabit easternmost parts of range given for *philomelos* from Sayan range to Lake Baikal; or may breed in northern Iran; wintering birds have been taken in southern Iran.

TURDUS MUPINENSIS

Turdus mupinensis Laubmann

> *Turdus auritus* Verreaux, 1871, Nouv. Arch. Mus. Hist. Nat. [Paris], 6 (1870), p. 34 — mountains of Chinese Tibet [= Mupin, *vide* Verreaux, 1872, *ibid.*, 7 (1871), p. 32]; *nec T. auritus* Gmelin, 1789.
>
> *Turdus mupinensis* Laubmann, 1920, Orn. Monatsb., 28, p. 17, new name for *T. auritus* Verreaux, preoccupied.
>
> *Turdus auritus conquisitus* Bangs, 1921, Bull. Amer. Mus. Nat. Hist., 44, p. 591 — Li-chiang, 10,000 ft., Snow Mountains, Yunnan.

China from Kansu, Shensi, and Hupeh, south in Szechwan, east-central Sikang, and northwestern Yunnan.

TURDUS VISCIVORUS

Turdus viscivorus viscivorus Linnaeus

> *Turdus viscivorus* Linnaeus, 1758, Syst. Nat., ed. 10, p. 168 — Europa; restricted to Essex, England, by Brit. Orn. Union List Comm., 1948, Ibis, p. 320; further restricted to Berechurch, near Colchester, Essex, southeastern England by Clancey, 1950, *ibid.*, p. 338.
>
> *Turdus viscivorus deichleri* Erlanger, 1897, Orn. Monatsb., 5, p. 192 — Ain-bou-Dries, Tunisia.
>
> *Turdus viscivorus reiseri* Schiebel, 1911, Orn. Monatsb., 19, p. 85 — Corsica.

Turdus viscivorus jubilaeus Lucanus and Zedlitz, 1917, Journ. f. Orn., 65, p. 511 — Slonim, Russia.

Turdus viscivorus uralensis Zarudny, 1918, Izvest. Turkest. Otd. Russk. Geogr. Obsht., 14, p. 125 — Orenburg [reference not verified].

Turdus viscivorus balticus Zarudny, 1918, *idem*, p. 125 — Pskov [reference not verified].

Turdus viscivorus neglectus Burg, 1921, Tierwelt, 31, p. 2 — Jura.

Turdus viscivorus theresae Meinertzhagen, 1939, Bull. Brit. Orn. Club, 59, p. 67 — Oulmes, Moyen Atlas, Morocco.

Turdus viscivorus bithynicus Keve-Kleiner, 1943, Aquila, 50, p. 360 — Mytilene, Aegean Islands.

Turdus viscivorus precentor Clancey, 1950, Limosa, 23, p. 337 — Darnley, East Renfrewshire, southwestern Scotland.

Turdus viscivorus hispaniae Jordans, 1950, Sylleg. Biol., Leipzig, p. 171 — Linares, Salamanca, western Spain.

T. (urdus) v. (iscivorus) tauricus Portenko, 1954, Keys Fauna USSR, no. 54, Birds, 3, p. 227 — Totakoi, near Simferopol, Crimea.

Breeds in Europe and the British Isles, north to lat. 68° N., east in western Siberia to the Ob River, intergrading beyond this area into *bonapartei*, south through Spain and Portugal east to the Crimea, Greece, northern Turkey (intergrading into *bonapartei* in the eastern Mediterranean), across the Caucasus to Iran east to the southeastern Caspian Sea area; northern population migrates from southern Scandinavia and central Europe south; less common on migration in Morocco, Tunisia, Egypt (?), Lebanon, and Israel; straggler to Azores.[1]

Turdus viscivorus bonapartei Cabanis

T. (urdus) Bonapartei Cabanis, 1860, Journ. f. Orn., 8, p. 183 — Himalaya.

[1] Specimens from western Europe show a continuous cline from northwest to south and southeast in color; more rich and buffy in the northwest, paler and grayer on the rump and head in the southeast. Size gradients parallel this; small, more slim-billed birds occur in the south of the range. It seems impossible to adjust this sliding scale to equal a valid segregation by geographical areas, recognized by given subspecific names. Even the eastern race *bonapartei* is at best a convenience.

Turdus pseudohodgsoni Kleinschmidt, 1909, Falco, 5, p. 20 — Tashkent.
Turdus viscivorus transcaspius Zarudny, 1918, Izvest. Turkest. Otd. Russk. Obsht., 14, p. 125 — Kopet Dagh.
T.(urdus) v.(iscivorus) expetibilis Portenko, 1954, Keys Fauna USSR, no. 54, Birds, p. 227 — Katon Karagai, Altai.

Breeds in Siberia from the Ob River east to the Angara, south to the Altai and Sayan ranges, western Sinkiang, Kirghizstan, Tadzhikistan and southern Turkmen and Uzbekistan to northern Afghanistan, northeastern Iran, southern Turkey, Syria (?), Lebanon (?), and West Pakistan and India from northern Baluchistan, North West Frontier Province and Chitral, Gilgit, Astor, and Hazara to Kashmir, east along the Himalayas to Kumaon and Nepal; migrates southward within this range to lower valleys, occasionally to southern Iran and the plains of the Punjab in West Pakistan and India.

TURDUS AURANTIUS

Turdus aurantius Gmelin

(Turdus) aurantius Gmelin, 1789, Syst. Nat., 1 (2), p. 832 — "in Jamaicae montibus."
Jamaica.

TURDUS RAVIDUS

Turdus ravidus (Cory)

Mimocichla ravida Cory, 1886, Auk, 3, p. 499 — Island of Grand Cayman, West Indies.
Grand Cayman.

TURDUS PLUMBEUS

Turdus plumbeus plumbeus Linnaeus

Turdus plumbeus Linnaeus, 1758, Syst. Nat., ed. 10, p. 169 — America [= Islands of Andros and Ilathera (Bahamas), in Catesby, 1731, Nat. Hist. Carolina, Florida, Bahama Islands, 1, p. 30, pl. 30].

Northern Bahama Islands: Grand Bahama, Abaco, Andros, New Providence, Eleuthera, Cat Island, and Exuma Cays; formerly on Great Exuma.

Turdus plumbeus schistaceus (Baird)

Mimocichla schistacea Baird, 1864, Rev. Amer. Birds, 1, p. 37 — Monte Verde, eastern Cuba.

Eastern Cuba.

Turdus plumbeus rubripes Temminck

Turdus rubripes Temminck, 1826, Pl. Col., livr. 69, pl. 409 — Cuba.

Mimocichla rubripes eremita Ridgway, 1905, Proc. Biol. Soc. Washington, 18, p. 213 — Swan Island, Caribbean Sea.

Central and western Cuba; Isle of Pines; Swan Island (extirpated?).

Turdus plumbeus coryi (Sharpe)

Mimocichla coryi Sharpe, 1902, in Seebohm, Monog. Turdidae, 2, p. 215 — Island of Cayman Brac.

Cayman Brac, Greater Antilles.

Turdus plumbeus ardosiaceus Vieillot

Turdus ardosiaceus Vieillot, 1823, Tabl. Encyc. Méth. Orn., 2, livr. 91, p. 646 — Santo Domingo.

Turdus ardosiaceus var. *portoricensis* Bryant, 1867, Proc. Boston Soc. Nat. Hist., 11, p. 93 — Puerto Rico.

Hispaniola, Gonave, Tortue, and Puerto Rico, Greater Antilles.

Turdus plumbeus albiventris (Sclater)

Mimocichla ardesiaca albiventris Sclater, 1889, Proc. Zool. Soc. London, p. 326 — Dominica.

Mimocichla verrillorum Allen, 1891, Auk, 8, p. 217 — Lasswa, Dominica.

Dominica, Lesser Antilles.

TURDUS CHIGUANCO

Turdus chiguanco chiguanco Lafresnaye and d'Orbigny

T.(urdus) chiguanco Lafresnaye and d'Orbigny, 1837, Synop. Av., 1, in Mag. Zool., 7, cl. 2, p. 16 — Tacua, Peru.

Coastal Peru, and northwestern Bolivia in northern La Paz.

Turdus chiguanco conradi Salvadori and Festa

Turdus conradi Salvadori and Festa, 1899, Boll. Mus. Zool. Anat. Comp. Torino, 14, p. 4 — Cuenca and Sigsig, Ecuador.

Andes of southern Ecuador and central Peru, above 6,000 feet.

Turdus chiguanco anthracinus Burmeister

Turdus anthracinus Burmeister, 1858, Journ. f. Orn., 6, p. 159 — Mendoza, Argentina.

Bolivia from central La Paz and Cochabamba south in the mountains to western Argentina in Mendoza, San Luis, and Córdoba, and to northeastern Chile in the Atacama region.

TURDUS NIGRESCENS

Turdus nigrescens Cabanis

Turdus nigrescens Cabanis, 1860, Journ. f. Orn., 8, p. 324 — Volcán de Irazú, Costa Rica.

Mountains of Costa Rica and western Panama (Volcán de Chiriquí).

TURDUS FUSCATER

Turdus fuscater cacozelus (Bangs)

Merula gigas cacozela Bangs, 1898, Proc. Biol. Soc. Washington, 12, p. 181 — Macotama, Sierra Nevada de Santa Marta, Colombia.

Sierra Nevada de Santa Marta, northern Colombia.

Turdus fuscater clarus Phelps and Phelps, Jr.

Turdus fuscater clarus Phelps and Phelps, Jr., 1953, Proc. Biol. Soc. Washington, 66, p. 9 — Cerro Tetaré, upper Río Negro, Sierra de Perijá, Zulia, Venezuela.

Known only from the Perijá Mountains on the Venezuela-Colombia border west of Lake Maracaibo.

Turdus fuscater quindio Chapman

Turdus fuscater quindio Chapman, 1925, Amer. Mus. Novit., no. 160, p. 1 — Laguneta, 10,300 ft., central Andes, Colombia.

Central and western Andes of Colombia south to southeastern Nariño and in northern Ecuador to the latitude of Baños.

Turdus fuscater gigas Fraser

Turdus gigas Fraser, 1841, Proc. Zool. Soc. London, (1840), p. 59 — Santa Fé de Bogotá.

Eastern Andes of Colombia and western Venezuela in Lara, Trujillo, Mérida, and Táchira.

Turdus fuscater gigantodes Cabanis

Turdus gigantodes Cabanis, 1873, Journ. f. Orn., 21, p. 315 — Maraynioc, Peru.

Southern Ecuador in Huigra District and temperate zone of Peru south to Junín.

Turdus fuscater ockendeni Hellmayr

Turdus fuscater ockendeni Hellmayr, 1906, Bull. Brit. Orn. Club. 16, p. 91 — Limbani, 9,500 ft., Carabaya, Peru.

Southeastern Peru in Cuzco and Puno.

Turdus fuscater fuscater Lafresnaye and d'Orbigny

Turdus Fuscater Lafresnaye and d'Orbigny, 1837, Synop. Av., 1, in Mag. Zool., 7, cl. 2, p. 16 — in Andibus (Bolivia) ; type from La Paz, *vide* Hellmayr, 1934, Field Mus. Nat. Hist. Publ., Zool. Ser., 13, pt. 7, p. 415.

Western Bolivia in La Paz and Cochabamba.

TURDUS SERRANUS

Turdus serranus infuscatus (Lafresnaye)

Mer.(ula) infuscata Lafresnaye, 1844, Rev. Zool. [Paris], 7, p. 41 — Mexico.

Mountains of southern and eastern Mexico in México, Guerrero, Oaxaca, Chiapas, eastern San Luis Potosí, southwestern Tamaulipas, and Veracruz; Guatemala, El Salvador, and central Honduras.

Turdus serranus cumanensis (Hellmayr)

Planesticus serranus cumanensis Hellmayr, 1919, Verh. Orn. Ges. Bayern, 14, p. 127 — mountains inland of Cumaná, "Bermúdez [= Sucre, *vide* Hellmayr, 1934, Field Mus. Nat. Hist. Publ., Zool. Ser., 13, pt. 7, p. 414].

Northeastern Venezuela in Anzoátegui, Sucre, and Monagas.

Turdus serranus atrosericeus (Lafresnaye)

Merula atro-sericea Lafresnaye, 1848, Rev. Zool. [Paris], 11, p. 3 — Caracas, Venezuela.

Northern Venezuela from the Andes of Mérida and Táchira northeast to Caracas; northeastern Colombia (Páramo de Tamá?).

Turdus serranus fuscobrunneus (Chapman)

Planesticus fuscobrunneus Chapman, 1912, Bull. Amer. Mus. Nat. Hist., 31, p. 158 — Cerro Munichique, 8,325 ft., western Andes west of Popayán, Cauca, Colombia.

Mountainous areas of Colombia, except Santa Marta and northeast, and Ecuador.

Turdus serranus serranus Tschudi

Turdus serranus Tschudi, 1844, Arch. f. Naturg., 10 (1), p. 280 — Peru.

Mountains of Peru (including the western slope of the Andes), and Bolivia (La Paz and Cochabamba).

TURDUS NIGRICEPS

Turdus nigriceps nigriceps Cabanis

Turdus nigriceps "Jelski" = Cabanis, 1874, Journ. f. Orn., 22, p. 97 — Peru.

Southeastern Ecuador, eastern Peru, eastern Bolivia, and western Argentina.

Turdus nigriceps subalaris (Seebohm)

Merula subalaris "Leverkühn" = Seebohm, 1887, Proc. Zool. Soc. London, p. 557 — Jutubá, presumably in the valley of the Rio Grande, Province of São Paulo, Brazil [= Jatubá, western Goiás, *vide* Hellmayr, 1934, Field Mus. Nat. Hist. Publ., Zool. Ser., 13, pt. 7, p. 411].

Southern Brazil in Goiás, Mato Grosso, and Paraná, in adjacent Argentina in Misiones, and in Paraguay.

TURDUS REEVEI

Turdus reevei Lawrence

Turdus reevei Lawrence, 1870, Ann. Lyc. Nat. Hist. New

York, 9, p. 234 — Puna Island, Gulf of Guayaquil, Ecuador.

Western Ecuador, south of the Chone River, and north-western Peru in Tumbes, Piura, and Lambayeque.

TURDUS OLIVATER

Turdus olivater sanctaemartae (Todd)

Planesticus olivater sanctae-martae Todd, 1913, Proc. Biol. Soc. Washington, 26, p. 170 — Cincinnati, Santa Marta, Colombia.

Santa Marta, Colombia.

Turdus olivater olivater (Lafresnaye)

Merula olivatra Lafresnaye, 1848, Rev. Zool. [Paris], 11, p. 2 — Caracas, Venezuela.

Norte de Santander, Colombia, and coastal mountains of Venezuela from Zulia, east in Lara, Yaracuy, Carabobo, Aragua, and Miranda.

Turdus olivater paraquensis Phelps and Phelps, Jr.

Turdus roraimae paraquensis Phelps and Phelps, Jr., 1946, Bol. Soc. Venezolana Cienc. Nat., 10, no. 67, p. 231 — Cerro Paraque, Territorio Amazonas, Venezuela.

Cerro Paraque, Amazonas, Venezuela.

Turdus olivater kemptoni Phelps and Phelps, Jr.

Turdus olivater kemptoni Phelps and Phelps, Jr., 1955, Proc. Biol. Soc. Washington, 68, p. 119 — Cerro de la Neblina, 1,800 m., headwaters of the Río Yatúa, Territorio Amazonas, Venezuela.

Cerro de la Neblina, Amazonas, Venezuela.

Turdus olivater duidae Chapman

Turdus roraimae duidae Chapman, 1929, Amer. Mus. Novit., no. 380, p. 23 — Mount Duida.

Mount Duida, Amazonas, Venezuela.

Turdus olivater roraimae Salvin and Godman

Turdus roraimae Salvin and Godman, 1884, Ibis, p. 443 — Roraima, "British Guiana" [= Bolívar, Venezuela, 1,600-1,800 m., *vide* Phelps and Phelps, 1950, Bol. Soc. Venezolana Cienc. Nat., 12, no. 75, p. 244].

Turdus roraimae ptaritepui Phelps and Phelps, Jr., 1946,

Bol. Soc. Venezolana Cienc. Nat., **10**, no. 67, p. 232 —
Cerro Ptari-tepuí, Gran Sabana, Bolívar, Venezuela.
Mountains of southern Bolívar, Venezuela and adjacent
British Guiana (Cerro Twek-quay).

Turdus olivater caucae (Chapman)

Planesticus caucae Chapman, 1914, Bull. Amer. Mus. Nat.
Hist., 33, p. 182 — La Sierra, 6,300 ft., central Andes,
Cauca, Colombia.
Central Andes, Colombia.

TURDUS MARANONICUS

Turdus maranonicus Taczanowski

Turdus maranonicus Taczanowski [ex Stolzman MS],
1880, Proc. Zool. Soc. London, p. 189, pl. 20 — Calla-
cate, northern Peru.
Northern Peru.

TURDUS FULVIVENTRIS

Turdus fulviventris Sclater

Turdus fulviventris Sclater [ex Verreaux MS], 1857,
Proc. Zool. Soc. London, p. 273 — Nova Grenada [=
Bogotá].
Eastern Andes of Colombia; Venezuela in Táchira, Méri-
da, and Trujillo; eastern Ecuador; extreme northern Peru.

TURDUS RUFIVENTRIS

Turdus rufiventris juensis (Cory)

Planesticus rufiventer juensis Cory, 1916, Field Mus. Nat.
Hist. Publ., Orn. Ser., 1, p. 344 — Juá, near Iguatu,
Ceará, Brazil.
Northeastern Brazil in Piauí, Ceará, Maranhão, Paraiba,
Pernambuco, and northern and western Bahia.

Turdus rufiventris rufiventris Vieillot

Turdus rufiventris Vieillot, 1818, Nouv. Dict. Hist. Nat.,
nouv. éd., 20, p. 226 — Brazil.
Turdus rufiventris chacoensis Brodkorb, 1939, Proc. Biol.
Soc. Washington, 52, p. 84 — Kilometer 195, west of
Puerto Casado, Paraguay.
Southern half of Brazil from southern Bahia, Minas

Gerais, Goiás, Mato Grosso, and Rio Grande do Sul; Uruguay; Paraguay; northern Argentina south to Córdoba and Buenos Aires; eastern Bolivia.

TURDUS FALCKLANDII

Turdus falcklandii falcklandii Quoy and Gaimard

Turdus falcklandii Quoy and Gaimard, 1824, in Freycinet, Voy. Uranie Physicienne, Zool., livr. 3, p. 104 — Falkland Islands.

Falkland Islands.

Turdus falcklandii magellanicus King

Turdus Magellanicus King, 1831, Proc. Comm. Sci. Corresp. Zool. Soc. London, 1, p. 14 — Fretu Magellanico.
Turdus falcklandii mochae Chapman, 1934, Amer. Mus. Novit., no. 762, p. 3 — Mocha Island, Chile.

Chile from Atacama to the Straits, Tierra del Fuego, Argentina in Tierra del Fuego, Más a Tierra, Más Afuera, and Patagonia.

Turdus falcklandii pembertoni Wetmore

Turdus magellanicus pembertoni Wetmore, 1923, Univ. California Publ. Zool., 21, p. 335 — Cerro Anecon Grande, Río Negro, Argentina.

Argentina in Río Negro and Neuquén.

TURDUS LEUCOMELAS

Turdus leucomelas albiventer Spix

Turdus albiventer Spix, 1824, Av. Spec. Nov. Brasil, 1, p. 70, pl. 69, fig. 2 — Brazil; restricted to Pará by Hellmayr, 1934, Field Mus. Nat. Hist. Publ., Zool. Ser., 13, pt. 7, p. 400.
Turdus ephippialis Sclater, 1862, Proc. Zool. Soc. London, p. 109 — Bogotá, Colombia.

Northern Colombia, from the upper Magdalena Valley east in the Santa Marta area and the eastern Andes to the Orinoco; Venzuela in Barinas, Cojedes, and Guárico south to Amazonas and Bolívar; the Guianas; and northeastern Brazil from Pará to Bahia west to the Tapajós River.

Turdus leucomelas cautor Wetmore

Turdus leucomelas cautor Wetmore, 1946, Smiths. Misc. Coll., 106, no. 16, p. 10 — Serrania de Macuire, 1,000 ft., above Nazaret, Guajira, Colombia.

Guajira Peninsula, Colombia.

Turdus leucomelas leucomelas Vieillot

Turdus leucomelas Vieillot, 1818, Nouv. Dict. Hist. Nat., nouv. éd., 20, p. 238, based on "Zorzal obscuro y blanco," Azara, no. 80, pt. ("male") — Paraguay.

Southern Brazil in Rio de Janiero, Minas Gerais, São Paulo, Goiás, and Mato Grosso; Paraguay; eastern Peru; and Bolivia (one specimen, Río Kaka).

TURDUS AMAUROCHALINUS

Turdus amaurochalinus Cabanis

T.(urdus) amaurochalinus Cabanis, 1851, Mus. Hein., 1, p. 5 — Brazil; restricted to Rio Grande do Sul by Pinto, 1944, Cat. Aves Brasil (Publ. Dept. Zool., São Paulo), pt. 2, p. 370.

Turdus ignobilis sandiae Carriker, 1933, Proc. Acad. Nat. Sci. Philadelphia, 85, p. 34 — Huacamayo, 1,500 ft., Prov. Sándia, Puno, Peru.

Brazil, southeastern Peru, Bolivia, Paraguay, northern and central Argentina south to the Río Negro, and Uruguay.

TURDUS PLEBEJUS

Turdus plebejus differens (Nelson)

Merula plebeia differens Nelson, 1901, Proc. Biol. Soc. Washington, 14, p. 175 — Pinabete, Chiapas, Mexico.

Mexico in highlands of southeastern Chiapas, and Guatemala.

Turdus plebejus rafaelensis Miller and Griscom

Turdus plebejus rafaelensis Miller and Griscom, 1925, Amer. Mus. Novit., no. 183, p. 4 — San Rafael del Norte, Nicaragua.

Nicaragua and El Salvador.

Turdus plebejus plebejus Cabanis

Turdus plebejus Cabanis, 1861, Journ. f. Orn., 8 (1860), p. 323 — Costa Rica.

Mountains of Costa Rica, and western Panama (Chiriquí).

TURDUS IGNOBILIS

Turdus ignobilis ignobilis Sclater

Turdus ignobilis Sclater, 1857, Proc. Zool. Soc. London, p. 273 — Nova Grenada [= Bogotá].

Eastern Colombia, on the slopes of the eastern and central Andes.

Turdus ignobilis goodfellowi Hartert and Hellmayr

Turdus ignobilis goodfellowi Hartert and Hellmayr, 1901, Novit. Zool., 8, p. 492 — Castilla, Cauca Valley, Colombia.

Cauca Valley and western slope of the western Andes, Colombia.

Turdus ignobilis debilis Hellmayr

Turdus ignobilis debilis Hellmayr, 1902, Journ. f. Orn., 50, p. 56 — Rio Madeira, western Brazil [= Salto Theotonio, vide Hellmayr, 1934, Field Mus. Nat. Hist. Publ., Zool. Ser., 13, pt. 7, p. 393].

Foothills of eastern Andes of Colombia; Venezuela in Táchira, Mérida, western Apure and Barinas; eastern Ecuador; northwestern Brazil as far east as the Rio Negro and Rio Madeira; eastern Peru; and northern Bolivia.

Turdus ignobilis murinus Salvin

Turdus murinus Salvin, 1885, Ibis, p. 197 — Merumé Mountains, Roraima (3,000-5,000 ft.), "British Guiana."

Amazonas (Duida) and Bolívar, Venezuela, and British Guiana in the foothills of Roraima and the Merumé Mountains.

Turdus ignobilis arthuri (Chubb)

Planesticus arthuri Chubb, 1914, Bull. Brit. Orn. Club, 33, p. 131 — Abary River, British Guiana.

Venezuela in the tropical lowlands near Duida, Amazonas; tropical British and French Guiana.

TURDUS LAWRENCII

Turdus lawrencii Coues /

Turdus brunneus Lawrence, 1878, Ibis, p. 57, pl. 1 — "Upper Amazons."

Turdus lawrencii Coues, 1880, Bull. U.S. Geol. Geog. Surveys Terr., 5, no. 4, p. 570, new name for *T. brunneus* Lawrence, 1878, *nec. T. brunneus* Boddaert, 1783.

Turdus altiloquus Todd, 1925, Proc. Biol. Soc. Washington, 38, p. 92 — Arima, Río Purús.

Upper Amazonia from Amazonas and Bolívar, Venezuela and western Brazil south to northern Mato Grosso, eastern Ecuador, and northeastern Peru (Pebas, Chamicuros); may occur in southeastern Colombia near the Río Negro.

TURDUS FUMIGATUS

Turdus fumigatus bondi Deignan

Turdus nigrirostris Lawrence, 1878, Ann. N. Y. Acad. Sci., 1, p. 146 — St. Vincent.

Turdus fumigatus bondi Deignan, 1951, Auk, 68, p. 379, new name for *Turdus nigrirostris* Lawrence, 1878, *nec Turdus nigrirostris* Karelin, 1875.

St. Vincent, Lesser Antilles.

Turdus fumigatus personus (Barbour)

Planesticus nigrirostris personus Barbour, 1911, Proc. Biol. Soc. Washington, 24, p. 58 — Grand Etang, 2,000 ft., Grenada.

Grenada, Lesser Antilles.

Turdus fumigatus aquilonalis (Cherrie)

Planesticus fumigatus aquilonalis Cherrie, 1909, Sci. Bull. Mus. Brooklyn Inst. Arts Sciences, 1 (16), p. 387 — heights of Aripo, Trinidad.

Trinidad and the northern coast of Venezuela from Perijá, Zulia, Lara, Yaracuy, Carabobo, and Distrito Federal to Sucre; northeastern Colombia in Norte de Santander (Petrólea).

Turdus fumigatus orinocensis Zimmer and Phelps

Turdus fumigatus orinocensis Zimmer and Phelps, 1955, Amer. Mus. Novit., no. 1709, p. 4 — Nericagua, 140 m., upper Orinoco River, Amazonas, Venezuela.

Venezuela in the upper Orinoco region in Apure, Táchira, and Barinas; Colombia in eastern Vichada (Maipures) and Meta.

Turdus fumigatus fumigatus Lichtenstein

Turdus fumigatus Lichtenstein, 1823, Verz. Doubl., p. 38 — Brazil; restricted to Rio Espirito Santo, by Hellmayr, 1934, Field Mus. Nat. Hist. Publ., Zool. Ser., 13, pt. 7, p. 385.

Planesticus bianchii Chrostovski, 1921, Ann. Zool. Mus. Polonici Hist. Nat. Warszawa, 1, p. 28 — Brazil.

British, Dutch and French Guiana; Brazil through the Amazon basin south to Rio de Janiero and west to Mato Grosso; eastern Bolivia (Chiquitos; no specimens; may refer to *hauxwelli?*).

Turdus fumigatus parambanus Hartert

Turdus obsoletus parambanus Hartert, 1920, Novit. Zool., 27, p. 475 — Paramba, northwestern Ecuador.

Pacific coast of Colombia and western Ecuador.

Turdus fumigatus obsoletus Lawrence

Turdus obsoletus Lawrence, 1862, Ann. Lyc. Nat. Hist. New York, 7, p. 470 — Atlantic side of the Isthmus of Panama, Panama Rail Road.

Caribbean slopes of Costa Rica and Panama; northwestern Colombia (Jiménez).

TURDUS HAUXWELLI

Turdus hauxwelli colombianus Hartert and Hellmayr 2

Turdus colombianus Hartert and Hellmayr, 1901, Novit. Zool., 8, p. 492 — Cali, Cauca Valley, western Colombia.

Eastern slopes of the western Andes, Colombia.

Turdus hauxwelli hauxwelli Lawrence

Turdus hauxwelli Lawrence, 1869, Ann. Lyc. Nat. Hist. New York, 9, p. 265 — Pebas, Peru.

Southeastern Colombia in the Orinoco region, adjacent Amazonas, Venezuela, eastern Peru, upper Amazonian Brazil from Marañon, Solimões, and Purús Rivers to the Rio Madeira, and south to northern Bolivia in the drainage area of the Río Beni.

TURDUS HAPLOCHROUS

Turdus haplochrous Todd

Turdus haplochrous Todd, 1931, Proc. Biol. Soc. Washington, 44, p. 54 — Palmarito, Río San Julian, Chiquitos, Bolivia.

Eastern Bolivia.

TURDUS GRAYI

Turdus grayi tamaulipensis (Nelson)

Merula tamaulipensis Nelson, 1899, Auk, 14, p. 75 — Ciudad Victoria, Tamaulipas, Mexico.

Eastern Mexico from southern Tamaulipas, central and southern Nuevo León, southeastern lowland San Luis Potosí, Veracruz, Tabasco, Chiapas (except the extreme south), Campeche, Yucatán, and Quintana Roo.

Turdus grayi microrhynchus Lowery and Newman

Turdus grayi microrhynchus Lowery and Newman, 1949, Occas. Papers Mus. Zool. Louisiana State Univ., no. 22, p. 5 — Santa María del Río, San Luis Potosí, Mexico.

Interior of south-central San Luis Potosí, Mexico.

Turdus grayi umbrinus Griscom

Turdus grayi umbrinus Griscom, 1930, Amer. Mus. Novit., no. 438, p. 5 — Finca El Cipres, 2,300 ft., near Mazatenango, Pacific slope, Guatemala.

Extreme southern Chiapas, Mexico, and Pacific lowlands of Guatemala.

Turdus grayi grayi Bonaparte

Turdus Grayi Bonaparte, 1838, Proc. Zool. Soc. London, (1837), p. 118 — Guatemala; restricted to Alta Vera Paz, Guatemala, by Griscom, 1930, *ibid.*, p. 6.

Turdus grayi megas Miller and Griscom, 1925, Amer. Mus. Novit., no. 183, p. 3 — Matagalpa, 2,200 ft., Nicaragua.

Southern Mexico (except the eastern coastal plain range of *tamaulipensis*) in the Sierra Madre range of eastern San Luis Potosí, Veracruz, Puebla, México, Guerrero, Hidalgo, Oaxaca, Tabasco, and Chiapas (except extreme south); eastern Guatemala; British Honduras; El Salvador; Honduras; and Nicaragua.

Turdus grayi casius (Bonaparte)

Planesticus casius Bonaparte, 1855, Compt. Rend. Acad. Sci. Paris, 41, p. 657 — Panama.

Costa Rica, Panama, and Colombia in northwestern Choco on the Gulf of Uraba.

Turdus grayi incomptus (Bangs)

Pl.(anesticus) luridus Bonaparte, 1854, Compt. Rend. Acad. Sci. Paris, 38, p. 4 — Nouvelle Grenade; *nec Turdus luridus* Hermann, 1804.

Merula incompta Bangs, 1898, Proc. Biol. Soc. Washington, 12, p. 144 — Santa Marta, Colombia.

Coastal region of northern Colombia from Barranquilla (sight record) to Santa Marta Peninsula.

TURDUS NUDIGENIS

Turdus nudigenis nudigenis Lafresnaye

Turdus nudigenis Lafresnaye, 1848, Rev. Zool. [Paris], 11, p. 4 — Caracas.

Lesser Antilles, from Martinique southward (excluding Barbados), Trinidad, Tobago, and Venezuela (except the northwest), from Portuguesa south to Barinas, Táchira, Apure, and Amazonas, east to Isla de Margarita, Sucre, and Bolívar; British, French and Dutch Guiana; northeastern Brazil north of the Amazon and east of the Rio Negro; extreme eastern Colombia east of the eastern Andes.

Turdus nudigenis extimus Todd

Turdus nudigenis extimus Todd, 1931, Proc. Biol. Soc. Washington, 44, p. 54 — Santarém, Amazon River, Brazil.

Northern Brazil, on the south bank of the lower Amazon (Santarém).

Turdus nudigenis maculirostris Berlepsch and Taczanowski

Turdus ignobilis maculirostris Berlepsch and Taczanowski, 1883, Proc. Zool. Soc. London, p. 538 — Chimbo, western Ecuador.

Western Ecuador and extreme northwestern Peru.

TURDUS JAMAICENSIS

Turdus jamaicensis Gmelin

(*Turdus*) *jamaicensis* Gmelin, 1789, Syst. Nat., 1 (2), p. 809, based on "Jamaican Thrush" Latham, 1783, Gen. Syn. Birds, 2, p. 20 — Jamaica.
Jamaica.

TURDUS ALBICOLLIS

Turdus albicollis calliphthongus Moore

Turdus assimilis calliphthongus Moore, 1937, Proc. Biol. Soc. Washington, 50, p. 204 — Baromicon, Sonora, Mexico.
Southeastern Sonora, northeastern Sinaloa, and adjacent Chihuahua, Mexico.

Turdus albicollis lygrus Oberholser

Merula tristis Swainson, 1827, Philos. Mag., n.s., 1, no. 5, p. 369 — Mexico; *nec Turdus tristis* Müller, 1776.
Turdus assimilis lygrus Oberholser, 1921, Proc. Biol. Soc. Washington, 34, p. 106, new name for *M. tristis* Swainson, preoccupied.
Turdus assimilis renominatus Miller and Griscom, 1925, Amer. Mus. Novit., no. 184, p. 10 — Juan Lisiarraga Mountain, 5,500 ft., southern Sinaloa.
Mexico in central and southern Sinaloa, Nayarit, Jalisco, Colima, Michoacán, Guerrero, western Oaxaca, southern Chihuahua, western Durango, western México, Morelos, and southwestern Chiapas (where intergrades with *leucauchen*).

Turdus albicollis assimilis Cabanis

Turdus assimilis Cabanis, 1850, Mus. Hein., 1, p. 4 — Jalapa, Veracruz.
Mexico in southern Tamaulipas, San Luis Potosí, Hidalgo, northern Oaxaca, eastern México, and western Veracruz.

Turdus albicollis leucauchen Sclater

Turdus leucauchen Sclater, 1858, Proc. Zool. Soc. London, p. 447 — Guatemala.
Turdus assimilis parcolor Austin, 1929, Bull. Mus. Comp. Zool., 69, p. 386 — Camp 6, Cayo District, British Honduras.

Mexico in southeastern Veracruz, northeastern Oaxaca, and Chiapas, intergrading with *lygrus* in southwestern Chiapas; Guatemala except along the Pacific slope; British Honduras; Honduras (San Pedro ?).

Turdus albicollis rubicundus (Dearborn)

Planesticus tristis rubicundus Dearborn, 1907, Field Mus. Nat. Hist. Publ., Orn. Ser., 1, p. 137 — Patulul, Sololá, Guatemala.

Guatemala, along the Pacific slope, and El Salvador.

Turdus albicollis atrotinctus Miller and Griscom

Turdus assimilis atrotinctus Miller and Griscom, 1925, Amer. Mus. Novit., no. 184, p. 12 — Tuma, Matagalpa, Nicaragua.

Caribbean highlands of Nicaragua.

Turdus albicollis oblitus Miller and Griscom

Turdus assimilis oblitus Miller and Griscom, 1925, Amer. Mus. Novit., no. 184, p. 12 — Tenorio, Costa Rica.

Costa Rica, grading into *cnephosus* in the southwest.

Turdus albicollis cnephosus (Bangs)

Merula leucauchen cnephosa Bangs, 1902, Proc. New England Zool. Club, 3, p. 92 — Boquete, Volcán de Chiriquí.

Southwestern Costa Rica, where intergrades with *oblitus*, and western Panama in Chiriquí and Veraguas.

Turdus albicollis coibensis Eisenmann

Turdus assimilis coibensis Eisenmann, 1950, Auk, 67, p. 366 — Coiba Island, Veraguas, Panama.

Coiba Island, Veraguas, Panama.

Turdus albicollis daguae Berlepsch

Turdus daguae Berlepsch, 1897, Orn. Monatsb., 5, p. 176 — San José, Río Dagua, Colombia.

Eastern Panama in Darién, Colombia west of the western Andes from the lower Atrato to Nariño, and northwestern Ecuador.

Turdus albicollis phaeopygoides Seebohm

Turdus phaeopygoides Seebohm, 1881, Cat. Birds Mus., 5, p. 404 — Tobago.

Merula phaeopyga minuscula Bangs, 1898, Proc. Biol. Soc. Washington, 12, p. 181 — Pueblo Viejo, Colombia.

Northeastern Colombia in Santa Marta and the eastern slope of the eastern Andes in Boyacá; northern Venezuela from Zulia east in Aragua, Carabobo, Distrito Federal, Anzoátegui, and Sucre; Trinidad and Tobago.

Turdus albicollis phaeopygus Cabanis

Turdus phaeopygus Cabanis, 1849, in Schomburgk, Reise, Brit. Guiana, 3 (1848), p. 666 — British Guiana.
Turdus phaeopygus cayennensis Todd, 1931, Proc. Biol. Soc. Washington, 44, p. 50 — Pied Saut, French Guiana.
Turdus phaeopygus coloratus Todd, 1931, Proc. Biol. Soc. Washington, 44, p. 51 — Colonia do Mojuy, Santarem, Brazil.

Extreme eastern Colombia along the Orinoco and Negro rivers, southern Venezuela in Barinas, Amazonas, and eastern Bolívar, the Guianas and northern Brazil from the northern side of the Rio Solimões to the coastal district of Maranhão, and south of the Amazon on the Rio Tapajós.

Turdus albicollis spodiolaemus Berlepsch and Stolzmann

Turdus phaeopygus spodiolaemus Berlepsch and Stolzmann, 1896, Proc. Zool. Soc. London, p. 326 — La Gloria, Chanchamayo, Junín, Peru.
Turdus phaeopygus berlepschi Todd, 1931, Proc. Biol. Soc. Washington, 44, p. 51 — Arimã, Rio Purús, Brazil.
Turdus albicollis purusensis Wolters, 1953, Bonn. Zool. Beitr., 3, p. 281, *nom. nov.* for *Turdus phaeopygus berlepschi* Todd (preoccupied if *Catharus* is merged with *Turdus*).

Eastern Ecuador and Peru east to western Brazil and south of Rio Solimões in the Rio Purús area; Beni, northern Bolivia.

Turdus albicollis contemptus Hellmayr

Turdus crotopezus contemptus Hellmayr, 1902, Journ. f. Orn., 50, p. 61 — Bueyes, Santa Cruz, Bolivia.
Bolivia in the yungas of La Paz, Santa Cruz, and Tarija.

Turdus albicollis crotopezus Lichtenstein

Turdus crotopezus Lichtenstein, 1823, Verz. Doubl., p. 38 — Bahia.
Eastern Brazil in Bahia, Espirito Santo, and Alagoas.

— **Turdus albicollis albicollis** Vieillot ♀

Turdus albicollis Vieillot, 1818, Nouv. Dict. Hist. Nat., nouv. éd., 20, p. 227 — Brazil [= Rio de Janiero, *vide* Hellmayr, 1934, Field Mus. Nat. Hist. Publ., Zool. Ser., 13, pt. 7, p. 366].

Southeastern Brazil from Rio de Janeiro to Rio Grande do Sul.

Turdus albicollis paraguayensis (Chubb)

Merula albicollis paraguayensis Chubb, 1910, Ibis, p. 608 — Sapucay, Paraguay.

Southwestern Brazil in Mato Grosso; Paraguay; northern Argentina in Misiones.

TURDUS RUFOPALLIATUS

Turdus rufopalliatus rufopalliatus Lafresnaye

Turdus rufo-palliatus Lafresnaye, 1840, Rev. Zool. [Paris], 3, p. 259 — Monterey, California [*errore;* Acapulco, Mexico, designated by Bangs and Penard, 1919, Bull. Mus. Comp. Zool., 63, p. 37].

Turdus rufo-palliatus grisior van Rossem, 1934, Bull. Mus. Comp. Zool., 77, p. 461 — Guirocoba, Sonora, Mexico.

Western Mexico from Sonora, Sinaloa, and Durango south to Nayarit, Jalisco, Michoacán, Guerrero, Distrito Federal, Morelos, and western Puebla.

Turdus rufopalliatus graysoni (Ridgway)

Merula flavirostris graysoni Ridgway, 1882, Proc. U.S. Nat. Mus., 5, p. 12 — Tres Marías.

Islas de las Tres Marías, Mexico.

TURDUS SWALESI

Turdus swalesi (Wetmore)

Haplocichla swalesi Wetmore, 1927, Proc. Biol. Soc. Washington, 40, p. 55 — Massif de la Selle, 6,000 ft.

La Selle ridge, Haiti.

TURDUS RUFITORQUES

Turdus rufitorques Hartlaub

Turdus (Merula) rufitorques Hartlaub, 1844, Rev. Zool. [Paris], 7, p. 214 — Guatemala.

Southeastern Mexico in the highlands of Chiapas; Guatemala; western El Salvador.

TURDUS MIGRATORIUS

Turdus migratorius migratorius Linnaeus

(*Turdus*) *migratorius* Linnaeus, 1766, Syst. Nat., ed. 12, 1, p. 292 — in America septentrionali.

Breeds from northern Alaska and northern Canada as far east as Thelon River, south in Manitoba, Ontario, Quebec, Prince Edward Island, Nova Scotia, and Alberta, west to central British Columbia and Saskatchewan, south in the west-central United States to western North Dakota, South Dakota, Nebraska to Kansas, Oklahoma, Missouri, Illinois, and Indiana, east through New England and coastal states south to Maryland and in the Appalachians to North Carolina and northwestern Virginia; winters in United States to the East coast and the Gulf, south in eastern Mexico to Durango, Nuevo León, and southern Veracruz; southern Florida, western Cuba (rare), and Bermuda; vagrant to Bahamas.

Turdus migratorius nigrideus Aldrich and Nutt

Turdus migratorius nigrideus Aldrich and Nutt, 1939, Sci. Publ. Cleveland Mus. Nat. Hist., 4, no. 2, p. 31 — Hodge Water, Avalon Peninsula, eastern Newfoundland.

Breeds in Canada from northern Quebec and Labrador to Newfoundland, and Miquelon and St. Pierre Islands; wintering and on migration from southeastern Canada west to Michigan, Illinois, and Ohio, south to the Gulf coast of Louisiana, in New England south to Maryland on the coast, and in the Appalachians.

Turdus migratorius achrusterus (Batchelder)

Merula migratoria achrustera Batchelder, 1900, Proc. New England Zool. Club, 1, p. 104 — Raleigh, North Carolina.

Breeds from southern Oklahoma east across southern Missouri to western Virginia, and Maryland south to Texas, and the Gulf and southern States to northern Florida; winters in the southern area of the breeding range and to southern Texas and Florida, to Mexico in Yucatán and Quintana Roo, and to western Cuba (one record).

Turdus migratorius caurinus (Grinnell)

Planesticus migratorius caurinus Grinnell, 1909, Univ.
California Publ. Zool., 5, no. 2, p. 241 — Windfall Harbor, Admiralty Island, Alaska.

Breeds form southeastern Alaska south along the coast
and adjacent islands to northwestern Oregon; winters from
southwestern British Columbia to west-central California;
casual in winter in Alaska.

Turdus migratorius propinquus Ridgway

T.(urdus) propinquus Ridgway, 1877, Bull. Nuttall Orn.
Club, 2, p. 9 — western region, including eastern base
of Rocky Mountains [= Laramie Peak, Wyoming, *vide*
Ridgway, 1907, Bull. U. S. Nat. Mus., 50, pt. 4, p. 103].

Breeds from southeastern Oregon and eastern Washing-
ton, southern and eastern British Columbia, southern Al-
berta and southwestern Saskatchewan to Montana, western
South Dakota, and western Nebraska, south to southern
California, southern Nevada, and Mexico from Sonora and
Chihuahua to Durango, Zacatecas, Guanajuato, México,
Morelos, Coahuila, Hidalgo, Puebla, and Neuvo León; in
Jalisco and Michoacán intergrades with *phillipsi;* winters
from southern part of breeding range south to Baja Cali-
fornia, Oaxaca and Guatemala highlands.

Turdus migratorius phillipsi Bangs

Turdus migratorius phillipsi Bangs, 1915, Proc. Biol. Soc.
Washington, 28, p. 125 — Las Viegas [= Las Vegas],
Veracruz, Mexico.
Turdus migratorius permixtus Griscom, 1934, Bull. Mus.
Comp. Zool., 75, p. 396 — Chilpancingo, 8,000 ft., Guer-
rero.

Mexico from southwestern Tamaulipas (where inter-
grades with *propinquus*) and western Veracruz south in
Jalisco and Michoacán, Distrito Federal, México, Morelos,
Hidalgo, Tlaxcala, and Puebla to Guerrero and Oaxaca.

Turdus migratorius confinis Baird

Turdus confinis Baird, 1864, Rev. Amer. Birds, 1, p. 29
— Todos Santos, Cape San Lucas, Baja California.
Mountains of Cape district, Baja California, Mexico.

SUBFAMILY **ORTHONYCHINAE**[1]

HERBERT G. DEIGNAN

GENUS **ORTHONYX** TEMMINCK

Orthonyx Temminck, 1820, Man. Orn., éd. 2, 1, p. 81. No species; generic details only. Type, by subsequent designation (Ranzani, 1822, Elem. Zool., 3, p. 19), *Orthonyx temminckii* Ranzani.

Macrorthonyx Mathews, 1912, Austral Avian Rec., 1 (1912-1913), p. 111. Type, by original designation and monotypy, *Orthonyx spaldingi* Ramsay.

Papuorthonyx Mathews, 1921, Birds Australia, 9, pt. 4, p. 177. Type, by original designation, *Orthonyx novaeguineae* A. B. Meyer.

ORTHONYX TEMMINCKII

Orthonyx temminckii novaeguineae Meyer

Orthonyx Novae Guineae A. B. Meyer, 1874, Sitzungsb. K. Akad. Wiss. Wien, Math.-Naturwiss. Cl., 69, p. 83 — Arfak Mountains, New Guinea.

Tamrau and Arfak Mountains, western New Guinea (Vogelkop).

Orthonyx temminckii dorsalis Rand

Orthonyx temminckii dorsalis Rand, 1940, Amer. Mus. Novit., no. 1074, p. 2 — Bele River, 18 km. north of Lake Habbema, Oranje Range, New Guinea.

Nassau and Oranje Ranges, western New Guinea.

Orthonyx temminckii victoriana van Oort

Orthonyx temminckii victoriana van Oort, 1909, Notes Leyden Mus., 30, p. 234 — Mount Victoria, Wharton Range, southeastern New Guinea.

Herzog Mountains and the Wharton Range, southeastern New Guinea.

Orthonyx temminckii temminckii Ranzani

Orthonyx Temminckii Ranzani, 1822, Elem. Zool., 3, p. 19 — Oceanica; restricted to Hat Hill, near Nowra, New South Wales, ex Vigors and Horsfield.

[1] MS read by J. Delacour, K. C. Parkes, A. L. Rand, and S. D. Ripley.

Or[thonyx]. *maculatus* Stephens, 1826, in Shaw, Gen. Zool., 14, p. 186 — Australia.

[*Orthonyx*] *Temminckii* Vigors and Horsfield, 1827 (Feb. 17), Trans. Linn. Soc. London, 15, p. 294 — near Hat Hill, New South Wales.

Orthonyx temminckii chandleri Mathews, 1912, Novit. Zool., 18, p. 329 — Richmond River, New South Wales.

Forests of southeastern Queensland and eastern New South Wales, south to the Williams River, Illawarra brushes (south of Sydney).

ORTHONYX SPALDINGII

— **Orthonyx spaldingii** Ramsay

Orthonix [sic] *spaldingii* Ramsay, 1868 (Mar. 21), Sydney *Morning Herald*, p. 4, bottom of column 6 — northeastern coast of Australia.

Orthonyx spaldingi Ramsay, 1868 (Oct.), Proc. Zool. Soc. London, p. 386 — Rockingham Bay, Queensland.

Macrothonyx spaldingi albiventer Mathews, 1915, Austral Avian Rec., 2, p. 130 — Atherton, northern Queensland.

Forests of coastal range of northern Queensland, from Trinity Bay southward to Rockingham Bay.

GENUS **ANDROPHOBUS** HARTERT AND PALUDAN

Androphobus Hartert and Paludan, 1934, Orn. Monatsb., 42, p. 46. Type, by original designation and monotypy, *Androphilus viridis* Rothschild and Hartert.

ANDROPHOBUS VIRIDIS

Androphobus viridis (Rothschild and Hartert)

Androphilus viridis Rothschild and Hartert, 1911, Bull. Brit. Orn. Club, 29, p. 33 — Mount Goliath, Oranje Range, New Guinea.

Weyland Mountains and the Nassau and Oranje Ranges, from 1,500 to 2,800 meters, western New Guinea.

GENUS **PSOPHODES** VIGORS AND HORSFIELD

Psophodes Vigors and Horsfield, 1827, Trans. Linn. Soc. London, 15, p. 328. Type by monotypy, *Muscicapa crepitans* Latham = *Corvus olivaceus* Latham.

PSOPHODES OLIVACEUS

Psophodes olivaceus lateralis North

> *Psophodes crepitans lateralis* North, 1897, Rec. Austr.
> Mus., 3, p. 13 — Boar Pocket, on the Barron River, *ca.*
> 32 miles from Cairns, northern Queensland.

Forests of northern Queensland (Cairns district).

Psophodes olivaceus magnirostris Mathews

> *Psophodes olivaceus magnirostris* Mathews, 1912, Austral Avian Rec., 1, p. 92 — Rockhampton, Queensland.

Forests of central Queensland (Rockhampton district).

Psophodes olivaceus olivaceus (Latham)

> C[*orvus*]. *olivaceus* Latham, 1801, Index Orn., Suppl., p.
> 26 — Australia; restricted to New South Wales, ex
> *Muscicapa crepitans* Latham, *ibid.,* p. 51.
> *Psophodes olivaceus scrymgeouri* Mathews, 1912, Novit.
> Zool., 18, p. 333 — Victoria.
> *Psophodes olivaceus sublateralis* Mathews, 1912, Novit.
> Zool., 18, p. 334 — Tweed River, New South Wales.

Forests of eastern Australia from southeastern Queensland southward, through eastern New South Wales, to southern Victoria (east of Melbourne).

PSOPHODES NIGROGULARIS

Psophodes nigrogularis leucogaster Howe and Ross

> *Psophodes nigrogularis leucogaster* Howe and Ross, 1933,
> Emu, 32, p. 147, pl. 22 — Manya, northwestern Victoria.

Mallee of northwestern Victoria and southeastern South Australia.

Psophodes nigrogularis nigrogularis Gould

> *Psophodes nigrogularis* Gould, 1844, Birds Australia, pt.
> 15 — Western Australia.
> *Psophodes nigrogularis pallida* [sic] Mathews, 1916,
> Austral Avian Rec., 3, p. 60 — Cape Mentelle, Western
> Australia.

Mallee of southwestern Western Australia (in recent years reported only from vicinity of Gnowangerup, Albany, and Borden).

GENUS **SPHENOSTOMA** GOULD

Sphenostoma Gould, 1838, Synops. Birds Australia, pt. 4, pl. [8]. Type, by monotypy, *Sphenostoma cristatum* Gould.

SPHENOSTOMA CRISTATUM

— **Sphenostoma cristatum** Gould

Sphenostoma cristatum Gould, 1838, Synops. Birds Australia, pt. 4, pl. [8] — New South Wales.

Sphenostoma cristatum pallidum Mathews, 1912, Novit. Zool., 18, p. 378 — Leigh's Creek, South Australia.

Sphenostoma cristatum occidentale Mathews, 1912, Novit. Zool., 18, p. 378 — Day Dawn, Western Australia.

Sphenostoma cristatum tanami Mathews, 1912, Novit. Zool., 18, p. 379 — Tanami, Northern Territory.

Widespread in interior of Australia, particularly the mulga country, north to Godfrey's Tank and Tanami. Reaches the coast only in western Australia, between Shark Bay and Exmouth Gulf.

GENUS **CINCLOSOMA** VIGORS AND HORSFIELD

Cinclosoma Vigors and Horsfield, 1827, Trans. Linn. Soc. London, 15 (1826), p. 219. Type, by monotypy, *Turdus punctatus* Latham, 1801 = *Turdus punctatus* Shaw, 1794.

Ajax Lesson, 1837, Hist. Nat. Mamm. Ois. découverts depuis 1788, 8, p. 435. Type, by tautonymy and monotypy, *Eupetes ajax* Temminck.

Samuela Mathews, 1912, Austral Avian Rec., 1, p. 112. Type, by original designation, *Cinclosoma cinnamomeum* Gould.

cf. Condon, 1962, Rec. South Austr. Mus., 14, pp. 337-370 (revision).

CINCLOSOMA PUNCTATUM

— **Cinclosoma punctatum punctatum** (Shaw)

Turdus punctatus Shaw, 1794, Zool. New Holland, 1, p. 25, pl. 9 — New South Wales.

Cinclosoma punctatum neglectum Mathews, 1912, Novit. Zool., 18, p. 330 — Frankston, Victoria.

Southeastern Queensland, eastern New South Wales, Victoria, and southeastern South Australia (now confined to Mount Lofty Ranges).

Cinclosoma punctatum dovei Mathews

Cinclosoma punctatum dovei Mathews, 1912, Novit. Zool., 18, p. 330 — Tasmania.

Tasmania.

CINCLOSOMA CASTANOTUM

Cinclosoma castanotum castanotum Gould

[*Cinclosoma*] *castanotus* [sic] Gould, 1840, Ann. Mag. Nat. Hist., 5, p. 117 — Australia; type from the Belts of the Murray, South Australia, *fide* Gould, 1841, Proc. Zool. Soc. London, pt. 8 (1840), p. 113.

Mallee districts of southeastern South Australia, north to Leigh Creek, and adjacent New South Wales and northwestern Victoria.

Cinclosoma castanotum mayri Condon

Cinclosoma castanotum mayri Condon, 1962, Rec. South Austr. Mus., 14, p. 355 — 20 miles south of Rankin Springs, New South Wales.

Mallee scrub in Murrumbidgee Irrigation area, New South Wales.

Cinclosoma castanotum morgani Condon

Cinclosoma castanotum morgani Condon, 1951, South Australian Orn., 20, p. 42 — eighteen miles northwest of Kimba, South Australia.

Eyre Peninsula and Gawler Ranges, South Australia.

Cinclosoma castanotum clarum Morgan

Cinclosoma castanotum clarum Morgan, 1926, South Australian Orn., 8 (1925-1926), p. 138 — Wipipippee, *ca.* five miles east of the southern end of Lake Gairdner, South Australia.

Southern portion of Northern Territory (north to MacDonnell Ranges), eastern Western Australia (northwest to Separation Well and southeast to Menzies and Kalgoorlie), and Musgrave and Everard Ranges in South Australia (south to Lake Gairdner).

Cinclosoma castanotum dundasi Mathews

Cinclosomá castanotum dundasi Mathews, 1912, Novit.
Zool., 18, p. 330 — Lake Dundas, Western Australia.
Southwestern Western Australia (except heavily forested
area of extreme southwest), north to the mulga line.

CINCLOSOMA CINNAMOMEUM

Cinclosoma cinnamomeum castaneothorax Gould[1]

Cinclosoma castaneothorax Gould, 1849, Proc. Zool. Soc.
London, pt. 16 (1848), p. 139 — Darling Downs [= upper Dawson River], Queensland.
[*Cinclosoma*] *erythrothorax* Sharpe, 1881, Ibis, p. 605.
New name for *Cinclosoma castaneothorax* Gould, on
grounds of purism.
Interior of southern Queensland, from the Darling Downs
westward to valley of the Warrego and Thompson River,
and adjacent area of New South Wales (near Bourke).

Cinclosoma cinnamomeum cinnamomeum Gould

Cinclosoma cinnamomeus [sic] Gould, 1846, Proc. Zool.
Soc. London, pt. 14, p. 68 — Sturt's Depot [lat. 29° 40′
S.], northwestern New South Wales.
Samuela cinnamomea todmordeni Mathews, 1923, Aus-
tral Avian Rec., 5, p. 35 — Todmorden, South Australia.
Desert regions of southeastern Northern Territory (north
to about Tropic of Capricorn), extreme southwestern
Queensland, extreme northwestern New South Wales, and
northern South Australia (south to vicinity of Leigh Creek
and Lake Frome).

Cinclosoma cinnamomea samueli (Mathews)

Samuela cinnamomea samueli Mathews, 1916, Austral
Avian Rec., 3, p. 60 — Gawler Ranges, South Australia.
South Australia, from southwest to Lake Eyre through
Stuart Range to Ooldea and the Gawler Ranges.

Cinclosoma cinnamomeum marginatum Sharpe[2]

Cinclosoma marginatum Sharpe, 1883, Cat. Birds Brit.
Mus. 7, p. 336 — northwestern Australia; error, type
locality "corrected" to northwestern New South Wales,

---

[1] Condon (1962, *op. cit.*, p. 365) considers this a separate species.
[2] Condon (1962, *op. cit.*, p. 364) considers this a separate species.

by Mathews, 1930, Syst. Avium Australasian, pt. 2, p. 558.

Cinclosoma castaneothorax nea [sic] Mathews, 1912, Novit. Zool., 18, p. 331 — Day Dawn, Western Australia.

Southwestern Australia from about lat. 30° S. (Yalgoo, Mount Kenneth, Mount Ida) north to slightly beyond the Tropic of Capricorn (Wanery River, Jigalong), east to Lake Darlot and Canning Stock Route.

Cinclosoma cinnamomeum alisteri Mathews[1]

Cinclosoma alisteri Mathews, 1910, Bull. Brit. Orn. Club, 27, p. 16 — Western Australia; type from Waddilinia, Nullarbor Plain, *fide* Hartert, 1920, Novit. Zool., 27, p. 488.

[*Cinclosoma*] *nullarborensis* [sic] Campbell, 1922, Emu, 21, p. 161, pl. 32 — Haig and Naretha, Nullarbor Plain, Western Australia.

Nullarbor Plain of southwestern South Australia and southeastern Western Australia.

CINCLOSOMA AJAX

Cinclosoma ajax ajax (Temminck)

Eupetes ajax Temminck, 1835, Pl. col., livr. 97, pl. 573 — Lobo [lat. 3° 45′ S., long. 134° 05′ E.], New Guinea.

Western shores of Geelvink Bay and Triton Bay, western New Guinea.

Cinclosoma ajax muscale Rand

Cinclosoma ajax muscalis [sic] Rand, 1940, Amer. Mus. Novit., no. 1074, p. 2 — five miles below Palmer Junction, Upper Fly River, New Guinea.

Valley of the Upper Fly, south-central New Guinea.

Cinclosoma ajax alare Mayr and Rand

Cinclosoma ajax alaris [sic] Mayr and Rand, 1935, Amer. Mus. Novit., no. 814, p. 6 — Wuroi [lat. 8° 50′ S., long. 143° 07′ E.], New Guinea.

Valleys of the Oriomo and Lower Fly, south-central New Guinea.

---

[1] Condon (1962, *op. cit.*, p. 359) considers this a separate species.

Cinclosoma ajax goldiei (Ramsay)

Eupetes goldiei Ramsay, 1879, Proc. Linn. Soc. New South Wales, 3, p. 302 — sixty miles inland from Port Moresby, New Guinea.
Hall Sound to Milne Bay, southeastern New Guinea.

GENUS **PTILORRHOA** PETERS

Ptilorrhoa Peters, 1940, Auk, 57, p. 94. Type, by original designation, *Eupetes caerulescens* Temminck.
Mollitor Iredale, 1956 (subgenus of *Ptilorrhoa* Peters), Birds New Guinea, 2, p. 86. Type, by original designation, *Eupetes leucostictus* Sclater.

PTILORRHOA LEUCOSTICTA

Ptilorrhoa leucosticta leucosticta (Sclater)

Eupetes leucostictus Sclater, 1874, Proc. Zool. Soc. London, (1873), p. 690, pl. 52 — Hatam, Arfak Mountains, New Guinea.
Tamrau and Arfak Mountains, western New Guinea (Vogelkop).

Ptilorrhoa leucosticta mayri (Hartert)

Eupetes leucostictus mayri Hartert, 1930, Novit. Zool., 36, p. 87 — Mount Wondiwoi, Wandammen Mountains, New Guinea.
Wandammen Mountains, western New Guinea.

Ptilorrhoa leucosticta centralis (Mayr)

Eupetes leucostictus centralis Mayr, 1936, Amer. Mus. Novit., no. 869, p. 1 — Weyland Mountains, New Guinea.
Weyland Mountains, Nassau and Oranje Ranges, western New Guinea.

Ptilorrhoa leucosticta sibilans (Mayr)

Eupetes leucostictus sibilans Mayr, 1931, Mitt. Zool. Mus. Berlin, 17, p. 691 — Cyclops Mountains, northern New Guinea.
Known only from type locality.

— **Ptilorrhoa leucosticta amabilis** (Mayr)

Eupetes leucostictus amabilis Mayr, 1931, Mitt. Zool. Mus. Berlin, 17, p. 691 — Junzaing [lat. 6° 23′ S., long. 147° 37′ E.], New Guinea.

Saruwaged Mountains, eastern New Guinea.

— **Ptilorrhoa leucosticta loriae** (Salvadori)

Eupetes loriae Salvadori, 1896, Ann. Mus. Civ. Genova, ser. 2, 16, p. 102 — Moroka [lat. 9° 24′ S., long. 147° 32′ E.], New Guinea.

Mountains of southeastern New Guinea, westward to Mount Hagen and (subsp. ?) Schraderberg.

PTILORRHOA CAERULESCENS

— **Ptilorrhoa caerulescens caerulescens** (Temminck) / o ʌ 'ʏ ʟ̶ ꜀)

Eupetes caerulescens Temminck, 1835, Pl. col., livr. 97, pl. 574 — Lobo [lat. 3° 45′ S., long. 134° 05′ E.], New Guinea.

Eupetes caerulescens occidentalis Neumann, 1924, Orn. Monatsb., 32, p. 39 — Waigama, Misol.

Misol; western New Guinea from the Vogelkop (Sorong district) eastward to Etna Bay and the Wanggar River.

Ptilorrhoa caerulescens neumanni (Mayr and de Schauensee)

Eupetes caerulescens neumanni Mayr and de Schauensee, 1939, Proc. Acad. Nat. Sci. Philadelphia, 91, p. 122 — Cyclops Mountains, northern New Guinea.

Northern New Guinea from the Mamberamo River southeastward to Astrolabe Bay.

— **Ptilorrhoa caerulescens nigricrissa** (Salvadori)

Eupetes nigricrissus Salvadori, 1876, Ann. Mus. Civ. Genova, 9, p. 36 — Naiabui, Hall Sound, and Fly River, southern New Guinea.

Southern New Guinea from Etna Bay and the Wanggar River eastward to Milne Bay.

— **Ptilorrhoa caerulescens geislerorum** (Meyer)

Eupetes geislerorum A. B. Meyer, 1892, Journ. f. Orn., 40, p. 259 — Butaueng [lat. 6° 36′ S., long. 147° 51′ E.], New Guinea.

Eastern New Guinea from the Huon Gulf (Finschhafen) southeastward to Collingwood Bay.

PTILORRHOA CASTANONOTA

Ptilorrhoa castanonota castanonota (Salvadori)

Eupetes castanonotus Salvadori, 1875, Ann. Mus. Civ.
Genova, 7, p. 966 — Mount Morait [lat. 0° 45′ S., long.
131° 30′ E.], New Guinea.

Mountains of the Vogelkop, western New Guinea.

Ptilorrhoa castanonota saturata (Rothschild and Hartert)

Eupetes castanonotus saturatus Rothschild and Hartert,
1911, Orn. Monatsb., 19, p. 157 — Snow Mountains,
New Guinea; type specimen from the Setekwa River,
Nassau Range, *fide* Mayr, 1941, List New Guinea Birds,
p. 111.

Nassau Range, western New Guinea.

Ptilorrhoa castanonota uropygialis (Rand)

Eupetes castanonotus uropygialis Rand, 1940, Amer. Mus.
Novit., no. 1074, p. 2 — six kilometers southwest of
Bernhard Camp, Idenburg River, New Guinea.

Northern slopes of the Oranje Range, western New
Guinea.

Ptilorrhoa castanonota buergersi (Mayr)

Eupetes castanonotus bürgersi Mayr, 1931, Mitt. Zool.
Mus. Berlin, 17, p. 691 — Lordberg, Sepik Mountains,
New Guinea.

Sepik Mountains (Lordberg and Etappenberg), central
New Guinea.

Ptilorrhoa castanonota par (Meise)

Eupetes castanonotus par Meise, 1930, Orn. Monatsb., 38,
p. 17 — Sattelberg [lat. 6° 30′ S., long. 147° 49′ E.],
New Guinea.

Saruwaged Range, eastern New Guinea.

Ptilorrhoa castanonota pulchra (Sharpe)

Eupetes pulcher Sharpe, 1882, Journ. Linn. Soc. London,
16 (1883), p. 319 — the back of the Astrolabe Range,
southeastern New Guinea.

Mountains of southeastern New Guinea from Herzog
Mountains southeastward to Owen Stanley Range.

GENUS **EUPETES** TEMMINCK

Eupetes Temminck, 1831, Pl. col., livr. 87, text to pl. 516.
Type, by original designation, *Eupetes macrocerus*
Temminck.

EUPETES MACROCERUS

Eupetes macrocerus macrocerus Temminck[1]

Eupetes macrocerus Temminck, 1831 (Jan.), Pl. col.,
livr. 87, pl. 516 — Padang [lat. 0° 58′ S., long. 100° 21′
E.], Sumatra.

Eupetes macrocercus Lesson, 1831 (June), Traité Orn.,
livr. 8, p. 649. *Nomen emendatum.*

Eupetes macrourus Temminck, 1835, Pl. col., livr. 97, text
to pl. 573. *Nomen emendatum.*

Eupetes macrocercus griseiventris Baker, 1917, Bull. Brit.
Orn. Club, 38, p. 8 — "Tang, Song Paa" [= Sathani
Thung Song (lat. 8° 10′ N., long. 99° 40′ E.)], Thailand.

Malay Peninsula from province of Surat Thani, Thailand,
southward probably to Johore; Sumatra; North Natuna Islands (Pulau Bunguran).

Eupetes macrocerus borneensis Robinson and Kloss

Eupetes macrocerus borneensis Robinson and Kloss, 1921,
Journ. Fed. Malay States Mus., 10, p. 204 — Batang
Samarahan, southern Sarawak.

Eupetes macrocercus subrufus Hachisuka, 1926, Bull.
Brit. Orn. Club, 47, p. 54 — Mount Dulit, Sarawak.
Borneo.

---

[1] The specific name *macrocerus* Temminck, 1831, seems to have re-
sulted from a printer's misreading of the word *macrourus* in the
author's hand-written copy. If there is to be any emendation, it must
be taken from Temminck's own corrected version of 1835.

Unfortunately, for more than a century the spelling has been un-
justifiably altered to *macrocercus* by almost all authors. Inasmuch as
the original spelling *macrocerus* forms a pronounceable (although in-
appropriate) name, carries a visual resemblance to *macrocercus*, and
has been used by several well-known ornithologists during the past
thirty years, I retain the solecistic name of 1831 and treat the corrected name of 1835 as a *nomen emendatum.*

Genus MELAMPITTA Schlegel

Melampitta Schlegel, 1873, Ned. Tijdsch. Dierk., 4, afd. 2 (1871), p. 47. Type, by monotypy, *Melampitta lugubris* Schlegel.
Mellopitta Stejneger, 1885, in Kingsley, Stand. Nat. Hist., 4, Birds, p. 466. Alternative name.

MELAMPITTA LUGUBRIS

Melampitta lugubris lugubris Schlegel
Melampitta lugubris ["Rosenberg"] Schlegel, 1873, Ned. Tijdsch. Dierk., 4, afd. 2 (1871), p. 47 — Arfak Mountains, New Guinea.
Arfak Mountains, western New Guinea (Vogelkop).

Melampitta lugubris rostrata (Ogilvie-Grant)
Mellopitta lugubris rostrata Ogilvie-Grant, 1913, Bull. Brit. Orn. Club, 31, p. 104 — Utakwa River, western New Guinea.
Weyland Mountains and Nassau Range, western New Guinea.

Melampitta lugubris longicauda Mayr and Gilliard
Melampitta lugubris longicauda Mayr and Gilliard, 1952, Amer. Mus. Novit., no. 1577, p. 1 — Mount Tafa, near the Wharton Range, southeastern New Guinea.
Mountains of north-central and eastern New Guinea from Oranje Range eastward to Owen Stanley Range; Huon Peninsula.

MELAMPITTA GIGANTEA

Melampitta gigantea (Rothschild)
Mellopitta gigantea Rothschild, 1899, Orn. Monatsb., 7, p. 137 — Mount Moari, Arfak Mountains, New Guinea.
Arfak Mountains and Nassau Range, western New Guinea.

Genus IFRITA Rothschild

Ifrita Rothschild, 1898, Bull. Brit. Orn. Club, 7, p. 53. Type, by monotypy, *Ifrita coronata* Rothschild.

IFRITA KOWALDI

Ifrita kowaldi kowaldi (De Vis)
Todopsis kowaldi De Vis, 1890, Ann. Rept. Brit. New

Guinea, 1888-1889, p. 59 — British New Guinea; type from Owen Stanley Mountains, *fide* Mayr, 1941, List New Guinea Birds, p. 113.

Ifrita coronata Rothschild, 1898, Bull. Brit. Orn. Club, 7, p. 54 — low country east of Port Moresby, British New Guinea; error, corrected to Owen Stanley Mountains, by Hartert, 1920, Novit. Zool., 27, p. 483.

Ifrita kowaldi schalowiana Stresemann, 1922, Orn. Monatsb., 30, p. 8 — Schraderberg, Sepik Mountains, New Guinea.

Highlands of eastern and central New Guinea.

Ifrita kowaldi brunnea Rand

Ifrita kowaldi brunnea Rand, 1940, Amer. Mus. Novit., no. 1074, p. 2 — Mount Kunupi, Weyland Mountains, New Guinea.

Westernmost part of central range of New Guinea (Nassau and Weyland Mountains).

SUBFAMILY TIMALIINAE[1]

HERBERT G. DEIGNAN

cf. Delacour, 1946, Oiseau Rev. Franç. Orn., 16, pp. 7-36.

GENUS PELLORNEUM SWAINSON

Pellorneum Swainson, 1832, in Swainson and Richardson, Fauna Bor.-Amer., 2 (1831), p. 487. Type, by original designation and monotypy, *Pellorneum ruficeps* Swainson.

Drymocataphus Blyth, 1849, Journ. Asiat. Soc. Bengal, 18, p. 815. Type, by original designation, *Brachypteryx nigrocapitata* Eyton.

Scotocichla Sharpe, 1883, Cat. Birds. Brit. Mus., 7, pp. 505 (in key), 522. Type, by monotypy, *Drymocataphus fuscocapillus* Blyth.

cf. Deignan, 1947, Smiths. Misc. Coll., 107 (14), pp. 1-20 (races of *ruficeps*).

Ripley, 1949, Ibis, pp. 414-421 (races of *albiventre*).

---

[1] MS read by J. Delacour, B. P. Hall (African forms), K. C. Parkes, A. L. Rand, and S. D. Ripley.

Delacour, 1951, Oiseau Rev. Franç. Orn., 21, pp. 88-90
(races of *ruficeps*, in part).
Ripley and Hall, 1954, Ibis, pp. 486-487 (races of *albiventre*, in part).

PELLORNEUM RUFICEPS

Pellorneum ruficeps olivaceum Jerdon

P[*ellorneum*]. *olivaceum* Jerdon, 1839, Madras Journ.
Lit. Sci., 10, p. 255 — "jungles of Trichoor, Wurgun-
cherry and Manantoddy"; restricted to Trichur, Tra-
vancore-Cochin, by Deignan, 1947, Smiths. Misc. Coll.,
107 (14), p. 3.
Pellorneum ruficeps granti Harington, 1913, Bull. Brit.
Orn. Club, 33, p. 81 — Mynall, Travancore-Cochin.
Southwestern India (Kerala).

Pellorneum ruficeps ruficeps Swainson

Pellorneum ruficeps Swainson, 1832, in Swainson and
Richardson, Fauna Bor.-Amer., 2 (1831), p. 487 —
India; type locality restricted to Coonoor, Madras, by
Deignan, 1947, Smiths. Misc. Coll., 107 (14), p. 3.
Hills and coastal lowlands of western India from Malabar
to Saurashtra; Eastern Ghats and adjacent lowlands from
southern Madras and Mysore to northern Andhra; isolated
hill tracts of Orissa and Bihar.

Pellorneum ruficeps punctatum (Gould)

Cinclidia punctata Gould, 1838, Proc. Zool. Soc. London,
5, p. 137 — "the Himalaya Mountains"; type locality
restricted to Kalka, Himachal Pradesh, by Deignan,
1947, Smiths. Misc. Coll., 107 (14), p. 5.
Pellorneum ruficeps jonesi Baker, 1920, Bull. Brit. Orn.
Club, 41, p. 9 — Kalka, Himachal Pradesh.
Western Himalayas from Kangra eastward to Garhwal.

Pellorneum ruficeps mandellii Blanford

Pellorneum Mandellii Blanford, 1871, Proc. Asiat. Soc.
Bengal, p. 216 — Sikkim.
Nepal, Sikkim, Bhutan, and West Bengal (Darjeeling
District).

—**Pellorneum ruficeps chamelum** Deignan

Pellorneum ruficeps chamelum Deignan, 1947, Smiths.
Misc. Coll., 107 (14), p. 6 — Gunjong, Cachar, Assam.
Assam (south of the Brahmaputra) from Garo Hills east-
ward to Naga Hills.

Pellorneum ruficeps pectorale Godwin-Austen

Pellorneum pectoralis [sic] Godwin-Austen, 1877, Journ.
Asiat. Soc. Bengal, 46, p. 41 — Sadiya, Assam.
Northeastern Assam (Mishmi Hills).

Pellorneum ruficeps ripleyi Deignan

Pellorneum ruficeps ripleyi Deignan, 1947, Smiths. Misc.
Coll., 107 (14), p. 7 — Margherita, Lakhimpur, Assam.
Northeastern Assam (Lakhimpur District, south of the
Brahmaputra).

Pellorneum ruficeps vocale Deignan

Pellorneum ruficeps vocale Deignan, 1951, Postilla, Yale
Univ., no. 7, p. 2 — Kanglatongbi, Manipur.
Valley of central Manipur.

Pellorneum ruficeps stageri Deignan

Pellorneum ruficeps stageri Deignan, 1947, Smiths. Misc.
Coll., 107 (14), p. 8 — N'Pon Village (on left bank of
the Irrawaddy, 15 miles north of Myitkyina), Kachin
State, Upper Burma.
Northeastern Burma (Myitkyina and Bhamo Districts).

Pellorneum ruficeps shanense Deignan

Pellorneum ruficeps shanense Deignan, 1947, Smiths.
Misc. Coll., 107 (14), p. 9 — Ma-li-pa [lat. 23° 41′ N.,
long. 98° 46′ E.], Northern Shan State, Upper Burma.
Northern and Southern Shan States and southwestern
Yunnan, so far as these lie between the Mekong and Sal-
ween Rivers.

Pellorneum ruficeps hilarum Deignan

Pellorneum ruficeps hilarum Deignan, 1947, Smiths. Misc.
Coll., 107 (14), p. 10 — Kyundaw, Pakokku District,
Magwe Division, Upper Burma.
Dry zone of central Burma.

Pellorneum ruficeps victoriae Deignan

Pellorneum ruficeps victoriae Deignan, 1947, Smiths. Misc. Coll., 107 (14), p. 10 — Mount Victoria, Southern Chin Hills, Upper Burma.

Burma (Chin Hills).

Pellorneum ruficeps minus Hume

Pellorneum minor [sic] Hume, 1873, Stray Feathers, 1, p. 298 — Thayetmyo District, Magwe Division, Upper Burma.

Burma (valley of the lower Irrawaddy from Thayetmyo District to its mouths).

Pellorneum ruficeps subochraceum Swinhoe

Pellorneum subochraceum Swinhoe, 1871, Ann. Mag. Nat. Hist., ser. 4, 7, p. 257 — "the Tenasserim provinces"; restricted to Moulmein, Amherst District, by Deignan, 1947, Smiths. Misc. Coll., 107 (14), p. 12.

Tenasserim Division of Lower Burma and evergreen forest of southwestern Thailand from southern Tak Province to Prachuap Khiri Khan Province.

Pellorneum ruficeps insularum Deignan

Pellorneum ruficeps insularum Deignan, 1947, Smiths. Misc. Coll., 107 (14), p. 12 — Domel Island [lat. 11° 37′ N., long. 98° 16′ E.], Mergui Archipelago.

Lower Burma (Mergui Archipelago).

Pellorneum ruficeps acrum Deignan

Pellorneum ruficeps acrum Deignan, 1947, Smiths. Misc. Coll., 107 (14), p. 13 — Yala [lat. 6° 30′ N., long. 101° 15′ E.], Thailand.

Open forests of central plains of Thailand (west of the Chao Phaya) and of Malay Peninsula southward to central Malaya.

Pellorneum ruficeps chthonium Deignan

Pellorneum ruficeps chthonium Deignan, 1947, Smiths. Misc. Coll., 107 (14), p. 14 — Doi Suthep [lat. 18° 50′ N., long. 98° 55′ E.], Thailand.

Northern plateau of Thailand (except area occupied by *indistinctum*).

Pellorneum ruficeps indistinctum Deignan

Pellorneum ruficeps indistinctum Deignan, 1947, Smiths. Misc. Coll., 107 (14), p. 16 — King Chiang Saen [lat. 20° 15′ N., long. 100° 05′ E.], Thailand.
Mekong drainage of the northern plateau of Thailand.

Pellorneum ruficeps oreum Deignan

Pellorneum ruficeps oreum Deignan, 1947, Smiths. Misc. Coll., 107 (14), p. 16 — Muong Moun [lat. 21° 42′ N., long. 103° 21′ E.], Tongking.
Hilly regions of southern Yunnan, northern Laos, and Tongking, so far as these lie between the Mekong River and the Black River-Red River divide.

Pellorneum ruficeps vividum La Touche *types*

Pellorneum nipalense vividum La Touche, 1921, Bull. Brit. Orn. Club, 42, p. 17 — Hokow, southeastern Yunnan.
Valley of the Red River of Tongking from Chinese frontier to its mouths and thence southward along coast of the Gulf of Tongking to central Annam.

Pellorneum ruficeps elbeli Deignan

Pellorneum ruficeps elbeli Deignan, 1956, Proc. Biol. Soc. Washington, 69, p. 208 — Ban Na Muang, near Muang Daen Sai [lat. 17° 15′ N., long. 101° 05′ E.], Thailand.
Northwestern portion of eastern plateau of Thailand.

Pellorneum ruficeps ubonense Deignan

Pellorneum ruficeps ubonense Deignan, 1947, Smiths. Misc. Coll., 107 (14), p. 18 — Ban Chanuman [lat. 16° 15′ N., long. 105° 00′ E.], Thailand.
Easternmost portion of eastern plateau of Thailand and adjacent region of southern Laos.

Pellorneum ruficeps deignani Delacour

Pellorneum ruficeps deignani Delacour, 1951, Oiseau Rev. Franç. Orn., 21, p. 89 — Da Ban [lat. 12° 38′ N., long. 109° 06′ E.], Annam.
Southern Annam.

Pellorneum ruficeps dilloni Delacour

Pellorneum ruficeps dilloni Delacour, 1951, Oiseau Rev. Franç. Orn., 21, p. 90 — Trang Bom [lat. 10° 56′ N., long. 107° 00′ E.], Cochin China.
Forests of northern Cochin China.

Pellorneum ruficeps euroum Deignan

Pellorneum ruficeps euroum Deignan, 1947, Smiths. Misc.
Coll., **107** (14), p. 19 — Chanthaburi [lat. 12° 35′ N.,
long. 102° 05′ E.], Thailand.

Central plains of Thailand (east of the Chao Phaya),
southeastern Thailand, and western Cambodia.

Pellorneum ruficeps smithi Riley

Pellorneum smithi Riley, 1924, Proc. Biol. Soc. Washington, 37, p. 129 — Ko Chang [lat. 12° 00′ N., long.
102° 30′ E.], Thailand.

Islets off coast of southeastern Thailand and of Cambodia.

PELLORNEUM PALUSTRE

Pellorneum palustre Gould

Pellorneum palustre "Jerd." = Gould, 1872, Birds Asia, 3,
pt. 24, pl. 65 and text — Cachar, Assam.

Assam (valley of the Brahmaputra).

PELLORNEUM FUSCOCAPILLUM

Pellorneum fuscocapillum babaulti (Wells)

Scotocichla fuscicapilla babaulti Wells, 1919, Bull. Brit.
Orn. Club, 39, p. 69 — Trincomalee, Ceylon.

Low-country dry zone of northern and eastern Ceylon.

Pellorneum fuscocapillum fuscocapillum (Blyth)

Dr[ymocataphus]. fuscocapillus Blyth, 1849, Journ.
Asiat. Soc. Bengal, **18**, p. 815 — Ceylon; inferentially
restricted to "S. W. Ceylon," by Baker, 1921, Journ.
Bombay Nat. Hist. Soc., **27** (3), p. 26, and here further restricted to Colombo.

Wet zone of southwestern Ceylon (except area occupied
by *scortillum*).

Pellorneum fuscocapillum scortillum Ripley

Pellorneum fuscocapillum scortillum Ripley, 1946, Spolia
Zeylanica, 24, p. 226 — Depedene Estate, Rakwana,
Sabaragamuwa, Ceylon.

Wettest parts of the wet zone of southwestern Ceylon
(from Hiniduma and Sinharadja Forest eastward through
the Pelmadulla area to Haputale).

PELLORNEUM CAPISTRATUM

Pellorneum capistratum nigrocapitatum (Eyton)

Brachypteryx nigrocapitata Eyton, 1839, Proc. Zool. Soc. London, pt. 7, p. 103 — Malaya.

Malay Peninsula from southern Tenasserim (Mergui District) and the Isthmus of Kra southward to Singapore; North Natuna Islands; Billiton.

Pellorneum capistratum nyctilampis (Oberholser)

Drymocataphus nigrocapitatus nyctilampis Oberholser, 1922, Smiths. Misc. Coll., 74 (2), p. 10 — Bukit Parmasang, Bangka.

Sumatra; Bangka.

Pellorneum capistratum capistratoides (Strickland)

Goldana capistratoides Strickland, 1849, in Jardine, Contrib. Orn., p. 128, pl. 36 — Borneo; here restricted to Pontianak [lat. 0° 02′ S., long. 109° 22′ E.].

Western and southern Borneo.

Pellorneum capistratum morrelli Chasen and Kloss

Pellorneum capistratum morrelli Chasen and Kloss, 1929, Journ. f. Orn., Ergänzungsb. 2, p. 118 — Kudat, North Borneo.

Northern and eastern Borneo; Banggai Island.

Pellorneum capistratum capistratum (Temminck)

Myiothera capistrata Temminck, 1823, Pl. col., livr. 31, pl. 185, fig. 1 — Java.

Java.

PELLORNEUM ALBIVENTRE

Pellorneum albiventre ignotum Hume

Pellorneum ignotum Hume, 1877 (Aug.), Stray Feathers, 5, p. 334 — "Dollah," near Sadiya, Assam.

Northeastern Assam (Mishmi Hills).

Pellorneum albiventre albiventre (Godwin-Austen)

Neornis albiventris Godwin-Austen, 1877 (Apr. 19), Journ. Asiat. Soc. Bengal, 45 (1876), p. 199 — Sengmai, Manipur.

Turdinus nagaënsis Godwin-Austen, 1877 (Dec.), Ann.

Mag. Nat. Hist., ser. 4, 20, p. 519 — eastern Naga Hills, Assam.

Bhutan-Assam border (Diwangiri); Assam (south of the Brahmaputra) from Cachar eastward to Naga Hills and Manipur, thence southward into western Burma (Chin Hills).

Pellorneum albiventre cinnamomeum (Rippon) / ., 4. ,

Drymocataphus cinnamomeus Rippon, 1900, Bull. Brit. Orn. Club, 11, p. 12 — Loi Mai [lat. 20° 25′ N., long. 97° 26′ E.], Southern Shan State, Upper Burma.

Drymocataphus albiventer vicinus Riley, 1940, Proc. Biol. Soc. Washington, 53, p. 132 — Fimnom [lat. 11° 47′ N., long. 108° 24′ E.], Annam.

Northern and Southern Shan States, Karenni State, and northwestern Thailand; southern Laos (Boloven Plateau); southern Annam (Lang Bian Plateau).

Pellorneum albiventre pusillum (Delacour)

Drymocataphus pusillus Delacour, 1927, Bull. Brit. Orn. Club, 47, p. 161 — Tam Dao [lat. 21° 27′ N., long. 105° 40′ E.], Tongking.

Eastern regions of northern Laos and western Tongking.

Genus **TRICHASTOMA** Blyth

Trichastoma Blyth, 1842, Journ. Asiat. Soc. Bengal, 11, p. 795. Type, by subsequent designation (G. R. Gray, 1855, Cat. Gen. Subgen. Birds, p. 41), *Trichastoma rostratum* Blyth. Not preoccupied by *Trichostoma* Pictet, 1834, Trichoptera.

Malacocincla Blyth, 1845, Journ. Asiat. Soc. Bengal, 14, p. 600. Type, by original designation and monotypy, *Malacocincla abbotti* Blyth.

Trichostoma Strickland, 1849, in Jardine, Contrib. Orn., p. 126. *Nomen emendatum.*

Illadopsis Heine, 1860, Journ. f. Orn., 7 (1859), p. 430. Type, by original designation and monotypy, *Turdirostris fulvescens* Cassin.

Nannothera Sundevall, 1872, Tentamen, pt. 1, p. 11. Type, by original designation and monotypy, *Brachypteryx sepiaria* Blyth, *i.e.*, Horsfield.

Erythrocichla Sharpe, 1883, Cat. Birds Brit. Mus., 7, pp.

506 (in key), 551. Type, by monotypy, *Brachypteryx bicolor* Lesson.

Ptilopyga Sharpe, 1883, Cat. Birds. Brit. Mus., 7, pp. 507 (in key), 585. Type, by subsequent designation (Anon., 1883, Ibis, p. 573), *Malacocincla rufiventris* Salvadori.

Anuropsis Sharpe, 1883, Cat. Birds. Brit. Mus., 7, pp. 507 (in key), 588. Type, by subsequent designation (Anon., 1883, Ibis, p. 573), *Brachypteryx malaccensis* Hartlaub.

Aethostoma Sharpe, 1902, Bull. Brit. Orn. Club, 12, p. 54. New name for *Trichostoma* Blyth, *i.e.*, Strickland (*nom. emend.*), preoccupied.

Elocincla Riley, 1939, Journ. Washington Acad. Sci., 29, p. 39. Type, by original designation and monotypy, *Elocincla aenigma* Riley = *Malacocincla rufiventris* Salvadori.

cf. Oberholser, 1932, Bull. U. S. Nat. Mus., 159, pp. 61-63 (races of *malaccense, sub nom. Anuropsis*).

Stresemann, 1940, Journ. f. Orn., 88, pp. 109-111 (races of *celebense, sub nom. Malacocincla*).

Delacour, 1946, Zoologica [New York], 31, p. 3.

Deignan, 1948, Journ. Washington Acad. Sci., 38, pp. 184-185 (races of *abbotti*, in part, *sub nom. Malacocincla*).

Chapin, 1953, Bull. Amer. Mus. Nat. Hist., 75A, pp. 206-221 (African forms, in part, *sub nom. Malacocincla*).

TRICHASTOMA TICKELLI

Trichastoma tickelli assamense (Sharpe)

Drymocataphus assamensis Sharpe, 1883, Cat. Birds Brit. Mus., 7, pp. 552 (in key), 557 — Dikrang valley, Dhollah, northeastern Assam.

Assam (both north and south of the Brahmaputra) and northwestern Burma (Upper Chindwin).

Trichastoma tickelli grisescens (Ticehurst)

Drymocataphus tickelli grisescens Ticehurst, 1932, Bull. Brit. Orn. Club, 53, p. 18 — "Nyaunggyo stream, Taungup-Prome cart-road, Arakan Yoma, 2,500 feet," Lower Burma.

Southwestern Burma (Arakan Yoma).

— **Trichastoma tickelli fulvum** (Walden)

Drymocataphus fulvus Walden, 1875, Ann. Mag. Nat. Hist., ser. 4, 15, p. 401 — Karenni State, Upper Burma.

Drymocataphus tickelli olivaceus Kinnear, 1924, Bull. Brit. Orn. Club, 45, p. 11 — Bao Ha [lat. 22° 10′ N., long. 104° 21′ E.], Tongking. Not *Malacopteron olivaceum* Strickland, 1847 = *Trichastoma abbotti olivaceum* (Strickland), nor *Mixornis olivaceus* Tickell, 1859 = *Pellornium* [sic] *Tickelli* Blyth = *Trichastoma tickelli tickelli* (Blyth).

Drymocataphus tickelli ochraceus Kinnear, 1934, Bull. Brit. Orn. Club., 55, p. 53. New name for *Drymocataphus tickelli olivaceus* Kinnear, preoccupied.

Northeastern Burma, Shan and Karenni States, northern Thailand, northern Laos, northern Annam, western Tongking, and southwestern Yunnan.

Trichastoma tickelli annamense (Delacour)

Drymocataphus tickelli annamensis Delacour, 1926, Bull. Brit. Orn. Club, 47, p. 17 — Col des Nuages [lat. 16° 11′ N., long. 108° 08′ E.], Annam.

Central Annam southward to northern Cochin China; southern Laos (Boloven Plateau).

Trichastoma tickelli tickelli (Blyth)

Pellornium [sic] *Tickelli* Blyth, 1859, Journ. Asiat. Soc. Bengal, 28, p. 414 — mountainous interior of Tenasserim; type from "Woods of Teewap'hado, 1,100 to 1,500 feet," Amherst District, *fide* Tickell, 1860, *ibid.* (1859), p. 449.

Drymocataphus tickelli australis Robinson and Kloss, 1921, Journ. Federated Malay States Mus., 10, p. 205 — Ginting Bidei [lat. 3° 18′ N., long. 101° 50′ E.], Selangor, Malaya.

From central Tenasserim (Amherst District) and adjacent areas of Thailand (Tak Province) southward over Malay Peninsula to southern Selangor.

— **Trichastoma tickelli buettikoferi** Vorderman

Trichostoma büttikoferi Vorderman, 1892, Natuurk. Tijdschr. Nederl.-Indië, 51, p. 230 — Lampung district, southern Sumatra.

Sumatra.

TRICHASTOMA PYRROGENYS

Trichastoma pyrrogenys pyrrogenys (Temminck)

Myiothera pyrrogenys Temminck, 1827, Pl. col., livr. 74,
pl. 442, fig. 2 — Java; type locality restricted to Bantam Province, by Kloss, 1931, Treubia, 13, p. 346.
Western Java.

Trichastoma pyrrogenys besuki (Kloss)

Aethostoma pyrrogenys besuki Kloss, 1931, Treubia, 13,
p. 346 — Tamansari [lat. 8° 16′ S., long. 113° 44′ E.],
near Banyuwangi, Java.
Eastern Java.

Trichastoma pyrrogenys erythrote (Sharpe)

Malacopterum erythrote Sharpe, 1883, Cat. Birds. Brit.
Mus., 7, pp. 564 (in key), 567, pl. 13, fig. 2 — Borneo;
type characteristic of population of western Sarawak.
Mounts Poi and Penrissen, western Sarawak.

Trichastoma pyrrogenys longstaffi (Harrisson and Hartley)

Malacocincla canicapillus longstaffi Harrisson and Hartley, 1934, Bull. Brit. Orn. Club, 54, p. 152 — Mount
Dulit, Sarawak.
Highlands of Sarawak (except area occupied by *erythrote*).

Trichastoma pyrrogenys canicapillum (Sharpe)

Turdinus canicapillus Sharpe, 1887, Ibis, p. 450 — Kina
Balu, North Borneo.
Highlands of North Borneo.

TRICHASTOMA MALACCENSE

Trichastoma malaccense malaccense (Hartlaub)

B[rachypteryx]. malaccensis Hartlaub, 1844, Syst. Verz.
Naturh. Samml. Ges. Mus. (Bremen), Abth. 1, p. 40
— Malacca.
Anuropsis malaccensis nesitis Oberholser, 1912, Smiths.
Misc. Coll., 60 (7), p. 8 — Pulau Tanah Masa, Batu
Islands.
Anuropsis malaccensis exsanguis Oberholser, 1912,
Smiths. Misc. Coll., 60 (7), p. 8 — Pulau Tuangku,
Banyak Islands.

Anuropsis malaccensis drymodrama Oberholser, 1922, Smiths. Misc. Coll., 74 (2), p. 9 — Sungei Mandau, eastern Sumatra.

Anuropsis malaccensis driophila Oberholser, 1922, Smiths. Misc. Coll., 74 (2), p. 9 — Khao Soi Dao [lat. 7° 20′ N., long. 99° 50′ E.], Thailand.

From southern Tenasserim (Mergui District) and Isthmus of Kra southward over Malay Peninsula to Singapore; Riouw and Lingga Archipelagos; Sumatra and islands off western coast (Banyak and Batu Groups) ; Anamba Islands and North Natuna Islands.

Trichastoma malaccense saturatum (Robinson and Kloss)

Anuropsis malaccensis saturata Robinson and Kloss, 1920, Bull. Brit. Orn. Club, 40, p. 68 — Sungei Tinjar, Baram district, Sarawak.

Anuropsis malaccensis docima Oberholser, 1922, Smiths. Misc. Coll., 74 (2), p. 10 — Tanjong Tedung, Bangka.

Bangka; Billiton; western half of Borneo.

Trichastoma malaccense poliogene (Strickland)

Brachypteryx poliogenis Strickland, 1849, in Jardine, Contrib. Orn., p. 93, pl. 31 — Borneo, ex Boie; type from the Sungei Karau, southeastern Borneo, *fide* Büttikofer, 1895, Notes Leyden Mus., 17, p. 84.

Anuropsis malaccensis sordidus [sic] Chasen and Kloss, 1929, Journ. f. Orn., Ergänzungsb., 2, p. 119 — Bettotan, near Sandakan, North Borneo.

Eastern half of Borneo.

Trichastoma (malaccense ?) feriatum (Chasen and Kloss)

Anuropsis malaccensis feriatus [sic] Chasen and Kloss, 1931, Novit. Zool., 36, p. 279 — Gunong Mulu [lat. 4° 04′ N., long. 114° 57′ E.], Sarawak.

Known only from unique type.

TRICHASTOMA CINEREICEPS[1]

Trichastoma cinereiceps (Tweeddale)

Drymocataphus cinereiceps Tweeddale, 1878, Proc. Zool. Soc. London, p. 617 — Puerto Princesa, Palawan.

Palawan and Balabac, Philippine Islands.

[1] *Trichastoma cinereiceps* and *malaccense* comprise a superspecies.

TRICHASTOMA ROSTRATUM

Trichastoma rostratum rostratum Blyth

Tr[ichastoma]. rostratum Blyth, 1842, Journ. Asiat. Soc. Bengal, 11, p. 795 — Singapore.

Ptilocichla leucogastra Davison, 1892, Ibis, p. 100 — Malay Peninsula.

Aethostoma rostrata [sic] *aethalea* [sic] Oberholser, 1922, Smiths. Misc. Coll., 74 (2), p. 12 — Pulau Karimun Anak, Riouw Archipelago.

Aethostoma rostrata [sic] *paganica* [sic] Oberholser, 1922, Smiths. Misc. Coll., 74 (2), p. 12 — Upper Siak River, eastern Sumatra.

Southern Tenasserim (Mergui District) and Isthmus of Kra southward over Malay Peninsula to Singapore; Riouw and Lingga Archipelagos; Sumatra; Billiton.

Trichastoma rostratum macropterum (Salvadori)

Brachypteryx macroptera Salvadori, 1868, Atti R. Accad. Sci. Torino, 3, p. 528 — Sarawak.

Brachypteryx umbratilis "(Temm.)" Salvadori, 1874, Ann. Mus. Civ. Genova, 5, p. 220 — Borneo. Misapplication of *Trichostoma umbratile* Strickland, 1849 = *Rhinomyias umbratilis* (Strickland).

[Aethostoma] umbratile "(Strickl.)" Sharpe, 1903, Handlist Gen. Spec. Birds, 4, p. 38 — Borneo. Misapplication of *Trichostoma umbratile* Strickland, 1849.

A[ëthostoma]. witmeri Sharpe, 1903, Hand-list Gen. Spec. Birds, 4, p. 358. New name for *[Aethostoma] umbratile* "(Strickl.)" Sharpe, considered preoccupied.

Borneo and Banggai.

TRICHASTOMA BICOLOR

Trichastoma bicolor (Lesson)

Brachypteryx bicolor Lesson, 1839, Rev. Zool., 2, p. 138 — Sumatra.

Erythrocichla bicolor whiteheadi Hartert, 1915, Bull. Brit. Orn. Club, 36, p. 36 — Bengkoka River, North Borneo.

Erythrocichla bicolor bankana Riley, 1938, Proc. Biol. Soc. Washington, 51, p. 96 — Klabat Bay, Bangka.

Southern Tenasserim (Mergui District) and Isthmus of

Kra southward over Malay Peninsula to Johore; eastern
Sumatra; Bangka; Borneo.

TRICHASTOMA SEPIARIUM

Trichastoma sepiarium tardinatum (Hartert)

Malacocincla sepiaria tardinata Hartert, 1915, Bull. Brit.
Orn. Club, 36, p. 35 — Gunong Tahan [lat. 4° 38′ N.,
long. 102° 14′ E.], Pahang.

Malay Peninsula from Siamese province of Pattani southward to Selangor and Pahang.

Trichastoma sepiarium liberale (Chasen)

Malacocincla sepiaria liberalis Chasen, 1941, in Chasen
and Hoogerwerf, Treubia, 18, suppl., p. 80 — Pulau
Munteh, near Pendeng [lat. 4° 08′ N., long. 97° 36′ E.],
Acheh, Sumatra.

Highlands of northwestern Sumatra.

Trichastoma sepiarium barussanum (Robinson and Kloss)

Malacocincla sepiaria barussana Robinson and Kloss,
1921, Journ. Federated Malay States Mus., 10, pt. 3,
p. 205 — Siulak Deras [lat. 1° 54′ S., long. 101° 18′ E.],
Sumatra.

Highlands of southwestern Sumatra.

Trichastoma sepiarium sepiarium (Horsfield)

Brachypteryx sepiaria Horsfield, 1821, Trans. Linn. Soc.
London, 13, p. 158 — Java; here restricted to Bogor
[lat. 6° 36′ S., long. 106° 48′ E.].

Western and central Java.

Trichastoma sepiarum minus (Meyer)

Turdinus sepiarius (Horsf.) var. *minor* A. B. Meyer,
1884, Zeitschr. ges. Orn., 1, p. 210 — Java; here restricted to Lawang [lat. 7° 50′ S., long. 112° 42′ E.].
Myiothera pyca "Boie" = Büttikofer, 1895, Notes Leyden
Mus., 17, p. 82 — Java; here restricted to Lawang.

Eastern Java; Bali.

Trichastoma sepiarium rufiventre (Salvadori)

Malacocincla rufiventris Salvadori, 1874, Ann. Mus. Civ.
Genova, 5, p. 229 — Sarawak.

Turdinus tephrops Sharpe, 1893, Bull. Brit. Orn. Club,
1, p. 54 — Mount Kalulong, Sarawak.
[1]*Elocincla aenigma* Riley, 1939, Journ. Washington Acad.
Sci., 29, p. 39 — Klumpang Bay, southeastern Borneo.
Western and southern Borneo.

Trichastoma sepiarium harterti (Chasen and Kloss)

Malacocincla sepiaria harterti Chasen and Kloss, 1929,
Journ. f. Orn., Ergänzungsb., 2, p. 116 — Bettotan, near
Sandakan, North Borneo.
Northern and eastern Borneo.

TRICHASTOMA CELEBENSE

Trichastoma celebense celebense (Strickland)

Trichostoma [sic] *celebense* Strickland, 1849, in Jardine,
Contrib. Orn., p. 127, pl. 35, ant. fig. — Celebes; type
specimen characteristic of population of northern peninsula.
Turdinus castaneus Büttikofer, 1893, Notes Leyden Mus.,
15, p. 261 — Minahassa Peninsula, Celebes.
Northern peninsula (Minahassa) of Celebes.

Trichastoma celebense connectens (Mayr)

Aethostoma celebense connectens Mayr, 1938, Orn. Monatsb., 46, p. 157 — Pinedapa, north-central Celebes.
North-central Celebes.

Trichastoma celebense rufofuscum (Stresemann)

Aethostoma celebense rufofuscum Stresemann, 1931, Orn.
Monatsb., 39, p. 45 — Uru [lat. 3° 30′ S., long. 119° 56′
E.], Celebes.
Western foreland of the Latimojong Range, Celebes.

Trichastoma celebense finschi (Walden)

Trichostoma [sic] *finschi* Walden, 1876, Ibis, p. 378, pl.
11, fig. 1 — Makassar District, Celebes.
Southern part of the southwestern peninsula, Celebes.

[1] A topotypical series may show that this name should be synony-
mized with *harterti*.

Trichastoma celebense improbatum Deignan, nom. nov.[1]

Aethostoma celebense sordidum Stresemann, 1938, Orn.
Monatsb., 46, p. 147 — Lalolai [lat. 4° 03′ S., long.
121° 53′ E.], Celebes. Not Anuropsis malaccensis sordida Chasen and Kloss, 1929 = Trichastoma malaccense poliogene (Strickland).

Eastern and southeastern peninsulas, Celebes; Pulau Buton.

Trichastoma celebense togianense (Voous)

Malacocincla celebensis togianensis Voous, 1952, Ardea,
40, p. 74 — Malenge, Togian Islands, Gulf of Tomini.

Known only from type locality.

TRICHASTOMA ABBOTTI

Trichastoma abbotti amabile (Koelz)

Malacocincla abbotti amabilis Koelz, 1952, Journ. Zool.
Soc. India, 4, p. 39 — Nichuguard, Naga Hills, Assam.

Low elevations from Sikkim and eastern Nepal eastward, along valley of the Brahmaputra, to eastern Assam.

Trichastoma abbotti alterum Sims

Trichastoma abbotti alterum Sims, 1957, Bull. Brit. Orn.
Club, 77, p. 154 — Thua Luu [lat. 16° 16′ N., long.
108° 00′ E.], Annam.

Central Laos (Wiang Chan or Vientiane) and central Annam.

Trichastoma abbotti williamsoni (Deignan)

Malacocincla abbotti williamsoni Deignan, 1948, Journ.
Washington Acad. Sci., 38, p. 185 — Ban Pak Chong
[lat. 14° 40′ N., long. 101° 25′ E.], Thailand.

Southwestern portion of the eastern plateau of Thailand and northwestern Cambodia.

Trichastoma abbotti obscurius (Deignan)

Malacocincla abbotti obscurior Deignan, 1948, Journ.
Washington Acad. Sci., 38, p. 185 — Khao Sa Bap [lat.
12° 35′ N., long. 102° 15′ E.], Thailand.

Coastal areas of southeastern Thailand from Chon Buri Province to Trat; the islet Ko Kut, off the coast of Trat.

[1] The homonymy of sordidum in Trichastoma has been brought to the attention of its author, who has disclaimed responsibility for its renaming.

Trichastoma abbotti abbotti (Blyth)

M[alacocincla]. Abbotti Blyth, 1845, Journ. Asiat. Soc. Bengal, 14, p. 601 — Ramree Island, Arakan, Lower Burma.

Malacocincla abbotti rufescentior Deignan, 1948, Journ. Washington Acad. Sci., 38, p. 184 — Ban Tha Lo [ca. lat. 9° 05′ N., long. 99° 15′ E.], Thailand.

Southern Burma from Arakan southeastward to southernmost Tenasserim (including Mergui Archipelago) ; Thailand from central plains west of Chao Phaya River southward, through southwestern and peninsular provinces (except area occupied by *olivaceum*), to extreme south; northwestern Malaya (Langkawi Islands, Perlis, and Kedah).

Trichastoma abbotti olivaceum (Strickland)

Malacopteron olivaceum Strickland, 1847, Proc. Zool. Soc. London, pt. 14 (1846), p. 102 — Malacca.

Peninsular Thailand (Pattani, Yala, and Narathiwat Provinces)and Malaya (except area occupied by *abbotti*) ; eastern Sumatra.

Trichastoma abbotti sirense (Oberholser)

Malacocincla Büttikoferi Finsch, 1901, Notes Leyden Mus., 22, p. 218 — Kahajan and Karau Rivers, southeastern Borneo. Not *Trichostoma* [sic] *büttikoferi* Vorderman, 1892.

Malacocincla abbotti sirensis Oberholser, 1917, Proc. U.S. Nat. Mus., 54, p. 195 — Pulau Mata Siri [lat. 4° 48′ S., long. 115° 48′ E.], Java Sea.

Malacocincla abbotti eritora Oberholser, 1922, Smiths. Misc. Coll., 74 (2), p. 11 — Telok Buding, Billiton.

Malacocincla abotti [sic] *voousi* Wynne, 1955, North Western Naturalist [Arbroath, Scotland], new ser., 3, p. 120. New name for *Malacocincla Büttikoferi* Finsch, preoccupied.

Billiton; Borneo; Pulau Mata Siri.

Trichastoma abbotti baweanum (Oberholser)

Malacocincla abbotti baweana Oberholser, 1917, Proc. U.S. Nat. Mus., 52, p. 194 — Pulau Bawean [lat. 5° 48′ S., long. 112° 39′ E.], Java Sea.

Bawean.

TRICHASTOMA PERSPICILLATUM

Trichastoma perspicillatum (Bonaparte)

[*Cacopitta*] *perspicillata* "Temm. Mus. Lugd." Bonaparte, 1850, Consp. Av., 1, p. 257 — "Java," error; corrected to Borneo, by Büttikofer, 1895, Notes Leyden Mus., 17, p. 83.

Known only from unique type (described by Büttikofer, 1895, *loc. cit. supra,* and depicted in Smythies, 1960, Birds Borneo, pl. 45, fig. 7).

TRICHASTOMA VANDERBILTI

Trichastoma vanderbilti (de Schauense and Ripley)

Malacocincla vanderbilti de Schauensee and Ripley, 1940, Proc. Acad. Nat. Sci. Philadelphia, 91, p. 351, pl. 20 — Koengke, 3,100 ft., Atjeh, north Sumatra.

Known only from type locality. (Possibly a geographical representative of *Tr. perspicillatum.*)

TRICHASTOMA PYRRHOPTERUM

Trichastoma pyrrhopterum (Reichenow and Neumann)

Callene pyrrhoptera Reichenow and Neumann, 1895, Orn. Monatsb., 3, p. 75 — Mau [lat. 0° 10′ S., long. 35° 41′ E.], Kenya.

Turdinus jacksoni Sharpe, 1900, Bull. Brit. Orn. Club, 11, p. 29 — Nandi District, Kenya.

Turdinus pyrrhopterus kivuensis Neumann, 1908, Bull. Brit. Orn. Club, 21, p. 55 — Mount Sabinyo [lat. 1° 22′ S., long. 29° 36′ E.], western Kivu Range, Belgian Congo.

Turdinus tanganjicae Reichenow, 1917, Journ. f. Orn., 65, p. 391 — primeval forest westward of Lake Tanganyika, Belgian Congo.

Turdinus pyrrhopterus elgonensis Granvik, 1923, Journ. f. Orn., 71, Sonderheft, p. 256 — Mount Elgon, Kenya.

Pseudoalcippe pyrrhoptera nyasæ Benson, 1939, Bull. Brit. Orn. Club, 59, p. 43 — Nyakhowa Mountain (Mount Laws), near Livingstonia, Nyasaland.

Highlands of eastern Congo, Uganda, Kenya (west of the Rift Valley), western Tanganyika, and northern Nyasaland.

TRICHASTOMA CLEAVERI

Trichastoma cleaveri johnsoni (Büttikofer)
Drymocataphus johnsoni Büttikofer, 1889, Notes Leyden
Mus., 11, p. 97 — Hill Town, Dukwia, Liberia.
Forests of Sierra Leone and Liberia.

Trichastoma cleaveri cleaveri (Shelley)
Drymocataphus cleaveri Shelley, 1874, Ibis, p. 89 —
Fanti region, Ghana.
Forests of Ghana.

Trichastoma cleaveri marchanti (Serle)
Illadopsis cleaveri marchanti Serle, 1956, Bull. Brit. Orn.
Club, 76, p. 22 — Omanelu [lat. 5° 12′ N., long. 6° 52′
E.], Nigeria.
Forests of southern Nigeria (Rivers and Owerri Prov-
inces).

Trichastoma cleaveri batesi (Sharpe)
Turdinus batesi Sharpe, 1901, Bull. Brit. Orn. Club, 12,
p. 2 — Efulen [lat. 2° 42′ N., long. 10° 30′ E.], French
Cameroons.
Forests of southeastern Nigeria, the Cameroons, Gabon,
and Moyen Congo.

Trichastoma cleaveri poense (Bannerman)
Illadopsis cleaveri poensis Bannerman, 1934, Bull. Brit.
Orn. Club, 54, p. 107 — near Bakaki, Fernando Po.
Fernando Po.

TRICHASTOMA ALBIPECTUS

Trichastoma albipectus albipectus (Reichenow)
Turdinus albipectus Reichenow, 1887, Journ. f. Orn., 35,
p. 307 — Stanley Falls, Belgian Congo.
Forests of northern Angola and the northern Congo eastward to the Ituri.

Trichastoma albipectus barakae (Jackson)
Turdinus barakæ Jackson, 1906, Bull. Brit. Orn. Club, 16,
p. 90 — Kibiro [lat. 1° 40′ N., long. 31° 15′ E.], Uganda.
Forests of the northern Congo east of the Ituri, Uganda,
the southern Sudan, and southwestern Kenya.

TRICHASTOMA RUFESCENS

Trichastoma rufescens (Reichenow)

Turdirostris rufescens Reichenow, 1878, Journ. f. Orn., 26, p. 209 — Liberia.

Forests of western Africa from Sierra Leone eastward to Ghana.

TRICHASTOMA RUFIPENNE

Trichastoma rufipenne extremum (Bates)

Malacocincla rufipennis extrema Bates, 1930, Bull. Brit. Orn. Club, 51, p. 49 — near Nzerekoré, French Guinea.

Forests of western Africa from eastern Sierra Leone eastward, through French Guinea and the Ivory Coast, to Ghana.

Trichastoma rufipenne rufipenne Sharpe

Trichastoma rufipennis [sic] Sharpe, 1872, Ann. Mag. Nat. Hist., ser. 4, 10, p. 451 — Gabon.

Turdinus pumilus Reichenow, 1904, Orn. Monatsb., 12, p. 28 — Bipindi, Cameroons.

Alethe polioparea Reichenow, 1912, Journ. f. Orn., 60, p. 321 — Angu [lat. 3° 30′ N., long. 24° 27′ E.], Belgian Congo.

Turdinus albipectus minutus van Someren, 1915, Bull. Brit. Orn. Club, 35, p. 126 — Mabira Forest, Uganda.

Forests of southern Nigeria, the Cameroons, Gabon, the Congo, Uganda, and southwestern Kenya.

Trichastoma rufipenne bocagei (Salvadori)

Turdinus bocagei Salvadori, 1903, Boll. Mus. Zool. Torino, 18, p. 1 — Fernando Po.

Fernando Po.

Trichastoma rufipenne distans (Friedmann)

Turdinus rufipennis distans Friedmann, 1928, Proc. New England Zool. Club, 10, p. 48 — Amani, Usambara Mountains, northeastern Tanganyika.

Illadopsis rufipennis puguensis Grant and Mackworth-Praed, 1940, Bull. Brit. Orn. Club, 60, p. 61 — Pugu Hills, inland from Dar-es-Salaam, Tanganyika.

Northeastern Tanganyika and Zanzibar.

TRICHASTOMA FULVESCENS

Trichastoma fulvescens gulare (Sharpe)

Illadopsis gularis Sharpe, 1870, Ibis, p. 474 — Elmina, Ghana.

Forests from French Guinea, through Sierra Leone, Liberia, and the Ivory Coast, to western Ghana.

Trichastoma fulvescens moloneyanum (Sharpe)

Turdinus moloneyanus Sharpe, 1892, Proc. Zool. Soc. London, p. 228, pl. 20, fig. 2 — Gold Coast [= Ghana].

Forests of eastern Ghana and Togo.

Trichastoma fulvescens iboense (Hartert)

Turdinus moloneyanus iboensis Hartert, 1907, Bull. Brit. Orn. Club, 19, p. 84 — Okuta, southwestern Nigeria.

Turdinus phœbei Kemp, 1908, Bull. Brit. Orn. Club, 21, p. 111 — Aguleri, southwestern Nigeria.

Forests of southwestern Nigeria.

Trichastoma fulvescens fulvescens (Cassin)

Turdirostris fulvescens Cassin, 1859, Proc. Acad. Nat. Sci. Philadelphia, 11, p. 54 — Cama River, Gabon.

Turdinus rufiventris Reichenow, 1893, Orn. Monatsb., 1, p. 177 — Yaunde, French Cameroons. Not *Malacocincla rufiventris* Salvadori, 1874.

Turdinus cerviniventris Sharpe, 1901, Bull. Brit. Orn. Club, 12, p. 3 — Conde, Congo [= Cabinda].

[*Turdinus*] *reichenowi* Sharpe, 1903, Hand-list Gen. Spec. Birds, 4, p. 33. New name for *Turdinus rufiventris* Reichenow, preoccupied.

Forests from the Cameroons southward to the western Congo.

Trichastoma fulvescens dilutius (White)

Malacocincla fulvescens dilutior White, 1953, Bull. Brit. Orn. Club, 73, p. 96 — Ndala Tando [= Vila Salazar], northern Angola.

Forests of northern Angola.

Trichastoma fulvescens ugandae (van Someren)

Turdinus ugandæ van Someren, 1915, Bull. Brit. Orn. Club, 35, p. 125 — Sezibwa River forest, Uganda.

Forests of the northern Congo and Uganda.

TRICHASTOMA PUVELI

Trichastoma puveli puveli (Salvadori)

Turdinus puveli Salvadori, 1901, Ann. Mus. Civ. Genova, ser. 2, 20 (1899-1901) p. 767 — Rio Cacine, Portuguese Guinea.

Forests of western Africa from Portuguese Guinea southward to Sierra Leone.

Trichastoma puveli strenuipes (Bannerman)

Turdinus strenuipes Bannerman, 1920, Bull. Brit. Orn. Club, 41, p. 5 — Iju Waterworks, near Lagos, southwestern Nigeria.

Ptyrticus puveli strenuiceps "(Bannerm.)" Sclater, 1930, Syst. Av. Aethiop., pt. 2, p. 360. *Lapsus.*

Forests of southern Nigeria, the Cameroons, and northeastern Congo.

TRICHASTOMA POLIOTHORAX

Trichastoma poliothorax (Reichenow)

Alethe poliothorax Reichenow, 1900, Orn. Monatsb., 8, p. 6 — "Bangwa," Cameroons [= "an elevated district at about 5° 15′ N., 10° 25′ E., with peaks rising to 5,000 ft.," *fide* Serle, 1951, in Bannerman, Birds Trop. West Afr., 8, p. 378].

Alethe moori Alexander, 1903, Bull. Brit. Orn. Club, 13, p. 37 — Bakaki, Fernando Po.

Highlands of Fernando Po and of southern (British) Cameroons (Cameroon Mountain, Kupé Mountain, Bamenda); highlands of the eastern Congo from west of Lake Albert to northwest of Lake Tanganyika; southern Ruanda; highlands of southwestern Kenya (Kavirondo District).

GENUS LEONARDINA MEARNS

Leonardia Mearns, 1905, Proc. Biol. Soc. Washington, 18, p. 1. Type, by original designation and monotypy, *Leonardia woodi* Mearns. Not *Leonardia* Tapparone-Canefri, 1890, Mollusca.

Leonardina Mearns, 1906, Proc. Biol. Soc. Washington, 18, p. 88. New name for *Leonardia* Mearns, preoccupied.

cf. Rand, 1950, Nat. Hist. Misc. [Chicago], no. 60, pp. 1-2.

LEONARDINA WOODI

Leonardina woodi (Mearns)

Leonardia woodi Mearns, 1905, Proc. Biol. Soc. Washington, 18, p. 2 — Todaya, 4,000 ft., Mount Apo, Mindanao.

Mindanao (Mount Apo and Mount Malindang), Philippine Islands.

GENUS PTYRTICUS HARTLAUB

Ptyrticus Hartlaub, 1883, Journ. f. Orn., 31, p. 425. Type, by monotypy, *Ptyrticus turdinus* Hartlaub.

PTYRTICUS TURDINUS

Ptyrticus turdinus harterti Stresemann

Ptyrticus turdinus harterti Stresemann, 1921, Anz. Orn. Ges. Bayern, 1, p. 38 — Upper Kadei region, Cameroons.

Grasslands of central Cameroons.

Ptyrticus turdinus turdinus Hartlaub

Ptyrticus turdinus Hartlaub, 1883, Journ. f. Orn., 31, p. 425 — Tomaya [lat. 4° 38′ N., long. 29° 50′ E.], Sudan.

Southwestern Sudan and northeastern Congo.

Ptyrticus turdinus upembae Verheyen

Ptyrticus turdinus upembæ Verheyen, 1951, Bull. Inst. Roy. Sci. Nat. Belgique, 21, p. 2 — Kabwe [lat. 8° 48′ S., long. 26° 50′ E.], Belgian Congo.

Southeastern Congo.

GENUS MALACOPTERON EYTON

Malacopteron Eyton, 1839, Proc. Zool. Soc. London, pt. 7, p. 102. Type, by subsequent designation (Strickland, 1844, Ann. Mag. Nat. Hist., 13, p. 417), *Malacopteron magnum* Eyton.

Setaria Blyth, 1844, Journ. Asiat. Soc. Bengal, 13, p. 385. Type, by monotypy, *Setaria albogularis* Blyth. Not *Setaria* Viborg, 1795, Vermes (Nematoda), nor *Setaria* Oken, 1815, Vermes (Gordiacea).

Malacornis Gistl, 1850, Isis (München), no. 6, p. 95. New name for *Malacopteron* Eyton, considered preoccupied by *Malacopterus* Audinet-Serville, 1834, Coleoptera.

Ophrydornis Büttikofer, 1895, Notes Leyden Mus., 17, p. 101. Type, by original designation and monotypy, *Setaria albogularis* Blyth.

Horizillas Oberholser, 1905, Smiths. Misc. Coll., 48 (1), p. 65. New name for *Malacopteron* Eyton, considered preoccupied by *Malacopterus* Audinet-Serville, 1834, Coleoptera.

cf. Büttikofer, 1895, Notes Leyden Mus., 17, pp. 101-106.
Voous, 1950, Sarawak Mus. Journ., 5, pp. 300-320.

MALACOPTERON MAGNIROSTRE

Malacopteron magnirostre magnirostre (Moore)

Alcippe magnirostris Moore, 1854, in Horsfield and Moore, Cat. Birds Mus. East India Co., 1, pp. xiv (in list), 407 — Malacca.

Malacopteron magnirostris flavum Chasen and Kloss, 1928, Journ. Malayan Branch Roy. Asiat. Soc., 6, pp. 47 (in list), 58 — Pulau Siantan, Anamba Islands.

Southern Tenasserim and peninsular Thailand (Prachuap Khiri Khan Province) southward over Malay Peninsula to Singapore; Pulau Tioman and Anamba Islands; Lingga Archipelago; Sumatra and Pulau Mansalar, off its western coast.

Malacopteron magnirostre cinereocapillum (Salvadori)

Alcippe cinereocapilla Salvadori, 1868, Atti R. Acad. Sci. Torino, 3, p. 530 — Sarawak.

Turdinus kalulongæ Sharpe, 1893, Bull. Brit. Orn. Club, 1, p. 54 — Mount Kalulong, Sarawak.

Borneo.

MALACOPTERON AFFINE

Malacopteron affine affine (Blyth)

Tr[ichastoma]. affine Blyth, 1842, Journ. Asiat. Soc. Bengal, 11, p. 795 — Singapore.

Alcippe Cantori Moore, 1854, in Horsfield and Moore, Cat. Birds Mus. East India Co., 1, pp. xiv (in list), 406 — Penang.

Malacopterum melanocephalum Davison, 1892, Ibis, p. 101 — mouth of the Temeling River, Pahang.

Southeastern peninsular Thailand (Pattani Province) southward through Malaya to Singapore; Sumatra.

Malacopteron affine notatum Richmond

Malacopteron notatum Richmond, 1902, Proc. Biol. Soc. Washington, 15, p. 190 — Pulau Bangkaru, Banyak Islands.

Known only from type locality.

Malacopteron affine phoeniceum Deignan

Malacopteron affine phoeniceum Deignan, 1950, Zoologica [New York], 35, p. 127 — Segah River [at *ca.* lat. 2° 12′ N., long. 117° 06′ E.], Borneo.

Borneo.

MALACOPTERON CINEREUM

Malacopteron cinereum indochinense (Robinson and Kloss)

Horizillas rufifrons indochinensis Robinson and Kloss, 1921, Journ. Fed. Malay States, 10, p. 205 — Trang Bom [lat. 10° 56′N., long. 107° 00′ E.], Cochin China.

Southwestern portion of eastern plateau and southeastern provinces of Thailand; Cambodia; southern Laos; Annam; Cochin China.

Malacopteron cinereum rufifrons Cabanis

M[alacopteron]. rufifrons Cabanis, 1850, Mus. Hein., 1 (1851), p. 65 — "Java oder Sumatra" [= Java].

Malacopterum lepidocephalum Sharpe, 1883, Cat. Birds Brit. Mus., 7, pp. 564 (in key), 567 — Sumatra and Java; type locality restricted to Java, by Büttikofer, 1895, Notes Leyden Mus., 17, p. 104.

Java.

Malacopteron cinereum cinereum Eyton

Malacopteron cinereus [sic] Eyton, 1839, Proc. Zool. Soc. London, pt. 7, p. 103 — Malaya.

Malacopterum cinereum bungurense Hartert, 1894, Novit. Zool., 1, p. 470 — Pulau Bunguran, North Natuna Islands.

Isthmus of Kra southward over Malay Peninsula to Johore; Riouw and Lingga Archipelagos; Sumatra; Bangka; North Natuna Islands; Borneo.

Malacopteron cinereum niasense (Riley)

Malacornis cinerea niasensis Riley, 1937, Proc. Biol. Soc. Washington, 50, p. 61 — Nias.

Nias.

MALACOPTERON MAGNUM

Malacopteron magnum magnum Eyton

> *Malacopteron magnum* Eyton, 1839, Proc. Zool. Soc. London, pt. 7, p. 103 — Malaya.
> *N[apothera]. pileata* "Müll. Mus. Lugd." Bonaparte, 1851, Consp. Av., 1 (1850), p. 359 — Sumatra and Borneo.

Southern Tenasserim (Mergui District) and Isthmus of Kra southward over Malay Peninsula to Johore; Sumatra; North Natuna Islands (Pulau Bunguran); Borneo (except area occupied by *saba*).

Malacopteron magnum saba Chasen and Kloss

> *Malacopteron magnum saba* Chasen and Kloss, 1930, Bull. Raffles Mus., 4, p. 75 — Samawang River, near Sandakan, North Borneo.

Northeastern Borneo.

MALACOPTERON PALAWANENSE

Malacopteron palawanense Büttikofer

> *Trichostoma rufifrons* Tweeddale, 1878, Proc. Zool. Soc. London, p. 616, pl. 38 — Puerto Princesa, Palawan. Not *M[alacopteron]. rufifrons* Cabanis, 1850.
> *Malacopteron palawanense* Büttikofer, 1895, Notes Leyden Mus., 17, p. 104. New name for *Trichostoma rufifrons* Tweeddale, preoccupied.

Palawan and Balabac, Philippine Islands.

MALACOPTERON ALBOGULARE

Malacopteron albogulare albogulare (Blyth)

> *S[etaria]. albogularis* Blyth, 1844, Journ. Asiat. Soc. Bengal, 13, p. 385 — "Singapore."

Malaya (Perak, Pahang, Selangor); northeastern Sumatra; Batu Islands (Pulau Pini).

Malacopteron albogulare moultoni (Robinson and Kloss)

> *Ophrydornis albogularis moultoni* Robinson and Kloss, 1919, Bull. Brit. Orn. Club, 40, p. 17 — Beton Saribas, Sarawak.
> *Setaria albigularis* [sic] *leucogastra* Hachisuka, 1926, Bull. Brit. Orn. Club, 47, p. 54 — Paku, Sarawak.

Northwestern Borneo.

GENUS **POMATORHINUS** HORSFIELD

Pomatorhinus Horsfield, 1821, Trans. Linn. Soc. London,
13, p. 164. Type, by monotypy, *Pomatorhinus montanus*
Horsfield.
Erythrogenys "Hodgs." = Baker, 1930, Fauna Brit. India,
Birds., ed. 2, 7, p. 39. Type, by original designation and
tautonymy, "*E*[*rythrogenys*]. *gouldii* = *P*[*omatorhinus*]. *erythrogenys* Vigors." Not *Erythrogenys* Brandt,
1841, *nomen emendatum* for *Erythrogonys* Gould, 1838.

POMATORHINUS HYPOLEUCOS

Pomatorhinus hypoleucos hypoleucos (Blyth)

O[*rthorhinus*]. *hypoleucos* Blyth, 1844, Journ. Asiat. Soc.
Bengal, 13, p. 371 — Arakan, Lower Burma.
P[*omatorhinus*]. (. . . *Orthorhinus*) *Inglisi* Hume, 1877,
Stray Feathers, 5, pp. 31, 32 — northeastern Cachar
District, Assam.
Assam (both north and south of the Brahmaputra) ; East
Pakistan; western and northern Burma.

Pomatorhinus hypoleucos tickelli Hume

Pomatorhinus (. . . *Orthorhinus*) *Tickelli* Hume, 1877,
Stray Feathers, 5, p. 32 — Mulayit Taung [lat. 16° 11'
N., long. 98° 32' E.], Tenasserim.
Pomatorhinus tickelli laotianus Delacour, 1926, Bull. Brit.
Orn. Club, 47, p. 16 — Chiang Khwang [lat. 19° 19' N.,
long. 103° 22' E.], Laos.
Pomatorhinus tickelli tonkinensis Delacour, 1927, Bull.
Brit. Orn. Club, 47, p. 159 — Babe Lakes, near Cho Ra
[lat. 22° 27' N., long. 105° 44' E.], Tongking.
Tenasserim (Amherst District) ; greater part of Thailand
(except central plains) south to Prachuap Khiri Khan
Province; northern Laos, Tongking, and northern Annam.

Pomatorhinus hypoleucos brevirostris Robinson and Kloss

Pomatorhinus tickelli brevirostris Robinson and Kloss,
1919, Ibis, p. 578 — Trang Bom [lat. 10° 56' N., long.
107° 00' E.], Cochin China.
Pomatorhinus tickelli friesi Delacour, 1927, Bull. Brit.
Orn. Club, 47, p. 160 — Thua Luu [lat. 16° 16' N., long.
108° 00' E.], Annam.

Southern Laos, central and southern Annam, and northern Cochin China.

Pomatorhinus hypoleucos hainanus Rothschild

Pomatorhinus tickelli hainanus Rothschild, 1903, Bull. Brit. Orn. Club., 14, p. 9 — No-tai [= Nata (lat. 19° 29' N., long. 109° 27' E.)], Hainan.

Hainan.

POMATORHINUS ERYTHROGENYS

Pomatorhinus erythrogenys erythrogenys Vigors

Pomatorhinus erythrogenys Vigors, 1832, Proc. Comm. Zool. Soc. London, pt. 1 (1830-1831), p. 173 — Himalayas; restricted to "the district Simla-Almora," by Ticehurst and Whistler, 1924, Ibis, p. 471.

Himalayas from northern West Pakistan southeastward to northwestern Uttar Pradesh.

Pomatorhinus erythrogenys ferrugilatus Hodgson

[*Pomatorhinus*] *Ferrugilatus* Hodgson, 1836, Asiat. Res., 19, p. 180 — Nepal.

Western and central Nepal.

Pomatorhinus erythrogenys haringtoni Baker

Pomatorhinus haringtoni Baker, 1914, Bull. Brit. Orn. Club, 33, p. 124 — Darjeeling, West Bengal.

Himalayas from Sikkim eastward into Bhutan.

Pomatorhinus erythrogenys mcclellandi Godwin-Austen

P[*omatorhinus*]. *McClellandi* "Jerdon" = Godwin-Austen, 1870, Journ. Asiat. Soc. Bengal, 39 (2), p. 103 — "at Nenglo beyond Asálu, under the Burrail range," Assam.

Pomatorhinus m'clellandi Jerdon, 1872, Ibis, p. 302 — Khasi Hills, Assam.

Pomatorhinus erythrogenys erythrotis Koelz, 1952, Journ. Zool. Soc. India, 4, p. 38 — Karong, Manipur.

Hill tracts of Assam south of the Brahmaputra and western Burma (Chin Hills).

Pomatorhinus erythrogenys imberbis Salvadori

Trichastoma rubiginosa [sic] Walden, 1875, Ann. Mag. Nat. Hist., ser. 4, 15, p. 402 — Karenni State, Burma. Not *Pomatorhinus rubiginosus* Blyth, 1855.

Pomatorhinus imberbis Salvadori, 1889, Ann. Mus. Civ.
Genova, ser. 2, 7, p. 410 — Yado, Karenni.
Karenni State, Burma.

Pomatorhinus erythrogenys celatus Deignan

Pomatorhinus erythrogenys celatus Deignan, 1941, Zoologica [New York], 26, p. 241 — Doi Luang Chiang Dao [lat. 19° 25′ N., long. 98° 55′ E.], Thailand.
Western portions of Northern and Southern Shan States of Burma and high mountains of northern portion of northwestern Thailand (Doi Pha Hom Pok, Doi Luang Chiang Dao).

Pomatorhinus erythrogenys odicus Bangs and Phillips

Pomatorhinus macclellandi odicus Bangs and Phillips, 1914, Bull. Mus. Comp. Zool., 58, p. 286 — Mengtsz, Yunnan.
Pomatorhinus erythrogenys minor Delacour and Jabouille, 1930, Oiseau Rev. Franç. Orn., 11, p. 400 — Pa Kha [lat. 22° 32′ N., long. 104° 18′ E.], Tongking.
Northeastern Burma (southern portion of Kachin State); eastern portion of Northern and Southern Shan States (east of the Salween); southwestern Yunnan; northern Laos; northwestern Tongking.

Pomatorhinus erythrogenys decarlei Deignan

Pomatorhinus erythrogenys decarlei Deignan, 1952, Proc. Biol. Soc. Washington, 65, p. 121 — near Yangtza [*ca*. kiang, northwestern Yunnan.
Northwestern Yunnan (except area occupied by *dedekeni*), southeastern Hsikang, and southwesternmost Szechwan.

Pomatorhinus erythrogenys dedekeni Oustalet

Pomatorhinus Dedekensi [sic] Oustalet, 1892, Ann. Sci. Nat., Zool., sér. 7, 12 (1891), p. 276 — "Tioungeu," error; type locality corrected to Tsonghai [lat. 29° 56′ N., long. 98° 40′ E.], Hsikang, by Oustalet, 1893, Nouv. Arch. Mus. Hist. Nat. [Paris], sér. 3, 5, p. 197, foot-note 1.
Pomatorhinus Armandi Oustalet, 1892, Ann. Sci. Nat., Zool., sér. 7, 12 (1891), p. 277 — "Aio," error; type locality corrected to Tatsienlu [= Kangting], Hsikang, by

Oustalet, 1893, Nouv. Arch. Mus. Hist. Nat. [Paris], sér. 3, 5, p. 199.

Pomatorhinus erythrogenys stoneae Deignan, 1952, Proc. Biol. Soc. Washington, 65, p. 121 — near Yangtza [*ca.* lat. 28° 15′ N., long. 98° 48′ E.], Yunnan.

East-central and south-central Hsikang, southward into northwestern Yunnan along the Mekong as far as *ca.* lat. 27° 12′ N.

Pomatorhinus erythrogenys gravivox David

Pomatorhinus gravivox David, 1873, Ann. Sci. Nat., Zool., sér. 5, 18, art. 5, p. 2 — Shensi; types from southern Shensi, *fide* David, 1874, *ibid.*, sér. 5, 19, art. 9, p. 2.

Northwestern Szechwan and southern portions of Kansu and Shensi (south of Chinling Mountains).

Pomatorhinus erythrogenys sowerbyi Deignan

Pomatorhinus erythrogenys sowerbyi Deignan, 1952, Proc. Biol. Soc. Washington, 65, p. 122 — twelve miles south of Fushih, northern Shensi.

Known only from type locality.

Pomatorhinus erythrogenys cowensae Deignan

Pomatorhinus erythrogenys cowensae Deignan, 1952, Proc. Biol. Soc. Washington, 65, p. 122 — Wanhsien, eastern Szechwan.

Eastern Szechwan, northern Kweichow, and southwestern Hupeh.

Pomatorhinus erythrogenys swinhoei David

Pomatorhinus Swinhoei David, 1874, Ann. Sci. Nat., Zool., sér. 5, 19, art. 9, p. 5 — western Fukien.

Anhwei, northeastern Kiangsi, and Fukien.

? Pomatorhinus erythrogenys abbreviatus Stresemann

Pomatorhinus swinhoei abbreviatus Stresemann, 1929, Journ. f. Orn., 77, p. 333 — Yao Shan, Kwangsi.

Southern Hunan, Kwangtung, and Kwangsi; doubtfully distinct from *swinhoei*.

Pomatorhinus erythrogenys erythrocnemis Gould

Pomatorhinus erythrocnemis Gould, 1863, Proc. Zool. Soc. London, (1862), p. 281 — Formosa.

Formosa.

POMATORHINUS HORSFIELDII

Pomatorhinus horsfieldii melanurus Blyth

P[omatorhinus]. melanurus Blyth, 1847, Journ. Asiat.
Soc. Bengal, 16, p. 451 — Ceylon; restricted to Uru-
gaha, south of Kalutara, Western Province, by Whist-
ler, 1942, Bull. Brit. Orn. Club, 62, p. 52.
? Pomatorhinus horsfieldii holdsworthi Whistler, 1942,
Bull. Brit. Orn. Club, 62, p. 52 — Ohiya, Uva Province,
Ceylon.[1]
Ceylon.

Pomatorhinus horsfieldii travancoreensis Harington

Pomatorhinus horsfieldi travancoreensis Harington, 1914,
Journ. Bombay Nat. Hist. Soc., 23, p. 333 — Peermade,
Travancore, Kerala.
Southwestern India, in Western Ghats and their outliers
from Kerala northward to North Kanara.

Pomatorhinus horsfieldii horsfieldii Sykes

Pomatorhinus Horsfieldii Sykes, 1832, Proc. Comm. Zool.
Soc. London, pt. 2, p. 89 — the Deccan [= Mahabalesh-
war, Satara North, fide Ali, 1953, Birds Travancore
Cochin, p. 35].
Western India, in Western Ghats from Goa northward to
the Satpura Range; Saurashtra.

Pomatorhinus horsfieldii obscurus Hume

Pomatorhinus Obscurus Hume, 1872, Stray Feathers, 1,
p. 7 — Mount Abu, Sirohi, Rajasthan.
Northwestern India, in Rajasthan (Aravalli Range).

Pomatorhinus horsfieldii maderaspatensis Whistler

Pomatorhinus horsfieldii maderaspatensis Whistler, 1936,
Journ. Bombay Nat. Hist. Soc., 38, p. 699 — Kurum-
bapatti, Salem District, Madras.
Eastern Ghats of India from the Salem District northward into Andhra.

[1] According to Whistler (loc. cit.), the rufescent melanurus is restricted to the low-country wet zone of southwestern Ceylon, while the
olivaceous holdsworthi occupies the rest of the island. Ripley (1946,
Spolia Zeylanica, 24, pp. 223-224) suggests rather that "an erythristic
rufous strain of mutants . . . appears at random, throughout areas of
densest population of this species." Material examined by me casts
grave doubt on the subspecific validity of holdsworthi.

POMATORHINUS SCHISTICEPS[1]

Pomatorhinus schisticeps leucogaster Gould

Pomatorhinus leucogaster Gould, 1838, Proc. Zool. Soc. London, 5, p. 137 — "the Himalaya Mountains"; restricted to Simla, *apud* Ripley, 1961, Synops. Birds India Pakistan, p. 349.
Pomatorhinus pinwilli Sharpe, 1883, Cat. Birds Brit. Mus., 7, p. 413 — northwestern Himalayas.

Himalayas from Himachal Pradesh eastward into northwestern Uttar Pradesh.

Pomatorhinus schisticeps schisticeps Hodgson

[*Pomatorhinus*] *Schisticeps* Hodgson, 1836, Asiat. Res., 19, p. 181 — Nepal.
P[*omatorhinus*]. *assamensis* Blyth (ex McClelland MS), 1847, Journ. Asiat. Soc. Bengal, 16, p. 451 — Assam; restricted to Khasi Hills, by Ripley, 1961, Synops. Birds India Pakistan, p. 349.

Himalayas from western Nepal eastward into Bhutan; hill tracts of Assam south of the Brahmaputra (Khasi Hills, Mikir Hills, Barail Range); northwestern Burma (Upper Chindwin District).

Pomatorhinus schisticeps salimalii Ripley

Pomatorhinus montanus sálimalii Ripley, 1948, Proc. Biol. Soc. Washington, 61, p. 101 — Tezu, Mishmi Hills, Assam.

Northeastern Assam (Mishmi Hills).

Pomatorhinus schisticeps cryptanthus Hartert

Pomatorhinus schisticeps cryptanthus Hartert, 1915, Bull. Brit. Orn. Club, 36, p. 35 — Margherita, Lakhimpur District, Assam.

Northeastern Assam (Lakhimpur District); northeastern Burma (?).

Pomatorhinus schisticeps mearsi Ogilvie-Grant

Pomatorhinus mearsi Ogilvie-Grant, 1905, Bull. Brit. Orn. Club, 15, p. 39 — Taungdwin Chaung [lat. 22° 52′ N., long. 94° 20′ E.], Upper Burma.

[1] Some authors consider *P. schisticeps*, and even *P. montanus*, conspecific with *P. horsfieldii*, but I prefer to treat the first two, possibly all three, as comprising a superspecies.

Hill tracts of western Burma from Lower Chindwin District southward to the Arakan Yomas.

Pomatorhinus schisticeps ripponi Harington

> *Pomatorhinus ripponi* Harington, 1910, Bull. Brit. Orn. Club, 27, p. 9 — Shan States; type specimen from Pyaunggaung [lat 23° 00′ N., long. 96° 28′ E.], Upper Burma, *fide* Kinnear, *in epistola.*

From Northern Shan State southward, through eastern portion of Southern Shan State, to northernmost Thailand and adjacent portion of northwestern Laos.

Pomatorhinus schisticeps nuchalis Ramsay

> *P[omatorhinus]. nuchalis* Ramsay (ex Tweeddale MS), 1877 (Oct.), Ibis, p. 465 — Karen Hills, Upper Burma.
> *Pomatorhinus nuchalis* Tweeddale, 1877 (Dec.), Ann. Mag. Nat. Hist., ser. 4, 20, p. 535 — "Thayetmyo, the Yoma and Karen hills, and Karen-nee"; types from the Karen Hills, *fide* Ramsay, 1881, Orn. Works Arthur, Ninth Marquis Tweeddale, p. 669.

Western portion of Southern Shan State and Karenni State of Upper Burma.

Pomatorhinus schisticeps difficilis Deignan

> *Pomatorhinus schisticeps difficilis* Deignan, 1956, Proc. Biol. Soc. Washington, 69, p. 208 — Doi Luang Chiang Dao [lat. 19° 25′ N., long. 98° 55′ E.], Thailand.

Mountains of northwestern Thailand from type locality southward into northern Tak Province; Tenasserim (Amherst District).

Pomatorhinus schisticeps olivaceus Blyth

> *P[omatorhinus]. olivaceus* Blyth, 1847, Journ. Asiat. Soc. Bengal, 16, p. 451 — Ye, Amherst District, Tenasserim.
> *Pomatorhinus olivaceus siamensis* Baker, 1917, Bull. Brit. Orn. Club, 38, p. 9 — Sathani Map Ammarit [lat. 10° 50′ N., long. 99° 20′ E.], Thailand.

Lowlands of Tenasserim from Amherst District southward into Mergui District; evergreen forest of southwestern and peninsular Thailand from southern Tak Province southward to Isthmus of Kra.

Pomatorhinus schisticeps fastidiosus Hartert

Pomatorhinus schisticeps fastidiosus Hartert, 1916, Bull.
Brit. Orn. Club, 36, p. 81 — Ban Khok Khan [lat. 7° 34′
N., long. 99° 38′ E.], Thailand.

Malay Peninsula from southernmost Tenasserim and
Isthmus of Kra southward to Siamese province of Trang.

Pomatorhinus schisticeps humilis Delacour

Pomatorhinus schisticeps humilis Delacour, 1932, Oiseau
Rev. Franç. Orn., nouv. sér., 2, p. 424 — Thateng [lat.
15° 31′ N., long. 106° 22′ E.], Laos.

Eastern portion of northern plateau of Thailand (Nan
Province) and provinces of eastern plateau bordering the
Mekong; southern Laos and adjacent portion of central
Annam.

Pomatorhinus schisticeps annamensis Robinson and Kloss

Pomatorhinus olivaceus annamensis Robinson and Kloss,
1919, Ibis, p. 577 — Dran [lat. 11° 49′ N., long. 108° 38′
E.], Annam.

Southern Annam (Lang Bian Plateau) and northeastern
Cochin China.

Pomatorhinus schisticeps klossi Baker

Pomatorhinus nuchalis klossi Baker, 1917, Bull. Brit. Orn.
Club, 38, p. 9 — Ban Khlong Manao [= Ban Huang Som
(lat. 11° 50′ N., long. 102° 50′ E.) and "Ban Sam
Khok," error = Ban Si Racha (lat. 13° 10′ N., long.
100° 55′ E.)], Thailand.

Southeastern Thailand and southwestern Cambodia.

POMATORHINUS MONTANUS

Pomatorhinus montanus occidentalis Robinson and Kloss

Pomatorhinus montanus occidentalis Robinson and Kloss,
1923, Journ. Fed. Malay States Mus., 11, p. 54 — Ginting Bidei, Selangor.

Malay Peninsula from northern Perak southward to Johore; Sumatra.

Pomatorhinus montanus montanus Horsfield

Pomatorhinus montanus Horsfield, 1821, Trans. Linn.
Soc. London, 13, p. 165 — Java; Gunong Merbabu [lat.
7° 29′ S., long. 110° 30′ E.] and Gunong Prahu [lat.

7° 12′ S., long. 109° 54′ E.] mentioned by Horsfield,
1824, Zool. Res. Java).
Western and central Java.

Pomatorhinus montanus ottolanderi Robinson

Pomatorhinus montanus ottolanderi Robinson, 1918,
Journ. Fed. Malay States Mus., 7, p. 235 — Sodong
Gerok, Ijen Mountains, near Banyuwangi [lat. 8° 12′
S., long. 114° 23′ E.], Java.
Eastern Java and Bali.

Pomatorhinus montanus bornensis Cabanis

P[*omatorhinus*]. *bornensis* Cabanis, 1851, Mus. Hein, 1
(1850), p. 84 — Borneo.
Borneo.

POMATORHINUS RUFICOLLIS

Pomatorhinus ruficollis ruficollis Hodgson

[*Pomatorhinus*] *Ruficollis* Hodgson, 1836, Asiat. Res.,
19, p. 182 — Nepal.
Western and central Nepal.

Pomatorhinus ruficollis godwini Kinnear

Pomatorhinus ruficollis godwini Kinnear, 1944, in Ludlow, Ibis, p. 79 — Chungkar, 6,500 ft., Trashigong-
Diwangiri Road, southeastern Bhutan.
Himalayas from eastern Nepal (east of the Arun Kosi)
eastward, through Sikkim, Bhutan, and southeastern Tibet,
into Assam (north of the Brahmaputra).

Pomatorhinus ruficollis bakeri Harington

Pomatorhinus ruficollis bakeri Harington, 1914, Journ.
Bombay Nat. Hist. Soc., 23, p. 336 — Shillong, Khasi
Hills, Assam.
Pomatorhinus ruficollis recter Koelz, 1954, Contrib. Inst.
Regional Explor., no. 1, p. 4 — Blue Mountain, Lushai
Hills, Assam.
Hill tracts of Assam (south of the Brahmaputra) and of
western Burma (Upper Chindwin, Chin Hills).

Pomatorhinus ruficollis similis Rothschild

Pomatorhinus ruficollis similis Rothschild, 1926, Novit.
Zool., 33, p. 261 — Hills round Tengyueh [= Teng-
chung], western Yunnan.

Northeastern Burma (south to Myitkyina District) and northwestern Yunnan.

? Pomatorhinus ruficollis bhamoensis Mayr

Pomatorhinus ruficollis bhamoensis Mayr, 1941, Ibis, p. 67 — Sinlumkaba, Bhamo District, Kachin State, Upper Burma.

Known only from type locality; doubtfully distinct from *albipectus*.

Pomatorhinus ruficollis albipectus La Touche

Pomatorhinus ruficollis albipectus La Touche, 1923, Bull. Brit. Orn. Club, 43, p. 173 — Szemao, southwestern Yunnan.

Southwestern Yunnan and northern portion of northern Laos.

Pomatorhinus ruficollis beaulieui Delacour and Greenway

Pomatorhinus schisticeps beaulieui Delacour and Greenway, 1940, Oiseau Rev. Franç. Orn., nouv. sér., 10, p. 63 — Chiang Khwang [lat. 19° 19′ N., long. 103° 22′ E.], Laos.

Northern Laos (except area occupied by *albipectus*).

Pomatorhinus ruficollis laurentei La Touche

Pomatorhinus ruficollis laurentei La Touche, 1921, Bull. Brit. Orn. Club, 42, p. 16 — Kopaotsun, near Kunming, Yunnan.

Known only from type locality.

Pomatorhinus ruficollis reconditus Bangs and Phillips

Pomatorhinus ruficollis reconditus Bangs and Phillips, 1914, Bull. Mus. Comp. Zool., 58, p. 286 — Mengtsz, southeastern Yunnan.

Pomatorhinus ruficollis saturatus Delacour, 1927, Bull. Brit. Orn. Club, 47, p. 159 — Tam Dao [lat. 21° 27′ N., long. 105° 40′ E.], Tongking.

Southeastern Yunnan, Tongking, and northernmost Annam.

Pomatorhinus ruficollis stridulus Swinhoe

Pomatorhinus stridulus Swinhoe, 1861, Ibis, p. 265 — Pehling Hills near Foochow, Fukien.

Pomatorhinus styani Seebohm, 1884, Ibis, p. 263 — Lu-shan, near Kiukiang, Kiangsi.
Hills of southeastern China in provinces of Kwangtung, Fukien, and Kiangsi.

Pomatorhinus ruficollis intermedius Cheng

Pomatorhinus ruficollis intermedius T. H. Cheng, 1962, Acta Zool. Sinica, 14, pp. 210, 217 — Yuanling, Hunan.
Southwestern Hupeh, Hunan (except extreme northern portion), Kwangsi, and Kweichow (?).

Pomatorhinus ruficollis eidos Bangs

Pomatorhinus ruficollis eidos Bangs, 1930, Occ. Papers Boston Soc. Nat. Hist., 5, p. 293 — Omei Shan, Szech-wan.
Pomatorhinus ruficollis usheri Hall, 1954, Bull. Brit. Orn. Club, 74, p. 43 — Chuan Chi Mine, near Pei Pei [lat. 29° 47′ N., long. 106° 25′ E.], Szechwan.
Southern Szechwan.

Pomatorhinus ruficollis musicus Swinhoe

Pomatorrhinus (sic) *musicus* Swinhoe, 1859, Journ. North China Branch Roy. Asiat. Soc., 1, p. 228 — For-mosa.
Formosa.

Pomatorhinus ruficollis nigrostellatus Swinhoe

Pomatorhinus nigrostellatus Swinhoe, 1870, Ibis, p. 250 — Central Hainan.
Hainan.

POMATORHINUS OCHRACEICEPS

Pomatorhinus ochraceiceps stenorhynchus Godwin-Austen

Pomatorhinus stenorhynchus Godwin-Austen, 1877, Journ. Asiat. Soc. Bengal, 46, pt. 2, p. 43 — "on Man-búm Tila, on Tenga Pani River, near Saddya at 800 ft.," Assam.
Northeastern Assam (Mishmi Hills) and northern Burma (Upper Chindwin District and Kachin State).

Pomatorhinus ochraceiceps austeni Hume

Pomatorhinus austeni Hume, 1881, Stray Feathers, 10, p. 152 — eastern Manipur.

Hill tracts of eastern Assam from Naga Hills southward to Barail Range and Manipur.

Pomatorhinus ochraceiceps ochraceiceps Walden

Pomatorhinus ochraceiceps Walden, 1873, Ann. Mag. Nat. Hist., ser. 4, 12, p. 487 — Karen Hills, Upper Burma.

Southern Shan and Karenni States, Burma; mountains of northern Thailand and Amherst District of Tenasserim (Mulayit Taung); Tongking and eastern portion of northern Laos (?).

Pomatorhinus ochraceiceps alius Riley

Pomatorhinus ochraceiceps alius Riley, 1940, Proc. Biol. Soc. Washington, 53, p. 47 — Dran [lat. 11° 49′ N., long. 108° 38′ E.], Annam.

Northwestern portion of eastern plateau of Thailand, southern Laos, and southern Annam.

POMATORHINUS FERRUGINOSUS

Pomatorhinus ferruginosus ferruginosus Blyth

P[omatorhinus]. ferruginosus Blyth, 1845, Journ. Asiat. Soc. Bengal, 14, p. 597 — Darjeeling, West Bengal, and Arakan; inferentially restricted to Darjeeling, by Blyth, 1847, Journ. Asiat. Soc. Bengal, 16, p. 452.

P[omatorhinus]. rubiginosus "nobis, XIV., 597" Blyth, 1847, Journ. Asiat. Soc. Bengal, 16, p. 542 — Darjeeling, West Bengal. *Lapsus.*

Along Himalayas from eastern Nepal eastward to eastern Assam (north of the Brahmaputra).

Pomatorhinus ferruginosus formosus Koelz

Pomatorhinus ferruginosus formosus Koelz, 1952, Journ. Zool. Soc. India, 4, p. 39 — Tura Mountain, Garo Hills, Assam.

Hill tracts of Assam south of the Brahmaputra and Manipur.

Pomatorhinus ferruginosus phayrei Blyth

P[omatorhinus]. Phayrei Blyth, 1847, Journ. Asiat. Soc. Bengal, 16, p. 452 — Arakan.

Hill tracts of southwestern Burma (Chin Hills and Arakan Yomas).

Pomatorhinus ferruginosus stanfordi Ticehurst

Pomatorhinus ferruginosus stanfordi Ticehurst, 1935,
Bull. Brit. Orn. Club, 55, p. 178 — Sadon-Sima Road,
east of Myitkyina, Kachin State, Upper Burma.
Northeastern Burma (Kachin State).

Pomatorhinus ferruginosus albogularis Blyth

P[omatorhinus]. albogularis Blyth, 1855, Journ. Asiat.
Soc. Bengal, 24, p. 274 — Mulayit Taung, Amherst Dis-
trict, Tenasserim.
Pomatorhinus Mariæ Walden, 1875, Ann. Mag. Nat. Hist.,
ser. 4, 15, p. 403 — Hills near Toungoo, Pegu, Lower
Burma.
Eastern Burma from Northern Shan State southward to
Karenni; northwestern Thailand and Amherst District of
Tenasserim.

Pomatorhinus ferruginosus orientalis Delacour

Pomatorhinus ferruginosus orientalis Delacour, 1927,
Bull. Brit. Orn. Club, 47, p. 159 — Tam Dao [lat. 21° 27′
N., long. 105° 40′ E.], Tongking.
Tongking and eastern portion of northern Laos.

GENUS **GARRITORNIS** IREDALE

Garritornis Iredale, 1956, Birds New Guinea, 2, p. 79.
Type, by original designation, *Pomatorhinus isidorei*
Lesson.

GARRITORNIS ISIDOREI

Garritornis isidorei calidus (Rothschild)

Pomatorhinus isidori calidus Rothschild, 1931, Novit.
Zool., 36, p. 266 — Siriwo River, 45 miles above mouth,
south of Geelvink Bay, New Guinea.
Northern New Guinea from head of Geelvink Bay eastward to Astrolabe Bay.

Garritornis isidorei isidorei (Lesson)

Pomatorhinus Isidorei Lesson, 1827, Dict. Sci. Nat. (éd.
Levrault), 50, p. 37 — Dorei Harbor [= Manokwari],
northwestern New Guinea.
Misol and all New Guinea (except area occupied by *calidus*).

Genus POMATOSTOMUS Cabanis

Pomatostomus Cabanis, 1851, in Cabanis and Heine, Mus. Hein., 1 (1850), p. 83. Type, by subsequent designation (G. R. Gray, 1855, Cat. Gen. Subgen. Birds., p. 45), *Pomatorhinus temporalis* Vigors and Horsfield.

Morganornis Mathews, 1912, Austral Avian Rec., 1, p. 112. Type, by original designation and monotypy, *Pomatorhinus superciliosus* Vigors and Horsfield.

POMATOSTOMUS TEMPORALIS

Pomatostomus temporalis tregellasi (Mathews)

Pomatorhinus temporalis tregellasi Mathews, 1912, Novit. Zool., 18, p. 334 — Victoria; restricted to Frankston, by Mathews, 1922, Birds Austr., 9 (1921-1922), p. 255.

Southeastern South Australia, Victoria, and southeastern New South Wales.

Pomatostomus temporalis trivirgatus (Temminck)

Pomatorhinus trivirgatus Temminck, 1827, Pl. col., livr. 75, pl. 443 — interior of New Holland [= Blue Mountains, New South Wales, *fide* Mathews, 1912, Novit. Zool., 18, p. 334].

Coastal region of New South Wales (except area occupied by *tregellasi*) and southern Queensland.

Pomatostomus temporalis temporalis (Vigors and Horsfield)

[*Pomatorhinus*] *Temporalis* Vigors and Horsfield, 1827, Trans. Linn. Soc. London, 15 (1826), p. 330 — Shoalwater Bay, Queensland.

Coastal region of central Queensland.

? Pomatostomus temporalis cornwalli (Mathews)

Pomatorhinus temporalis cornwalli Mathews, 1912, Novit. Zool., 18, p. 335 — Cairns, northern Queensland.

Coastal region of northern Queensland from Cairns northward to extremity of Cape York; doubtfully distinct from *temporalis*.

Pomatostomus temporalis strepitans (Mayr and Rand)

Pomatorhinus temporalis strepitans Mayr and Rand, 1935, Amer. Mus. Novit., no. 814, p. 6 — Dogwa, Oriomo River, Western Division, Territory of Papua.

Lowland savannas of southern New Guinea from Oriomo River westward to Princess Marianne Straits.

Pomatostomus temporalis intermedius (Mathews)

Pomatorhinus temporalis intermedius Mathews, 1912, Novit. Zool., 18, p. 335 — Alexandra Station, Northern Territory.

Pomatostomus innominatus Mathews, 1924, Birds Austr., Suppl. Vol., Suppl. no. 3, p. 223 — Point Torment, West Kimberley, West Australia. "New name for the bird figured in my *Birds of Australia*, Vol. IX., pl. 432 (top figure), and described on p. 255 (pt. VI., Feb. 15th, 1922)."

Arid scrub of interior of Australia from Queensland and northern South Australia northwestward, through Northern Territory, to Kimberley Division of Western Australia.

Pomatostomus temporalis mountfordae Deignan

Pomatostomus temporalis mountfordae Deignan, 1950, Emu, 50, p. 19 — Groote Eylandt, Northern Territory (in Gulf of Carpentaria).

Known only from type locality.

Pomatostomus temporalis browni Deignan

Pomatostomus temporalis browni Deignan, 1950, Emu, 50, p. 20 — Cape Arnhem Peninsula, Northern Territory.

Known only from type locality.

Pomatostomus temporalis rubeculus (Gould)

Pomatorhinus rubeculus Gould, 1840, Proc. Zool. Soc. London, p. 144 — "North-west coast of Australia" [= Port Essington, Northern Territory].

Mainland of Northern Territory (except areas occupied by *mountfordae* and *intermedius*).

Pomatostomus temporalis bamba (Mathews)

Pomatorhinus temporalis bamba Mathews, 1912, Austral Avian Rec., 1, p. 43 — Melville Island, Northern Territory.

Known only from type locality.

Pomatostomus temporalis nigrescens (Mathews)

Pomatorhinus temporalis nigrescens Mathews, 1912, Novit. Zool., 18, p. 335 — Strelley River, Western Australia.

Western Australia from Pilbara Goldfield southward to the Wooramel River, the upper Murchison (Mileura, Karalundi) and Wiluna.

POMATOSTOMUS SUPERCILIOSUS

Pomatostomus superciliosus gilgandra (Mathews)

Pomatorhinus superciliosus gilgandra Mathews, 1912, Novit. Zool., 18, p. 336 — New South Wales; type locality restricted to Gilgandra (*ca.* 210 miles northwest of Sydney), by Mathews, 1922, Birds Austr., 9 (1921-1922), p. 265.

Mallee and arid scrub of interior of New South Wales, Victoria, and South Australia (including Eyre Peninsula).

Pomatostomus superciliosus superciliosus (Vigors and Horsfield)

[*Pomatorhinus*] *Superciliosus* Vigors and Horsfield, 1827, Trans. Linn. Soc. London, 15 (1826), p. 330 — south coast of Australia; here further restricted to Yorke Peninsula, South Australia.

Southeastern South Australia (Yorke and Fleurieu Peninsulas) and possibly Victoria (except area occupied by *gilgandra*).

Pomatostomus superciliosus ashbyi Mathews

Pomatostomus superciliosus ashbyi Mathews, 1911, Bull. Brit. Orn. Club, 27, p. 87 — southern West Australia; restricted to Broome Hill, by Mathews, 1922, Birds Austr., 9 (1921-1922), p. 265.

Southwestern West Australia (mainly inland), north to the Tropic of Capricorn.

? Pomatostomus superciliosus gwendolenae (Mathews)

Pomatorhinus superciliosus gwendolenae Mathews, 1912, Novit. Zool., 18, p. 336 — Carnarvon, West Australia.

West Australia (valley of the Gascoyne).

POMATOSTOMUS RUFICEPS

Pomatostomus ruficeps (Hartlaub)

P[omatorhinus]. ruficeps Hartlaub, 1852, Rev. Mag. Zool.
[Paris], sér. 2, 4, p. 316 — "Adelaide," South Aus-
tralia.[1]
Pomatostomus ruficeps bebba Mathews, 1916, Austral
Avian Rec., 3, p. 60 — southern Queensland.
Pomatostomus ruficeps parsonsi Mathews, 1918, Bull.
Brit. Orn. Club, 38, p. 48 — Pungonda, South Australia.
Mallee and arid scrub of southwestern Queensland, west-
ern New South Wales, northwestern Victoria, and north-
eastern South Australia.

GENUS XIPHIRHYNCHUS BLYTH

Xiphirhynchus Blyth, 1842, Journ. Asiat. Soc. Bengal,
11, p. 175. Type, by monotypy, Xiphirhynchus superciliaris Blyth.
Xiphorhamphus Blyth, 1843, Journ. Asiat. Soc. Bengal,
12, p. 947. New name for Xiphirhynchus Blyth, considered preoccupied by Xiphorhynchus Swainson, 1827.

XIPHIRHYNCHUS SUPERCILIARIS

Xiphirhynchus superciliaris superciliaris Blyth

X[iphirhynchus]. superciliaris Blyth, 1842, Journ. Asiat.
Soc. Bengal, 11, p. 176 — Darjeeling, Bengal.
Easternmost Nepal, Darjeeling District of West Bengal,
Sikkim, and Bhutan.

Xiphirhynchus superciliaris intextus Ripley

Xiphirhynchus superciliaris intextus Ripley, 1948, Proc.
Biol. Soc. Washington, 61, p. 105 — Dreyi, Mishmi
Hills, Assam.
Xiphorhamphus superciliaris arquatellus Koelz, 1954,
Contrib. Inst. Regional Explor., no. 1, p. 4 — Blue
Mountain, Lushai Hills, Assam.
Hill tracts of Assam south and east of the Brahmaputra.

[1] Mathews (1918, Bull. Brit. Orn. Club, 38, p. 49), asserting that
Hartlaub's type specimen came not from Adelaide, but from the interior, has altered the type locality to Broken Hill, New South Wales.
A specimen in the U. S. National Museum, collected in 1918, was taken
at Morgan, South Australia, only 85 miles northeast of Adelaide.

Xiphirhynchus superciliaris forresti Rothschild

Xiphirhynchus superciliaris forresti Rothschild, 1926, Novit. Zool., 33, p. 262 — Shweli-Salween Divide, northwestern Yunnan.

Mountains of northeastern Burma and northwestern Yunnan.

Xiphirhynchus superciliaris rothschildi Delacour and Jabouille

Xiphirhynchus superciliaris rothschildi Delacour and Jabouille, 1930, Oiseau Rev. Franç. Orn., 11, p. 613 — Loquiho, Fan Si Pan [lat. 22° 17′ N., long. 104° 47′ E.], 2,500 m., Tongking.

Known only from type locality.

Genus **JABOUILLEIA** Delacour

Jabouilleia Delacour, 1927, Bull. Brit. Orn. Club, 47, p. 160. Type, by original designation, *Rimator danjoui* Robinson and Kloss.

JABOUILLEIA DANJOUI

Jabouilleia danjoui parvirostris Delacour

Jabouilleia danjoui parvirostris Delacour, 1927, Bull. Brit. Orn. Club, 47, p. 160 — Bana, near the Col des Nuages [lat. 16° 11′ N., long. 108° 08′ E.], Annam.

Known only from type locality.

Jabouilleia danjoui danjoui (Robinson and Kloss)

Rimator danjoui Robinson and Kloss, 1919, Ibis, p. 579 — Lang Bian Peaks [*ca.* lat. 12° 02′ N., long. 108° 26′ E.], Annam.

Known only from type locality.

Genus **RIMATOR** Blyth

Rimator Blyth, 1847, Journ. Asiat. Soc. Bengal, 16, p. 154. Type, by original designation and monotypy, *Rimator malacoptilus* Blyth.

RIMATOR MALACOPTILUS

Rimator malacoptilus malacoptilus Blyth

R[imator]. malacoptilus Blyth, 1847, Journ. Asiat. Soc. Bengal, 16, p. 155 — Darjeeling, Bengal.

Rimator malacoptilus amadoni Koelz, 1954, Contrib. Inst. Regional Explor., no. 1, p. 5 — Maoflang, Khasi Hills, Assam.

Himalayas from Sikkim to eastern Assam; hill tracts of Assam south of the Brahmaputra and northeastern Burma.

Rimator malacoptilus pasquieri Delacour and Jabouille

Rimator pasquieri Delacour and Jabouille, 1930, Oiseau Rev. Franç. Orn., 11, p. 401 — Fan Si Pan [lat. 22° 17′ N., long. 104° 47′ E.], 2,500 m., Tongking.

Known only from type locality.

Rimator malacoptilus albostriatus Salvadori

Rimator albo-striatus Salvadori, 1879, Ann. Mus. Civ. Genova, 14, p. 224 — Mount Singgalang [lat. 0° 24′ S., long. 100° 20′ E.], Sumatra.

Highlands of western Sumatra.

GENUS **PTILOCICHLA** SHARPE

Ptilocichla Sharpe, 1877, Trans. Linn. Soc. London, ser. 2, 1, p. 332. Type, by original designation and monotypy, *Ptilocichla falcata* Sharpe.

PTILOCICHLA LEUCOGRAMMICA

Ptilocichla leucogrammica (Bonaparte)

[Cacopitta] leucogrammica "Temm. Mus. Lugd." Bonaparte, 1850, Consp. Av., 1, p. 257 — Borneo; type specimen from Pontianak [lat. 0° 02′ S., long. 109° 22′ E.], *fide* Büttikofer, 1895, Notes Leyden Mus., 17, p. 72.

Borneo.

PTILOCICHLA MINDANENSIS

Ptilocichla mindanensis minuta Bourns and Worcester

Ptilocichla minuta Bourns and Worcester, 1894, Occ. Papers Minnesota Acad. Nat. Sci., 1, p. 24 — Samar, Philippine Islands; type specimens from Catbalogan.

Leyte and Samar, Philippine Islands.

Ptilocichla mindanensis fortichi Rand and Rabor

Ptilocichla mindanensis fortichi Rand and Rabor, 1957, Fieldiana Zool. [Chicago], 42, p. 13 — Cantaub, Sierra Bullones, Bohol, Philippine Islands.

Bohol, Philippine Islands.

Ptilocichla mindanensis mindanensis (Blasius)

Ptilopyga mindanensis Blasius, 1890 (Apr. 15), Braunschweigische Anzeigen, no. 87, p. 877; *idem*, 1890 (Oct.), Journ. f. Orn., 38, p. 146 — Mindanao, Philippine Islands; type from Davao, *fide* Hartert, 1920, Novit. Zool., 27, p. 482.

Ptilocichla (?) *mindanensis* Steere, 1890 (July 14), List Birds Mammals Steere Exped., p. 18 — Mindanao, Philippine Islands; type from Ayala, *fide* Hachisuka, 1935, Birds Philippine Islands, pt. 4, p. 405.

Mindanao, Philippine Islands.

Ptilocichla mindanensis basilanica Steere

Ptiocichla [sic] (?) *Basilanica* Steere, 1890 (July 14), List Birds Mammals Steere Exped., p. 18 — Basilan, Philippine Islands.

Basilan, Philippine Islands.

PTILOCICHLA FALCATA

Ptilocichla falcata Sharpe

Ptilocichla falcata Sharpe, 1877, Trans. Linn. Soc. London, ser. 2, 1, p. 332, pl. 50, fig. 3 — Puerto Princesa, Palawan.

Palawan and Balabac, Philippine Islands.

GENUS KENOPIA GRAY

Kenopia "Bl. 1855" G. R. Gray, 1869 (subgenus of *Malacopteron* Eyton), Handl. Gen. Spec. Birds Brit. Mus., pt. 1, p. 317. Type, by subsequent designation (Salvadori, 1874, Ann. Mus. Civ. Genova, 5, p. 223), *Timalia striata* Blyth.

KENOPIA STRIATA

Kenopia striata (Blyth)

T[*imalia*]. *striata* Blyth, 1842, Journ. Asiat. Soc. Bengal, 11, p. "783" [= 793] — "Singapore."

Malay Peninsula from southern Thailand (Krabi Province) southward to Johore; eastern Sumatra; Borneo.

GENUS **NAPOTHERA** GRAY

Napothera "Boie (1835)" G. R. Gray, 1842, App. List. Gen. Birds, p. 8. Type, by original designation and monotypy, *Myiothera epilepidota* Temminck.

Turdinus Blyth, 1844, Journ. Asiat. Soc. Bengal, 13, p. 382. Type, by original designation and monotypy, *Malacopteron macrodactylum* Strickland.

Cacopitta Bonaparte, 1850, Consp. Av., 1, p. 257. Type, by subsequent designation (G. R. Gray, 1855, Cat. Gen. Subgen. Birds, p. 41), *Myiothera loricata* S. Müller = *Turdinus marmoratus* Ramsay.

Hadropezus Sundevall, 1872, Tentamen, pt. 1, p. 11. New name for *Turdinus* Blyth.

Turdinulus Hume, 1878, in Hume and Davison, Stray Feathers, 6, pp. 234, 235. Type, by original designation and monotypy, *Pnoepyga roberti* Godwin-Austen and Walden.

Gypsophila Oates, 1883, Handb. Birds Brit. Burmah, 1, p. 61. Type, by original designation and monotypy, *Turdinus crispifrons* Blyth.

Corythocichla Sharpe, 1883, Cat. Birds Brit. Mus., 7, pp. 507 (in key), 592. Type, by subsequent designation (Anonymous, 1883, Ibis, p. 573), *Turdinus brevicaudatus* Blyth.

Lanioturdinus Büttikofer, 1895, Notes Leyden Mus., 17, pp. 67 (in key), 72. Type, by original designation and monotypy, *Corythocichla crassa* Sharpe.

Curzonia Skinner, 1905, List Birds Brit. India (St. Mary Cray, Kent), pp. [1], 6. New name for *Gypsophila* Oates, considered preoccupied by *Gypsophila* Linnaeus, 1756, in Botany.

Cursonia "Skinner, 1898" = Baker, 1922, Fauna Brit. India, Birds, ed. 2, 1, p. 248. *Lapsus*.

Turdinus "Hume" Baker (ex Hume MS) 1930, Fauna Brit. India, Birds, ed. 2, 7, p. 48. *Lapsus* for *Turdinulus* Hume.

cf. Voous, 1949, Limosa, 22, pp. 347-352 (species of *Turdinus, sensu stricto*).

NAPOTHERA RUFIPECTUS

— Napothera rufipectus (Salvadori)

Turdinus rufipectus Salvadori, 1879, Ann. Mus. Civ. Genova, 14, p. 224 — Mount Singgalang [lat. 0° 24′ S., long. 100° 20′ E.], Sumatra.
Highlands of western Sumatra.

NAPOTHERA ATRIGULARIS

~ Napothera atrigularis (Bonaparte)

C[*acopitta*]. *atrigularis* Bonaparte, 1850, Consp. Av., 1, p. 257 — Borneo.
Borneo.

NAPOTHERA MACRODACTYLA

—Napothera macrodactyla macrodactyla (Strickland)

Malacopteron macrodactylum Strickland, 1844, Ann. Mag. Nat. Hist., 13, p. 417 — Malacca.
Turdinus macrodactylus bakeri Hachisuka, 1926, Bull. Brit. Orn. Club, 47, p. 54 — Sathani Lam Phura [lat. 7° 40′ N., long. 99° 35′ E.], Thailand. Not *Turdinulus epilepidotus bakeri* Harington, 1913 = *Napothera epilepidota bakeri* (Harington).
Malay Peninsula from Isthmus of Kra southward to Johore.

Napothera macrodactyla beauforti (Voous)

Turdinus macrodactylus beauforti Voous, 1949, Limosa, 22, p. 348 — Sungai Tasik, Langkat, northeastern Sumatra.
Northeastern Sumatra.

Napothera macrodactyla lepidopleura (Bonaparte)

[*Cacopitta*] *lepidopleura* Bonaparte, 1850, Consp. Av., 1, p. 257 — Java.
Java.

NAPOTHERA MARMORATA

Napothera marmorata grandior (Voous)

Turdinus marmoratus grandior Voous, 1949, Limosa, 22, p. 351 — Semangko Pass, Selangor-Pahang boundary, Malaya.

Highlands of central Malaya (Selangor-Pahang boundary).

Napothera marmorata marmorata (Ramsay)

Myiothera loricata S. Müller, 1836, Tijdschr. Natuur. Gesch. Phys., 2 (1835), p. 348, pl. "V. fig. 5" [= IX, fig. 5] — Sumatra. Not *M[yiothera]. loricata* Lichtenstein, 1823.

Turdinus marmoratus Ramsay, 1880, Proc. Zool. Soc. London, p. 15 — Padang Highlands, western Sumatra. Highlands of western Sumatra.

NAPOTHERA CRISPIFRONS

Napothera crispifrons annamensis (Delacour and Jabouille)

Corythocichla annamensis Delacour and Jabouille, 1928, Bull. Brit. Orn. Club, 48, p. 131 — Phuqui, northern Annam.

Cursonia crispifrons saxatilis Bangs and Van Tyne, 1930, Field Mus. Nat. Hist. Publ., Zool. Ser., 18, p. 3 — Bac Tan Trai, northwestern Tongking.

Northern Laos, western Tongking, and northern Annam (restricted to calcareous formations).

Napothera crispifrons calcicola Deignan

Napothera crispifrons calcicola Deignan, 1939, Journ. Washington Acad. Sci., 29, p. 177 — Sathani Hin Lap [lat. 14° 40′ N., long. 101° 10′ E.], Thailand. Known only from type locality.

Napothera crispifrons crispifrons (Blyth)

Turdinus crispifrons Blyth, 1855, Journ. Asiat. Soc. Bengal, 24, p. 269 — Tenasserim; type from Mulayit Taung, *fide* Sclater, 1892, Ibis, p. 76.

Calcareous formations of northern Thailand (Phrae Province), central Tenasserim, and southwestern Thailand (Kanchanaburi Province).

NAPOTHERA BREVICAUDATA

Napothera brevicaudata striata (Walden)

Turdinus striatus Walden, 1871, Ann. Mag. Nat. Hist., ser. 4, 7, p. 241 — near Cherrapunji, Khasi Hills, Assam.

Napothera brevicaudata naphaea Koelz, 1954, Contrib.
Inst. Regional Explor., no. 1, p. 5 — Sangua, Lushai
Hills, Assam.

Hill tracts of Assam (south of the Brahmaputra), Manipur, and southwestern Burma.

? Napothera brevicaudata venningi (Harington)

Turdinulus brevicaudatus venningi Harington, 1913, Bull.
Brit. Orn. Club, 33, p. 44 — Southern Shan State, Upper Burma.

Western Yunnan; northeastern Burma from southern
portion of Kachin State southward, through Northern and
Southern Shan States, to Karenni State; doubtfully distinct from *brevicaudata*.

Napothera brevicaudata brevicaudata (Blyth)

T[urdinus]. brevicaudatus Blyth, 1855, Journ. Asiat. Soc.
Bengal, 24, p. 272 — mountainous interior of the Tenasserim provinces; type from Mulayit Taung, Amherst District, *fide* Sclater, 1892, Ibis, p. 76.

Northern Thailand, on the west southward to Amherst
District of Tenasserim.

Napothera brevicaudata stevensi (Kinnear)

Turdinulus brevicaudatus stevensi Kinnear, 1925, Bull.
Brit. Orn. Club, 45, p. 74 — Ngai Tio [lat. 22° 40′ N.,
long. 103° 47′ E.], Tongking.

Corythocichla brevicaudata obscura Delacour, 1927, Bull.
Brit. Orn. Club, 47, p. 161 — Tam Dao [lat. 21° 27′ N.,
long. 105° 40′ E.], Tongking.

Eastern portion of northern Laos, Tongking, and northern Annam.

Napothera brevicaudata proxima Delacour

Napothera brevicaudata proxima Delacour, 1930, Oiseau
Rev. Franç. Orn., 11, p. 654 — Bana, between the Col
des Nuages and Tourane, central Annam.

Central Annam and southern Laos.

Napothera brevicaudata rufiventer (Delacour)

Corythocichla brevicaudata rufiventer Delacour, 1927,
Bull. Brit. Orn. Club, 47, p. 162 — Jiring [lat. 11° 31′
N., long. 108° 01′ E.], Annam.

Southern Annam (Lang Bian Plateau).

Napothera brevicaudata griseigularis (Delacour and Jabouille)

Corythocichla griseigularis Delacour and Jabouille, 1928, Bull. Brit. Orn. Club, 48, p. 131 — Le Boc Kor [lat. 10° 37′ N., long. 104° 03′ E.], Cambodia.

Corythocichla brevicaudata cognata Riley, 1933, Proc. Biol. Soc. Washington, 46, p. 156 — Khao Sa Bap [lat. 12° 35′ N., long. 102° 15′ E.], Thailand.

Southeastern Thailand and southwestern Cambodia.

Napothera brevicaudata leucosticta (Sharpe)

Corythocichla leucosticta Sharpe, 1887, Proc. Zool. Soc. London, p. 438 — Larut Range [ca. lat. 4° 48′ N., long. 100° 45′ E.], Perak.

Corythocichla brevicaudata herberti Baker, 1917, Bull. Brit. Orn. Club, 38, p. 10 — Sathani Thung Song [lat. 8° 10′ N., long. 99° 40′ E.], Thailand.

Malay Peninsula from Siamese province of Nakhon Si Thammarat southward to Negri Sembilan; Pulau Tioman.

NAPOTHERA CRASSA

Napothera crassa (Sharpe)

Corythocichla crassa Sharpe, 1888, Ibis, p. 391 — Kina Balu, North Borneo.

Highlands of North Borneo.

NAPOTHERA RABORI

Napothera rabori Rand

Napothera rabori Rand, 1960, Fieldiana Zool. [Chicago], 39, p. 377 — Tabbug, Pagudpud, Ilocos Norte, Luzon, Philippine Islands.

Northern Luzon (Ilocos Norte and Cagayán Provinces), Philippine Islands.

NAPOTHERA EPILEPIDOTA

Napothera epilepidota guttaticollis (Ogilvie-Grant)

Turdinulus guttaticollis Ogilvie-Grant, 1895, Ibis, p. 432 — Miri Hills, Assam.

Hill tracts of Assam north of the Brahmaputra.

Napothera epilepidota roberti (Godwin-Austen and Walden)

Pnoepyga roberti Godwin-Austen and Walden, 1875, Ibis, p. 252 — Chakha, Manipur.
Corythocichla squamata Baker, 1901, Journ. Bombay Nat. Hist. Soc., 13, p. 403, pl. H — Hangmai Peak, Cachar, Assam.

Hill tracts of Assam south of the Brahmaputra; north-western Burma.

? Napothera epilepidota bakeri (Harington)

Turdinulus epilepidotus bakeri Harington, 1913, Bull. Brit. Orn. Club, 33, p. 44 — Na Noi, Southern Shan State, Upper Burma.

Southern Shan and Karenni States of Burma; doubtfully distinct from davisoni.

Napothera epilepidota davisoni (Ogilvie-Grant)

Turdinulus davisoni Ogilvie-Grant, 1910, Bull. Brit. Orn. Club, 25, p. 97 — Thaungya Sakan and Mulayit Taung, Amherst District, Tenasserim.

Northern Thailand, on the west southward to Amherst District of Tenasserim.

Napothera epilepidota amyae (Kinnear)

Turdinulus epilepidotus amyæ Kinnear, 1925, Bull. Brit. Orn. Club, 45, p. 73 — Bao Ha [lat. 22° 10′ N., long. 104° 21′ E.], Tongking.
Turdinulus epilepidotus laotianus Delacour, 1926, Bull. Brit. Orn. Club, 47, p. 17 — Chiang Khwang [lat. 19° 19′ N., long. 103° 22′ E.], Laos.

Eastern portion of northern Laos, northern Annam, and Tongking.

Napothera epilepidota delacouri Yen

Napothera epilepidota Delacouri Yen, 1934, Oiseau Rev. Franç. Orn., nouv. sér., 4, p. 32 — Yao Shan, Kwangsi.
Known only from type locality.

Napothera epilepidota hainana (Hartert)

Turdinulus roberti hainanus Hartert, 1910, Novit. Zool., 17, p. 230 — Mount Wuchi, Hainan.
Hainan.

Napothera epilepidota clara (Robinson and Kloss)

> *Turdinulus epilepidotus clarus* Robinson and Kloss, 1919,
> Ibis, p. 582 — Dalat [lat. 11° 55′ N., long. 108° 26′ E.],
> Annam.

Southern Annam (Lang Bian Plateau).

Napothera epilepidota granti (Richmond)

> *Turdinulus granti* Richmond, 1900, Proc. U.S. Nat. Mus.,
> 22, p. 320 — Khao Soi Dao [lat. 7° 20′ N., long. 99° 50′
> E.], Thailand.
> *Turdinulus humei* Hartert, 1902, Novit. Zool., 9, p. 564
> — Gunong Tahan [lat. 4° 38′ N., long. 102° 14′ E.],
> Pahang.

Malay Peninsula from Siamese province of Nakhon Si
Thammarat southward to Johore.

Napothera epilepidota diluta (Robinson and Kloss)

> *Turdinulus epilepidotus dilutus* Robinson and Kloss, 1916,
> Journ. Straits Branch Roy. Asiat. Soc., no. 73, p. 276
> — Sungei Kumbang [lat. 1° 48′ S., long. 101° 20′ E.],
> Sumatra.
> *Napothera exsul lucilleae* de Schauensee and Ripley, 1940,
> Proc. Acad. Nat. Sci. Philadelphia, 91 (1939), p. 352,
> pl. 21, lower fig. — Meluwak, near Gunong Leuser,
> Acheh, Sumatra.[1]

Highlands of western Sumatra (except area occupied by
mendeni).

? Napothera epilepidota mendeni Neumann

> *Napothera epilepidota mendeni* Neumann, 1937, Bull.
> Brit. Orn. Club, 57, p. 152 — Gunong Dempu [lat.
> 4° 01′ S., long. 103° 07′ E.], Sumatra.

Highlands of southwestern Sumatra; doubtfully distinct
from *epilepidota*.

---

[1] The authors, reporting this form as virtually sympatric with
diluta in the Acheh district, divide the populations here treated as
races of *epilepidota* between two species: *N. epilepidota*, with *diluta*,
mendeni, and *epilepidota*, and *N. exsul*, with all the rest. I am unable
to appreciate the characters believed by them to have specific value,
and since *lucilleae* was separated from the very distinct *exsul* of
Borneo rather than from the sympatric *diluta*, I must consider it a
mere variant of *diluta*.

Napothera epilepidota epilepidota (Temminck)

Myiothera epilepidota Temminck, 1827, Pl. col., livr. 74,
pl. 448, fig. 2 — Java and Sumatra; restricted to Java,
by Hartert, 1902, Novit. Zool., 9, p. 565.
Highlands of western and central Java.

Napothera epilepidota exsul (Sharpe)

Turdinulus exsul Sharpe, 1888, Ibis, p. 479 — Kina Balu,
North Borneo.
Highlands of northern Borneo.

GENUS PNOEPYGA HODGSON

Microura Gould, 1837, Icones Avium, pt. 1, pl. [5]. Type,
by original designation and monotypy, *Microura squamata* Gould. *Nomen oblitum.*[1]

Pnoepyga Hodgson, 1844, in J. E. Gray, Zool. Misc., no. 3,
p. 82. Type, by subsequent designation (Zimmer and
Vaurie, 1954, Bull. Brit. Orn. Club, 74, p. 41), *Tesia
albiventer* Hodgson.

PNOEPYGA ALBIVENTER

Pnoepyga albiventer pallidior Kinnear

Pnoepyga albiventer pallidior Kinnear, 1924, Bull. Brit.
Orn. Club, 45, p. 10 — Dharmsala, Kangra, East Pun-
jab.
Himalayas from East Punjab eastward to western Nepal.

Pnoepyga albiventer albiventer (Hodgson)

[Tesia] Albiventer Hodgson, 1837 (Feb.), Journ. Asiat.
Soc. Bengal, 6, p. 102 — Nepal; restricted to eastern
Nepal by Rand and Fleming, 1957, Fieldiana Zool.
[Chicago], 41, p. 124; further restricted to Ilam Dis-
trict by Ripley, 1961, Synopsis Birds India Pakistan,
p. 357.

Microura squamata Gould, 1837 (Aug.), Icones Avium,
pt. 1, pl. [5] — "Himalaya Mountains" and Nepal.

Pnoepyga mutica Thayer and Bangs, 1912, Mem. Mus.
Comp. Zool., 40, p. 172, pl. 4, fig. 1 — Wa Shan, southwestern Szechwan.

[1] This name has been placed on the Official Index of Rejected Names
(Opinion 695).

Pnoepyga squamata magnirostris Rothschild, 1925, Novit.
Zool., 32, p. 297 — Shweli Valley, northwestern Yunnan.

Pnoepyga albiventer vegeta Koelz, 1954, Contrib. Inst.
Regional Explor., no. 1, p. 11 — Kohima, Naga Hills,
Assam.

Himalayas from central Nepal eastward, through Bhutan,
Assam (both north and south of the Brahmaputra), northern Burma, northwestern Yunnan, and southeastern Hsikang, to southwestern Szechwan; northwestern Tongking.

PNOEPYGA PUSILLA

Pnoepyga pusilla pusilla Hodgson

[*Pnoepyga*] *pusillus* [sic] Hodgson, 1845, Proc. Zool. Soc.
London, p. 25 — Nepal.

Pnœpyga pusilla tonkinensis Delacour and Jabouille, 1930,
Oiseau Rev. Franç. Orn., 11, p. 404 — Chapa [lat.
22° 20′ N., long. 103° 50′ E.], Tongking.

Pnœpyga pusilla pygmæa Koelz, 1952, Journ. Zool. Soc.
India, 4, p. 40 — Karong, Manipur.

Himalayas from Nepal eastward, through Bhutan, Assam
(both north and south of the Brahmaputra), northern Burma, northwestern Yunnan, and southeastern Hsikang, to
southwestern Szechwan, southward in western Burma to
the Chin Hills and in eastern Burma, through the Shan
States, to Karenni; northwestern Thailand and Amherst
District of Tenasserim (Mulayit Taung); northern Laos
and Tongking; southeastern Yunnan, Kwangsi, and Fukien.

Pnoepyga pusilla formosana Ingram

Pnoëpyga formosana Ingram, 1909, Bull. Brit. Orn. Club,
23, p. 97 — Arisan, central Formosa.

Highlands of Formosa.

Pnoepyga pusilla annamensis Robinson and Kloss

Pnoepyga pusilla annamensis Robinson and Kloss, 1919,
Ibis, p. 591 — Lang Bian Peaks, southern Annam.

Southern Annam (Lang Bian Plateau); population of
southern Laos (Boloven Plateau) may be referable here.

Pnoepyga pusilla harterti Robinson and Kloss

Pnoepyga pusilla . . . *harterti* Robinson and Kloss, 1918, Journ. Federated Malay States Mus., 8, p. 205 — Gunong Ijau, Larut Range, Perak.

Highlands of Malay Peninsula from Siamese province of Nakhon Si Thammarat southward to southern Selangor and northern Pahang.

Pnoepyga pusilla lepida Salvadori

Pnoepyga lepida Salvadori, 1879, Ann. Mus. Civ. Genova, 14, p. 227 — Mount Singgalang [lat. 0° 24' S., long. 100° 20' E.], Sumatra.

Highlands of western Sumatra.

Pnoepyga pusilla rufa Sharpe

Pnoepyga rufa Sharpe, 1882, Cat. Birds Brit. Mus., 6 (1881), pp. 302 (in key), 304 — Mount Gedeh [lat. 6° 47' S., long. 106° 59' E.], Java.

Highlands of Java.

Pnoepyga pusilla everetti Rothschild

Pnoepyga everetti Rothschild, 1897, Novit. Zool., 4, p. 168 — southern Flores.

Highlands of Flores.

Pnoepyga pusilla timorensis Mayr

Pnoepyga pusilla timorensis Mayr, 1944, Bull. Amer. Mus. Nat. Hist., 83, pp. 135 (in annotated list), 157 — Mount Mutis, Timor.

Highlands of Timor.

GENUS **SPELAEORNIS** DAVID AND OUSTALET

Spiloptera Jerdon (ex Blyth MS), 1862 (subgenus of *Troglodytes* Vieillot), Birds India, 1, p. 492. Type, by original designation and monotypy, *Troglodytes punctatus* Blyth. Not *Spiloptera* Zetterstedt, 1860, Diptera.

Spelæornis David and Oustalet, 1877, Oiseaux Chine, p. 228. Type, by subsequent designation (Sharpe, 1882, Cat. Birds Brit. Mus., 6 (1881), p. 264), *Pnoepyga troglodytoides* J. Verreaux.

Urocichla Sharpe, 1882, Cat. Birds Brit. Mus., 6 (1881),

pp. 181 (in key), 263. Type, by monotypy, *Pnoepyga longicaudata* Moore.

Elachura Oates, 1889, Fauna Brit. India, Birds, 1, pp. 328 (in key), 339. Type, by original designation and monotypy, *Troglodytes punctatus* Blyth. New name for *Spiloptera* "Blyth" Jerdon, preoccupied.

cf. Ripley, 1950, Auk, 67, pp. 390-391.
Ripley, 1954, Postilla, Yale Univ., no. 20, pp. [1]-4.

SPELAEORNIS CAUDATUS

Spelaeornis (caudatus) caudatus (Blyth)

T[esia]. caudata Blyth, 1845, Journ. Asiat. Soc. Bengal, 14, p. 588 — Darjeeling, West Bengal.
Sikkim and West Bengal (Darjeeling District) ; eastern Bhutan.

Spelaeornis (caudatus?) badeigularis Ripley

Spelaeornis badeigularis Ripley, 1948, Proc. Biol. Soc. Washington, 61, p. 103 — Dreyi, Mishmi Hills, northeastern Assam.
Known only from type locality.

SPELAEORNIS TROGLODYTOIDES

Spelaeornis troglodytoides sherriffi Kinnear

Spelæornis souliei sherriffi Kinnear, 1934, Bull. Brit. Orn. Club, 54, p. 107 — Donga La, between Lingste and Trashi Yangtsi, eastern Bhutan.
Eastern Bhutan.

Spelaeornis troglodytoides souliei Oustalet

Spelæornis Souliei Oustalet, 1898, Bull. Mus. Hist. Nat. Paris, 4, p. 257 — Tzeku, northwestern Yunnan.
Northeastern Burma and northwestern Yunnan (west of the Mekong).

Spelaeornis troglodytoides rocki Riley

Spelaeornis rocki Riley, 1929, Proc. Biol. Soc. Washington, 42, p. 214 — mountains of Ho-fu-ping, Mekong Valley, northwestern Yunnan.
Northwestern Yunnan (east of the Mekong).

Spelaeornis troglodytoides troglodytoides (Verreaux)

Pnoepyga troglodytoides J. Verreaux, 1870, Nouv. Arch.
Mus. Hist. Nat. [Paris], 6, p. 34; *idem*, 1871, *ibid.*, 7,
p. 30 — "les montagnes du Thibet chinois"; type from
Muping [= Paohing, Hsikang, *fide* Verreaux, 1871].
Hsikang (Kangting and Paohing) and northwestern
Szechwan (Wenchwan).

Spelaeornis troglodytoides halsueti (David)

Pnoepyga? Halsueti David, 1875, L'Institut, nouv. sér.,
3, p. 76 — Shensi; type from Chinling Mountains.
Shensi (Chinling Mountains).

SPELAEORNIS FORMOSUS

Spelaeornis formosus (Walden)

T[roglodytes]. punctatus Blyth, 1845, Journ. Asiat. Soc.
Bengal, 14, p. 589 — Darjeeling, West Bengal. Not *Troglodytes punctatus* C. L. Brehm, 1823, nor *Troglodites*
[sic] *punctatus* Boie, 1822.
T[roglodytes]. formosus Walden, 1874, Ibis, p. 91. New
name for *Troglodytes punctatus* Blyth, preoccupied.
Elachura laurentei La Touche, 1923, Bull. Brit. Orn. Club,
43, p. 172 — Mahuangpu, southeastern Yunnan.
Sikkim and West Bengal (Darjeeling District); Bhutan,
Assam north of the Brahmaputra (Dafla Hills); western
Burma (Mount Victoria); southeastern Yunnan; northwestern Fukien.

SPELAEORNIS CHOCOLATINUS

Spelaeornis chocolatinus chocolatinus (Godwin-Austen and
Walden)

Pnoepyga chocolatina Godwin-Austen and Walden, 1875,
Ibis, p. 252 — Kedimai, Manipur.
Elachura haplonota Baker, 1892, Ibis, p. 62, pl. 2, fig. 1 —
Hangrum Peak, North Cachar Hills, Assam.
Elachura immaculata Baker, 1893, Journ. Bombay Nat.
Hist. Soc., 7 (1892), pl. opp. p. 319 — North Cachar,
Assam. *Lapsus.*
Spelaeornis chocolatinus nagaensis Ripley, 1951, Postilla,
Yale Univ., no. 6, p. 4 — Mount Japvo, Naga Hills, Assam.

Hill tracts of Assam south of the Brahmaputra (North Cachar and Naga Hills) and Manipur.

Spelaeornis chocolatinus oatesi (Rippon)

Urocichla oatesi Rippon, 1904, Bull. Brit. Orn. Club, **14**, p. 83 — Mount Victoria, Southern Chin Hills, Upper Burma.

Known only from type locality.

Spelaeornis chocolatinus reptatus (Bingham)

Urocichla reptata Bingham, 1903, Bull. Brit. Orn. Club, **13**, p. 55 — Loi Pang Nao [lat. 21° 20′ N., long. 100° 20′ E.], Southern Shan State, Upper Burma.

Urocichla kauriensis Harington, 1908, Ann. Mag. Nat. Hist., ser. 8, 2, p. 246 — Watan, Bhamo District, Kachin State, Upper Burma.

Urocichla sinlumensis Harington, 1908, Ann. Mag. Nat. Hist., ser. 8, 2, p. 246 — Sinlumkaba [lat. 24° 15′ N., long. 97° 31′ E.], Kachin State, Upper Burma.

Northeastern Burma from southern portion of Kachin State southward to eastern portion of Southern Shan State; southwestern Yunnan (west of the Mekong).

Spelaeornis chocolatinus kinneari Delacour and Jabouille

Spelœornis [sic] *longicaudatus kinneari* Delacour and Jabouille, 1930, Oiseau Rev. Franç. Orn., **11**, p. 403 — Chapa [lat. 22° 20′ N., long. 103° 50′ E.], Tongking.

Northwestern Tongking.

SPELAEORNIS LONGICAUDATUS

Spelaeornis longicaudatus (Moore)

Pnoëpyga longicaudata Moore, 1854, in Horsfield and Moore, Cat. Birds Mus. East India Co., 1, p. 398 — "Afghanistan," error; corrected to northern India, by Moore, 1855, Proc. Zool. Soc. London, pt. 22 (1854), p. 74, and restricted to Khasi Hills, Assam, by Sharpe, 1882, Cat. Birds Brit. Mus., 6 (1881), p. 263.

Hill tracts of Assam (south of the Brahmaputra) and Manipur.

GENUS **SPHENOCICHLA** GODWIN-AUSTEN AND WALDEN

Sphenocichla Godwin-Austen and Walden, 1875, Ibis, p. 250. Type, by monotypy, *Sphenocichla roberti* Godwin-Austen and Walden.

SPHENOCICHLA HUMEI

Sphenocichla humei humei (Mandelli)

Heterorhynchus Humei Mandelli, 1873, Stray Feathers, 1, p. 415 — Sikkim.

Himalayas from Sikkim eastward into Assam (north of the Brahmaputra).

Sphenocichla humei roberti Godwin-Austen and Walden

Sphenocichla roberti Godwin-Austen and Walden, 1875, Ibis, p. 251 — North Cachar and Manipur Hills.

Hill tracts of Assam south of the Brahmaputra and northern Burma.

GENUS **NEOMIXIS** SHARPE

Eroessa Hartlaub, 1866, Proc. Zool. Soc. London, p. 218. Type, by monotypy, *Eroessa tenella* Hartlaub. Not *Eroessa* Doubleday, 1847, Lepidoptera.

Damia . . . "Pollen, *in litteris*" Schlegel, 1867, Proc. Zool. Soc. London, (1866), p. 422. Type, by monotypy, *Damia pusilla* "Pollen" Schlegel. Not *Damia* Lacordaire, 1848, Coleoptera.

Neomixis Sharpe, 1881, Proc. Zool. Soc. London, p. 195. Type, by original designation and monotypy, *Neomixis striatigula* Sharpe.

Dauria "Pollen" Oberholser, 1899, Proc. Acad. Nat. Sci. Philadelphia, p. 211. *Lapsus.*

Hartertula Stresemann, 1925, Orn. Monatsb., 33, p. 186. Type, by original designation and monotypy, *Neomixis flavoviridis* Hartert.

NEOMIXIS TENELLA

Neomixis tenella tenella (Hartlaub)

Eroessa tenella Hartlaub, 1866, Proc. Zool. Soc. London, p. 219, text figs. — Madagascar; inferentially restricted to northeastern Madagascar, by Milne Edwards and

Grandidier, 1882, Histoire Physique, Naturelle Poli-
tique Madagascar, 12 (1879), p. 322.
Damia pusilla "Pollen, *in litteris*" Schlegel, 1867, Proc.
Zool. Soc. London, (1866), p. 422. In synonymy with
Eroessa tenella Hartlaub.
Northern Savanna, northern portion of the Western
Savanna, and northern portion of the Humid East district,
Madagascar.

Neomixis tenella decaryi Delacour
Neomixis tenella decaryi Delacour, 1931, Oiseau Rev.
Franç. Orn., nouv. sér., 1, p. 482 — Tsiandro, Anka-
vandra, Ankoja, and Tsiroanomandidy, west-central
Madagascar; restricted to Tsiandro, by Salomonsen,
1934, Ann. Mag. Nat. Hist., ser. 10, 14, p. 63.
Southern portion of the Western Savanna, Madagascar.

Neomixis tenella orientalis Delacour
Neomixis tenella orientalis Delacour, 1931, Oiseau Rev.
Franç. Orn., nouv. sér., 1, p. 482 — Fanovana, east-
central Madagascar.
Central and southern portions of the Humid East district,
Madagascar.

Neomixis tenella debilis Delacour
Neomixis tenella debilis Delacour, 1931, Oiseau Rev.
Franç. Orn., nouv. sér., 1, p. 482 — Tabiky, southwest-
ern Madagascar.
Arid plains of southwestern Madagascar.

NEOMIXIS VIRIDIS

Neomixis viridis delacouri Salomonsen
Neomixis viridis delacouri Salomonsen, 1934, Ann. Mag.
Nat. Hist., ser. 10, 14, p. 63 — Maroantsetra, north-
eastern Madagascar.
Northern portion of the Humid East district, Madagascar.

Neomixis viridis viridis (Sharpe)
Eroessa viridis Sharpe, 1883, Cat. Birds Brit. Mus., 7, pp.
· 151 (in key), 152 — Ankafana forest, Madagascar.
Southern portion of the Humid East district, Madagascar.

NEOMIXIS STRIATIGULA

Neomixis striatigula sclateri Delacour

Neomixis striatigula sclateri Delacour, 1931, Oiseau Rev. Franç. Orn., nouv. sér., 1, p. 480 — Fanovana, eastern Madagascar.

Northern and central portions of the Humid East district, Madagascar.

Neomixis striatigula striatigula Sharpe

Neomixis striatigula Sharpe, 1881, Proc. Zool. Soc. London, p. 195, pl. 19 — Fianarantsoa, southeastern Madagascar.

Eroessa tenella, var. *major* Grandidier, 1882, in Milne Edwards and Grandidier, Histoire Physique, Naturelle Politique Madagascar, 12 (1879), p. 323, pls. 113 A and 113 B — Fianarantsoa, southeastern Madagascar.

Southern portion of the Humid East district, Madagascar.

Neomixis striatigula pallidior Salomonsen

Neomixis striatigula pallidior Salomonsen, 1934, Ann. Mag. Nat. Hist., ser. 10, 14, p. 65 — Ampotaka, southwestern Madagascar.

Arid plains of southwestern Madagascar.

NEOMIXIS FLAVOVIRIDIS

Neomixis flavoviridis Hartert

Neomixis flavoviridis Hartert, 1924, Bull. Brit. Orn. Club, 45, p. 35 — Analamazastra, Madagascar.

Central and southern portions of the Humid East district, from Vondrozo to Sianaka, Madagascar.

GENUS **STACHYRIS** HODGSON

Stachyris Hodgson, 1844, in Blyth, Journ. Asiat. Soc. Bengal, 13, p. 379, footnote. Type, by original designation and monotypy, *Stachyris nigriceps* Blyth.

Strachyris J. E. Gray, 1846, Cat. Mamm. Birds Nepal Thibet, pp. 74, 153. *Lapsus.*

Stachyrhis Agassiz, 1846, Nomencl. Zool. Index Univ. *Nomen emendatum.*

Cyanoderma Salvadori, 1874, Ann. Mus. Civ. Genova, 5,

p. 213. Type, by original designation, *Timalia erythrop-tera* Blyth.

Dasycrotapha Tweeddale, 1878, Proc. Zool. Soc. London, p. 114, pl. 9. Type, by original designation and mono-typy, *Dasycrotapha speciosa* Tweeddale.

Dasycrotopha Sharpe, 1883, Cat. Birds Brit. Mus., 7, pp. 506 (in key), 574. *Lapsus.*

Stachyridopsis "Sharpe" Oates, 1883, Handb. Birds Brit. Burmah, 1, p. 52. Type, by subsequent designation (Sharpe, 1883, Cat. Birds Brit. Mus., 7, p. 597), *Stachyris pyrrhops* Blyth.

Thringorhina Oates, 1889, Fauna Brit. India, Birds, 1, p. 155. Type, by original designation, *Turdinus guttatus* Blyth.

Stachyrhidopsis Oates, 1889, Fauna Brit. India, Birds, 1, p. 164. *Nomen emendatum.*

Zosterornis Ogilvie-Grant, 1894, Bull. Brit. Orn. Club, 3, p. 50. Type, by original designation and monotypy, *Zosterornis whiteheadi* Ogilvie-Grant.

Sterrhoptilus Oberholser, 1918, Journ. Washington Acad. Sci., 8, p. 394. Type, by original designation, *Mixornis* (?) *capitalis* Tweeddale.

Nigravis Baker, 1920, Bull. Brit. Orn. Club, 41, p. 10; *idem*, 1921, *ibid.*, 41, p. 110. Type, by monotypy, *Nigravis herberti* Baker.

Borisia Hachisuka, 1935, Birds Philippine Is., pt. 4, pp. 398 (in key), 416. Type, by original designation, *Zosterornis dennistouni* Ogilvie-Grant.

STACHYRIS RODOLPHEI

Stachyris rodolphei Deignan

Stachyris rodolphei Deignan, 1939, Field Mus. Nat. Hist. Publ., Zool. Ser., 24, p. 110 — Doi Luang Chiang Dao [lat. 19° 25′ N., long. 98° 55′ E.], Thailand.

Known only from type locality.

STACHYRIS RUFIFRONS

Stachyris rufifrons pallescens (Ticehurst)

Stachyridopsis rufifrons pallescens Ticehurst, 1932, Bull. Brit. Orn. Club, 53, p. 18 — Gamon Chaung, Sandoway District, Arakan.

Hill tracts of southwestern Burma (Chin Hills and the Arakan Yoma).

Stachyris rufifrons rufifrons Hume

Stachyris rufifrons Hume, 1873, Stray Feathers, 1, p. 479 — western slopes of the Pegu Hills, Lower Burma.

Southeastern Burma from Shan States southward, through Karenni, into Tenasserim; western Thailand (Chiang Mai and Kanchanaburi Provinces).

Stachyris rufifrons obscura (Baker)

Stachyridopsis rufifrons obscura Baker, 1917, Bull. Brit. Orn. Club, 38, p. 10 — Khlong Bang Lai, a stream near Sathani Map Ammarit [lat. 10° 50′ N., long. 99° 20′ E.], Thailand.

Mergui District, Tenasserim (?); central portion of peninsular Thailand (Chumphon and Surat Thani Provinces).

Stachyris rufifrons poliogaster Hume

Stachyris poliogaster Hume, 1880, Stray Feathers, 9, p. 116 — Gunong Pulai [lat. 1° 36′ N., long. 103° 33′ E.], Johore.

Western portion of Malaya from southern Perak southward to Johore; Sumatra.

? Stachyris rufifrons sarawacensis (Chasen)

Cyanoderma rufifrons sarawacensis Chasen, 1939, Treubia, 17, p. 205 — Mount Poi, western Sarawak.

Borneo; doubtfully distinct from *poliogaster*.

STACHYRIS AMBIGUA

Stachyris ambigua ambigua (Harington)

Stachyrhidopsis rufifrons ambigua Harington, 1915, Journ. Bombay Nat. Hist. Soc., 23, pp. 628 (in key), 631 — Gunjong, Barail Range, Assam.

Himalayas from Sikkim eastward, through Bhutan, into hill tracts of Assam, both north and south of the Brahmaputra.

Stachyris ambigua planicola Mayr

Stachyris ruficeps planicola Mayr, 1941, in Stanford and Mayr, Ibis, p. 70 — Shingaw [lat. 25° 39′ N., long. 97° 53′ E.], Upper Burma.

Northeastern Burma (Kachin State).

Stachyris ambigua adjuncta Deignan

Stachyris rufifrons adjuncta Deignan, 1939, Field Mus.
Nat. Hist. Publ., Zool. Ser., 24, p. 110 — Phong Saly
[lat. 21° 42′ N., long. 102° 07′ E.], Laos.

Northern Thailand and northwestern portion of the eastern plateau of Thailand; northern Laos and northwestern
Tongking.

Stachyris ambigua insuspecta Deignan

Stachyris rufifrons insuspecta Deignan, 1939, Field Mus.
Nat. Hist. Publ., Zool. Ser., 24, p. 111 — Thateng [lat.
15° 31′ N., long. 106° 22′ E.], Laos.

Southern Laos (Boloven Plateau).

STACHYRIS RUFICEPS

Stachyris ruficeps ruficeps Blyth

Stachyris ruficeps Blyth, 1847, Journ. Asiat. Soc. Bengal,
16, p. 452 — Darjeeling, West Bengal.

Along Himalayas from Sikkim eastward, through Bhutan, into Assam (north of the Brahmaputra).

Stachyris ruficeps rufipectus Koelz

Stachyris ruficeps rufipectus Koelz, 1954, Contrib. Inst.
Regional Explor., no. 1, p. 6 — Kohima, Naga Hills,
Assam.

Hill tracts of northeastern Assam from the Mishmis
southward to Manipur; northwestern Burma (Upper Chindwin District).

Stachyris ruficeps bhamoensis (Harington)

Stachyridopsis bhamoensis Harington, 1908, Ann. Mag.
Nat. Hist., ser. 8, 2, p. 245 — Sinlumkaba [lat. 24° 15′
N., long. 97° 31′ E.], Upper Burma.

Northeastern Burma (Kachin and Northern Shan States)
and northwestern Yunnan.

Stachyris ruficeps davidi (Oustalet)

Stachyridiopsis [sic] *Davidi* Oustalet, 1899, Bull. Mus.
Hist. Nat. Paris, 5, p. 119 — western Szechwan.
S[tachyridopsis]. sinensis Ogilvie-Grant and La Touche,
1907, Ibis, p. 184 — China; here restricted to Fukien.
Stachyridopsis ruficeps bangsi La Touche, 1923, Bull.

Brit. Orn. Club, 44, p. 32 — Milati, southeastern Yunnan.

Along Yangtze Valley from Szechwan eastward to Anhwei, thence southward and westward through coastal provinces to southeastern Yunnan; Tongking, northern Annam, and eastern portion of northern Laos.

Stachyris ruficeps praecognita Swinhoe

Stachyrhis [sic] præcognitus [sic] Swinhoe, 1866, Ibis, p. 310 — Formosa.

Formosa.

Stachyris ruficeps goodsoni (Rothschild)

Stachyridopsis ruficeps goodsoni Rothschild, 1930, Bull. Brit. Orn. Club, 14, p. 8 — Mount Wuchi, Hainan.

Hainan.

Stachyris ruficeps pagana (Riley)

Stachyridopsis ruficeps paganus Riley, 1940, Proc. Biol. Soc. Washington, 53, p. 132 — Fimnom [lat. 11° 47' N., long. 108° 24' E.], Annam.

Southern Annam.

STACHYRIS PYRRHOPS

Stachyris pyrrhops Blyth

St[achyris]. pyrrhops Blyth, 1844, Journ. Asiat. Soc. Bengal, 13, p. 379 — Nepal.

Stachyris pyrrhops ochrops Koelz, 1954, Contrib. Inst. Regional Explor., no. 1, p. 6 — Kotla, Kangra, East Punjab.

Himalayas from West Pakistan (Rawalpindi District) eastward to central Nepal.

STACHYRIS CHRYSAEA

Stachyris chrysaea chrysaea Blyth

St[achyris]. chrysæa Blyth, 1844, Journ. Asiat. Soc. Bengal, 13, p. 379 — Nepal.

Stachyris chrysaea chrysocoma Koelz, 1954, Contrib. Inst. Regional Explor., no. 1, p. 5 — Karong, Manipur.

Himalayas from Nepal eastward, through Sikkim, Bhutan, Assam (both north and south of the Brahmaputra, except area occupied by *binghami*), and northern Burma, to western Yunnan (west of the Salween).

Stachyris chrysaea binghami Rippon

> *Stachyris binghami* Rippon, 1904, Bull. Brit. Orn. Club, 14, p. 84 — Mount Victoria [lat. 21° 15′ N., long. 93° 55′ E.], Upper Burma.
>
> *Stachyris chrysaea crocina* Koelz, 1954, Contrib. Inst. Regional Explor., no. 1, p. 6 — Sangau, Lushai Hills, Assam.

Southeastern Assam (Lushai Hills) and southwestern Burma (Chin Hills and Arakan Yoma).

Stachyris chrysaea aurata de Schauensee

> *Stachyris chrysaea aurata* de Schauensee, 1938, Proc. Acad. Nat. Sci. Philadelphia, 90, p. 29 — Doi Pha Hom Pok [lat. 20° 05′ N., long. 99° 10′ E.], Thailand.

Southern Shan State (east of the Salween), northernmost Thailand, northern Laos, northern half of Annam, and Tongking.

Stachyris chrysaea assimilis (Walden)

> S[*trachyrhis*]. *assmilis* [sic] Walden, 1875, in Blyth, Journ. Asiat. Soc. Bengal, 43, extra no., p. 116 — Karenni State, Upper Burma.

Southern Shan State (west of the Salween) southward, through Karenni and hills of northwestern Thailand, to central Tenasserim.

Stachyris chrysaea chrysops Richmond

> *Stachyris chrysops* Richmond, 1902, Proc. Biol. Soc. Washington, 15, p. 157 — Khao Nam Pliu [lat. 7° 35′ N., long. 99° 50′ E.], Thailand.

Hills of Malay Peninsula from southern Tenasserim and Isthmus of Kra southward to southern Selangor and Pahang.

Stachyris chrysaea frigida (Hartlaub)

> H[*eleia*]. *frigida* "Müll." Hartlaub, 1865, Journ. f. Orn., 13, p. 27 — Sumatra.
>
> *Stachyris bocagei* Salvadori, 1879, Ann. Mus. Civ. Genova, 14, p. 223 — Mount Singgalang [lat. 0° 24′ S., long. 100° 20′ E.], Sumatra.

Highlands of western Sumatra.

STACHYRIS PLATENI

Stachyris plateni pygmaea (Ogilvie-Grant)

Mixornis plateni "Blas." Bourns and Worcester, 1894, Occ. Papers Minnesota Acad. Nat. Sci., 1, p. 58 — Samar, Philippine Islands. Not *Mixornis plateni* Blasius, 1890.

Zosterornis pygmæus Ogilvie-Grant, 1896, Bull. Brit. Orn. Club, 6, p. 18 — Samar, Philippine Islands.

Samar and Leyte, Philippine Islands.

Stachyris plateni plateni (Blasius)

Mixornis Plateni Blasius, 1890 (Apr. 15), Braunschweigische Anzeigen, no. 87, p. 877; *idem*, 1890 (Oct.), Journ. f. Orn., 38, p. 147 — Mindanao, Philippine Islands; type from near Dávao, *fide* Hachisuka, 1935, Birds Philippine Is., pt. 4, p. 420.

Mindanao, Philippine Islands.

STACHYRIS CAPITALIS

Stachyris (capitalis ?) dennistouni (Ogilvie-Grant)

Zosterornis dennistouni Ogilvie-Grant, 1895, Bull. Brit. Orn. Club, 5, p. 2 — Cape Engaño, northeastern Luzon, Philippine Islands.

Northeastern Luzon (Cagayan and Isabela Provinces), Philippine Islands.

Stachyris capitalis affinis (McGregor)

Zosterornis affinis McGregor, 1907, Philippine Journ. Sci., 2, p. 292 — Lamao, Bataan Province, Luzon, Philippine Islands.

Southern Luzon, Philippine Islands.

Stachyris capitalis nigrocapitata (Steere)

Mixornis nigrocapitatus Steere, 1890, List Birds Mammals Steere Exped., p. 17 — Samar and Leyte, Philippine Islands.

Samar and northern Leyte, Philippine Islands.

Stachyris capitalis boholensis Rand and Rabor

Stachyris nigrocapitata boholensis Rand and Rabor, 1957, Fieldiana: Zool. [Chicago], 42, p. 14 — Cantaub, Sierra Bullones, Bohol, Philippine Islands.

Bohol, Philippine Islands.

Stachyris capitalis capitalis (Tweeddale)

Mixornis (?) *capitalis* Tweeddale, 1877, Ann. Mag. Nat. Hist., ser. 4, 20, p. 535 — Dinagat, Philippine Islands.

Panaon, Dinagat, Mindanao, and Basilan, Philippine Islands.

STACHYRIS SPECIOSA

Stachyris speciosa (Tweeddale)

Dasycrotapha speciosa Tweeddale, 1878, Proc. Zool. Soc. London, p. 114, pl. 9 — Valencia, Negros, Philippine Islands.

Lowlands of Negros, Philippine Islands.

STACHYRIS WHITEHEADI

Stachyris whiteheadi (Ogilvie-Grant)

Zosterornis whiteheadi Ogilvie-Grant, 1894, Bull. Brit. Orn. Club, 3, p. 50 — mountains of northern Luzon, Philippine Islands.

Highlands of northern Luzon, and (subspecies ?) extreme southern Luzon, Philippine Islands.

STACHYRIS STRIATA

Stachyris striata (Ogilvie-Grant)

Zosterornis striatus Ogilvie-Grant, 1894, Bull. Brit. Orn. Club, 4, p. ii — Isabela Province, Luzon, Philippine Islands.

Highlands of northern Luzon, Philippine Islands.

STACHYRIS NIGRORUM[1]

Stachyris nigrorum Rand and Rabor

Stachyris nigrorum Rand and Rabor, 1952, Nat. Hist. Misc. [Chicago], no. 100, p. [1] — Cuernos de Negros, Negros Oriental Province, Negros, Philippine Islands.

Highlands of Negros, Philippine Islands.

STACHYRIS HYPOGRAMMICA

Stachyris hypogrammica Salomonsen

Stachyris hypogrammica Salomonsen, 1961, Dansk Orn.

[1] *Stachyris nigrorum* and *striata* comprise a superspecies.

For. Tidsskr., 55, p. 219 — Mount Mataling, Mantalingajan Range, Palawan, Philippine Islands.
Highlands of Palawan, Philippine Islands.

STACHYRIS GRAMMICEPS

Stachyris grammiceps (Temminck)

Myiothera grammiceps Temminck, 1827, Pl. col., livr. 74, pl. 448, fig. 3 — Java.
T[imalia]. grammicephala Bonaparte (ex Kuhl MS), 1850, Consp. Av., 1, p. 217 — Java.
Western Java.

STACHYRIS HERBERTI

Stachyris herberti (Baker)

Nigravis herberti Baker, 1920, Bull. Brit. Orn. Club, 41, p. 10 — "Ban Sao, Siam," error [= Ban Lak Sao (lat. 18° 11′ N., long. 104° 58′ E.), Laos].
Known only from type locality.

STACHYRIS NIGRICEPS

Stachyris nigriceps nigriceps Blyth

St[achyris]. nigriceps Blyth (ex Hodgson MS), Journ. Asiat. Soc. Bengal, 13, p. 378 — Nepal.
Himalayas from central Nepal, through Sikkim and Bhutan, to Assam (Abor Hills).

Stachyris nigriceps coei Ripley

Stachyris nigriceps coei Ripley, 1952, Postilla, Yale Univ., no. 14, p. [2] — Dreyi, Mishmi Hills, Assam.
Eastern Assam (Mishmi Hills).

Stachyris nigriceps coltarti Harington

Stachyris nigriceps coltarti Harington, 1913, Bull. Brit. Orn. Club, 33, p. 61 — Margherita, eastern Assam.
Eastern Assam (Naga Hills) and northern Burma.

Stachyris nigriceps spadix Ripley

Stachyris nigriceps spadix Ripley, 1948, Bull. Brit. Orn. Club, 68, p. 89 — Laisung, Cachar, Assam.
Stachyris nigriceps ravida Koelz, 1954, Contrib. Inst.

Regional Explor., no. 1, p. 5 — Sangau, Lushai Hills, Assam.

Hill tracts of Assam south of the Brahmaputra and of southern Burma (except Shan State) south and east to northern Tenasserim; northwestern Thailand.

Stachyris nigriceps yunnanensis La Touche

Stachyris nigriceps yunnanensis La Touche, 1921, Bull. Brit. Orn. Club, 42, p. 18 — Hokow [lat. 23° 15′ N., long. 103° 39′ E.], Yunnan.

Eastern Burma (Shan State); eastern portion of northern plateau of Thailand; northern Laos; northern Annam; western Tongking; southwestern Yunnan.

Stachyris nigriceps rileyi Chasen

Stachyris nigriceps dilutus [sic] Robinson and Kloss, 1919, Ibis, p. 584 — Dran [lat. 11° 49′ N., long. 108° 38′ E.], Annam. Not *Stachyris poliocephala diluta* Robinson and Kloss, 1918.

Stachyris nigriceps rileyi Chasen, 1936, Bull. Brit. Orn. Club, 56, p. 115. New name for *Stachyris nigriceps diluta* Robinson and Kloss, preoccupied.

Southern half of Annam.

Stachyris nigriceps dipora Oberholser

Stachyris nigriceps dipora Oberholser, 1922, Smiths. Misc. Coll., 74 (2), p. 7 — Khao Soi Dao [lat. 7° 20′ N., long. 99° 50′ E.], Thailand.

Malay Peninsula from southern Tenasserim (Mergui District) and Isthmus of Kra southward to Siamese province of Trang.

Stachyris nigriceps davisoni Sharpe

Stachyris davisoni Sharpe, 1892, Bull. Brit. Orn. Club, 1, p. 7 — Gunong Tahan, Pahang.

Malay Peninsula from Siamese province of Pattani southward to Negri Sembilan.

Stachyris nigriceps larvata (Bonaparte)

T[imalia]. larvata "Müll. Mus. Lugd." Bonaparte, 1850, Consp. Av., 1, p. 217 — Sumatra.

Sumatra and Lingga Archipelago.

? Stachyris nigriceps tionis Robinson and Kloss

Stachyris nigriceps tionis Robinson and Kloss, 1927,
Journ. Federated Malay States Mus., 13, p. 211 —
Juara Bay, Pulau Tioman, Johore-Pahang Archipelago.
Pulau Tioman; doubtfully distinct from *larvata*.

Stachyris nigriceps natunensis Hartert

Stachyris natunensis Hartert, 1894, Novit. Zool., 1, p. 470
— Pulau Bunguran, North Natuna Islands.
North Natuna Islands.

Stachyris nigriceps hartleyi Chasen

Stachyris nigriceps hartleyi Chasen, 1935, Bull. Raffles
Mus., no. 10, p. 43 — Mount Poi, Sarawak.
Highlands of western Sarawak (Mounts Poi and Penrissen).

Stachyris nigriceps borneensis Sharpe

Stachyris borneensis Sharpe, 1887, Ibis, p. 449 — Kina
Balu, North Borneo.
Stachyris larvata vermiculata Harrisson and Hartley,
1934, Bull. Brit. Orn. Club, 54, p. 153 — Mount Dulit,
Sarawak.
Borneo (except area occupied by *hartleyi*).

STACHYRIS POLIOCEPHALA

Stachyris poliocephala (Temminck)

Timalia poliocephala Temminck, 1836, Pl. col., livr. 100,
pl. 593, fig. 2 — Sumatra and Borneo; restricted to
Benkulen [lat. 3° 47′ S., long. 102° 15′ E.], Sumatra,
by Kloss, 1931, Treubia, 13, p. 348.
Stachyris poliocephala diluta Robinson and Kloss, 1918,
Ibis, p. 587 — Taiping, Perak.
Stachyris poliocephala pulla Kloss, 1931, Treubia, 13, p.
348 — Tuntungan, Deli District, northeastern Sumatra.
Malay Peninsula southward from Isthmus of Kra; Sumatra; Borneo.

STACHYRIS STRIOLATA

Stachyris striolata swinhoei Rothschild

Stachyris guttata swinhoei Rothschild, 1903, Bull. Brit.
Orn. Club, 14, p. 8 — Mount Wuchi, Hainan.
Hainan.

Stachyris striolata tonkinensis Kinnear

> *Thringorhina guttata diluta* Kinnear, 1924, Bull. Brit. Orn. Club, 45, p. 11 — Thai Nien [lat. 22° 23′ N., long. 104° 06′ E.], Tongking. Not *Stachyris poliocephala diluta* Robinson and Kloss, 1918, nor *Stachyris nigriceps diluta* Robinson and Kloss, 1919.
> *Thringorhina guttata sinensis* Stresemann, 1929, Orn. Monatsb., 37, p. 141 — Yao Shan, Kwangsi. Not *Stachyridopsis sinensis* Ogilvie-Grant and La Touche, 1907 = *Stachyris ruficeps davidi* (Oustalet).
> *Stachyris guttata tonkinensis* Kinnear, 1938, Bull. Brit. Orn. Club, 58, p. 82. New name for *Thringorhina guttata diluta* Kinnear, preoccupied.

Kwangsi, eastern Tongking, and northern Annam.

Stachyris striolata helenae Delacour and Greenway

> *Stachyris striolata helenæ* Delacour and Greenway, 1939, Bull. Brit. Orn. Club, 59, p. 130 — Ban Nam Khuang [lat. 20° 24′ N., long. 100° 14′ E.], Laos.

Northern Laos and eastern portion of northern plateau of Thailand (Nan Province).

Stachyris striolata guttata (Blyth)

> *Turdinus guttatus* "Tickell" Blyth, 1859, Journ. Asiat. Soc. Bengal, 28, p. 414 — Tenasserim; type from "Woods near Theethoungplee. 3000 ft.," Amherst District, *fide* Tickell, 1860, *ibid.*, 28, no. 5 (1859), p. 450.

Central Tenasserim and adjacent areas of western Thailand (Tak Province).

Stachyris striolata nigrescentior Deignan

> *Stachyris striolata nigrescentior* Deignan, 1947, Journ. Washington Acad. Sci., 37, p. 104 — Khao Nok Ra [lat. 7° 25′ N., long. 99° 55′ E.], Thailand.

Peninsular Thailand from Isthmus of Kra southward to Trang Province.

Stachyris striolata umbrosa (Kloss)

> *Thringorhina striolata umbrosa* Kloss, 1921, Journ. Federated Malay States Mus., 10, p. 212 — Bandar Bahru, Upper Deli District, northeastern Sumatra.

Northeastern Sumatra.

Stachyris striolata striolata (Müller)

Timalia strialata [sic] S. Müller, 1835, Tijdschr. Natuur. Gesch. Phys., 2, p. 345 — Sumatra; restricted to Padang Residencies, by Kloss, 1921, Journ. Federated Malay States Mus., 10, p. 213.

Highlands of western Sumatra.

STACHYRIS OGLEI[1]

Stachyris oglei (Godwin-Austen)

Actinura Oglei Godwin-Austen, 1877, Journ. Asiat. Soc. Bengal, 46, p. 42 — near Sadiya, Assam.

Eastern Assam.

STACHYRIS MACULATA

Stachyris maculata maculata (Temminck)

Timalia maculata Temminck, 1836, Pl. col., livr. 100, pl. 593, fig. 1 — Borneo and Sumatra; restricted to Borneo, by Robinson and Kloss, 1918, Ibis, p. 588.

Malay Peninsula from southern Thailand (Krabi Province) southward to Johore; Riouw Archipelago; Sumatra; Borneo.

Stachyris maculata banjakensis Richmond

Stachyris banjakensis Richmond, 1902, Proc. Biol. Soc. Washington, 15, p. 190 — Pulau Tuangku, Banyak Islands.

Known only from type locality.

Stachyris maculata hypopyrrha Oberholser

Stachyris maculata hypopyrrha Oberholser, 1912, Smiths. Misc. Coll., 60 (7), p. 9 — Pulau Pini, Batu Islands.

Pulau Pini and Pulau Tanah Masa, Batu Islands.

STACHYRIS LEUCOTIS

Stachyris leucotis leucotis (Strickland)

Timalia leucotis Strickland, 1848, in Jardine, Contr. Orn., p. 63, pl. 12 — Malacca.

Lowland forests of Malay Peninsula from Siamese province of Trang southward to Johore.

---

[1] *Stachyris oglei* and *striolata* comprise a superspecies.

Stachyris leucotis sumatrensis Chasen

Stachyris leucotis sumatrensis Chasen, 1939, Treubia, 17, p. 184 — between Lesten [lat. 4° 12′ N., long. 97° 40′ E.] and Pendeng [lat. 4° 08′ N., long. 97° 36′ E.], Sumatra.

Known only from type locality.

Stachyris leucotis obscurata Mayr

Stachyris leucotis goodsoni Hartert, 1915, Bull. Brit. Orn. Club, 36, p. 7 — Gunong Mulu [lat. 4° 04′ N., long. 114° 57′ E.], Sarawak. Not *Stachyridopsis ruficeps goodsoni* Rothschild, 1903 = *Stachyris ruficeps goodsoni* (Rothschild).

Stachyris leucotis obscurata Mayr, 1942, Auk, 59, p. 117. New name for *Stachyris leucotis goodsoni* Hartert, preoccupied.

Borneo.

STACHYRIS NIGRICOLLIS

Stachyris nigricollis (Temminck)

Timalia nigricollis Temminck, 1836, Pl. col., livr. 100, pl. 594, fig. 2 — Borneo.

Malay Peninsula from southern Thailand (Nakhon Si Thamarat Province) southward to Johore; eastern Sumatra; Borneo.

STACHYRIS THORACICA

Stachyris thoracica thoracica (Temminck)

Pitta thoracica Temminck, 1821, Pl. col., livr. 13, pl. 76 — Java; here restricted to Mount Salak [lat. 6° 42′ S., long. 106° 44′ E.].

Southern Sumatra; western and central Java.

Stachyris thoracica orientalis Robinson

Stachyris orientalis Robinson, 1918, Journ. Federated Malay States, 7, p. 236 — Sodong Gerok, near Banyuwangi [lat. 8° 12′ S., long. 114° 23′ E.], Java.

Eastern Java.

STACHYRIS ERYTHROPTERA

Stachyris erythroptera erythroptera (Blyth)

T[*imalia*]. *erythroptera* Blyth, 1842, Journ. Asiat. Soc. Bengal, 11, p. 794 — Singapore.

Cyanoderma erythropterum sordida [sic] Baker, 1917, Bull. Brit. Orn. Club, 38, p. 10 — Khlong Wang Hip, a stream near Sathani Thung Song [lat. 8° 10' N., long. 99° 40' E.], and Sathani Map Ammarit [lat. 10° 50' N., long. 99° 20' E.], Thailand.

Cyanoderma erythroptera [sic] *neocara* [sic] Oberholser, 1932, Bull. U. S. Nat. Mus., 159, pp. 6 (in list), 66 — Pulau Bunguran, Northern Natuna Islands.

Malay Peninsula from southern Tenasserim (Mergui District) and Isthmus of Kra southward to Singapore; North Natuna Islands.

Stachyris erythroptera pyrrhophaea (Hartlaub)

Timalia pyrrhophæa Hartlaub, 1844, Rev. Zool. [Paris], 7, p. 402 — Malacca and Sumatra; inferentially restricted to Sumatra, by Robinson and Kloss, 1920, Journ. Straits Branch Roy. Asiat. Soc., no. 81, p. 106.

Cyanoderma erythropterum pellum Oberholser, 1912, Smiths. Misc. Coll., 60 (7), p. 9 — Pulau Tanah Masa, Batu Islands.

Cyanoderma erythroptera [sic] *eripella* [sic] Oberholser, 1922, Smiths. Misc. Coll., 74 (2), p. 7 — Upper Sungei Siak [probably at *ca.* lat. 0° 46' N., long. 101° 46' E.], Sumatra.

Cyanoderma erythroptera [sic] *apega* [sic] Oberholser, 1922, Smiths. Misc. Coll., 74 (2), p. 8 — Tanjong Tedong, Bangka.

Sumatra; Batu Islands; Bangka; Billiton; birds from Pulau Tioman may be provisionally referred to this race.

Stachyris erythroptera fulviventris (Richmond)

Cyanoderma fulviventris Richmond, 1903, Proc. U. S. Nat. Mus., 26, p. 507 — Pulau Tuangku, Banyak Islands.
Known only from type locality.

Stachyris erythroptera bicolor (Blyth)

T[imalia]. bicolor Blyth, 1865, Ibis, p. 46 — no provenience, but Labuan inferentially suggested as type locality by Collin and Hartert, 1927 (*vide infra*).

Cyanoderma labuanensis [sic] Collin and Hartert, 1927, Novit. Zool., 34, p. 51. New name for *T[imalia]. bicolor*

Blyth, erroneously considered preoccupied by "Tim-[alie]. bicolor" Lafresnaye, 1835 (French vernacular). Northern and eastern Borneo; Banggai.

Stachyris erythroptera rufa (Chasen and Kloss)

Cyanoderma erythroptera [sic] *rufa* [sic] Chasen and Kloss, 1927, Bull. Brit. Orn. Club, 48, p. 47 — Sampit [lat. 2° 32′ S., long. 112° 59′ E.], Borneo.
Southwestern Borneo.

STACHYRIS MELANOTHORAX

Stachyris melanothorax melanothorax (Temminck)

Myiothera melanothorax Temminck, 1823, Pl. col., livr. 31, pl. 185, fig. 2 — Java; inferentially restricted to Mount Gedeh [lat. 6° 47′ S., long. 106° 59′ E.], by Stresemann, 1930, Orn. Monatsb., 38, p. 148.
Mount Gedeh and Mount Pangerango, western Java.

Stachyris melanothorax albigula (Stresemann)

Cyanoderma melanothorax albigula Stresemann, 1930, Orn. Monatsb., 38, p. 148 — Mount Papandayan [lat. 7° 20′ S., long. 107° 44′ E.], Java.
Known only from type locality.

Stachyris melanothorax mendeni (Neumann)

Cyanoderma melanothorax mendeni Neumann, 1935, Bull. Brit. Orn. Club, 55, p. 136 — Indramayu [lat. 6° 20′ S., long. 108° 19′ E.], Java.
Known only from type locality.

Stachyris melanothorax intermedia (Robinson)

Stachyridopsis melanothorax intermedia Robinson, 1918, Journ. Federated Malay States Mus., 7, p. 236 — Sodong Gerok, near Banyuwangi [lat. 8° 12′ S., long. 114° 23′ E.], Java.
Ijen Mountains, eastern Java.

Stachyris melanothorax baliensis (Hartert)

Cyanoderma melanothorax baliensis Hartert, 1915, Bull. Brit. Orn. Club, 36, p. 2 — Bali.
Lowlands of Bali.

Genus **DUMETIA** Blyth

Dumetia Blyth, 1849, Cat. Birds Mus. Asiat. Soc., p. 140.
Type, by subsequent designation (G. R. Gray, 1855,
Cat. Gen. Subgen. Birds, p. 45), *Timalia hyperythra*
Franklin.

DUMETIA HYPERYTHRA

Dumetia hyperythra hyperythra (Franklin)
Timalia hyperythra Franklin, 1831, Proc. Comm. Zool.
Soc. London, pt. 1 (1830-31), p. 118 — Ganges between
Calcutta and Benares.
Southwestern Nepal; northern and central India from
Uttar Pradesh southward, through Vindhya Pradesh and
western Madhya Pradesh, to southern Hyderabad.

Dumetia hyperythra albogularis (Blyth)
M[alacocercus]. (?) *albogularis* Blyth, 1847, Journ.
Asiat. Soc. Bengal, 16, p. 453 — southern India.
Dumetia albigularis abuensis Harington, 1915, Journ.
Bombay Nat. Hist. Soc., 23, p. 429 — Mount Abu, Ara-
valli Range, Rajasthan, India.
India from southern Rajasthan (Aravalli Range) south-
ward along Western Ghats to Kerala (where approaches
phillipsi), and thence northeastward through Eastern Ghats
to the lower Kistna (where approaches *hyperythra*).

Dumetia hyperythra phillipsi Whistler
Dumetia hyperythra phillipsi Whistler, 1941, Ibis, p. 319
— Kumbalgamna, Ceylon.
Ceylon.

Genus **RHOPOCICHLA** Oates

Rhopocichla Oates, 1889, Fauna British India, 1, pp. 131
(in key), 159. Type, by original designation, *Brachypteryx atriceps* Jerdon.

RHOPOCICHLA ATRICEPS

Rhopocichla atriceps atriceps (Jerdon)
B[rachypteryx]. atriceps Jerdon, 1839, Madras Journ.
Lit. Sci., 10, p. 250 — Trichur, Wadakkancheri, Coo-

noor, and the Wynaad, southwestern India; restricted
to the Wynaad, by Whistler, 1935, in Ali and Whistler,
Journ. Bombay Nat. Hist. Soc., 38, p. 82.
Western Ghats of India from the Belgaum District of
Bombay southward to Nilgiri Hills.

Rhopocichla atriceps bourdilloni (Hume)

Alcippe Bourdilloni Hume, 1876, Stray Feathers, 4, p.
485 — Mynall, Kerala, southwestern India.
Hill tracts of Kerala.

Rhopocichla atriceps siccata Whistler

Rhopocichla atriceps siccatus [sic] Whistler, 1941, Bull.
Brit. Orn. Club, 62, p. 38 — Kalawewa, North Central
Province, Ceylon.
Low-country dry zone of northern and eastern Ceylon
and the central hills.

Rhopocichla atriceps nigrifrons (Blyth)

Alcippe nigrifrons Blyth, 1849, Journ. Asiat. Soc. Bengal,
18, p. 815 — Ceylon; restricted to Uragaha, Southern
Province, by Whistler, 1941, Bull. Brit. Orn. Club, 62,
p. 38.
Low-country wet zone of southwestern Ceylon.

GENUS MACRONOUS JARDINE AND SELBY

Macronous[1] Jardine and Selby, 1835, Illustr. Orn., text to
pl. 150. Type, by monotypy, *Macronous ptilosus* Jardine
and Selby.
Mixornis Blyth, 1842, Journ. Asiat. Soc. Bengal, 11, p.
794, footnote. Type, by original designation, *Timalia
chloris* Blyth = *Macronous gularis rubicapillus* (Tickell).
Minodoria Hachisuka, 1934, Tori, 8, p. 220. Type, by
original designation, *Macronus striaticeps* Sharpe.

cf. Oberholser, 1932, Bull. U. S. Nat. Mus., 159, pp. 68-73
    (races of *gularis,* sub nom. *Mixornis*).
Delacour, 1936, Oiseau Rev. Franç. Orn., nouv. sér., 6,

---

[1] Spelling used by first revisor (Blyth, 1842, Journ. Asiat. Soc. Bengal, 11, p. 795). However, the plate is lettered *Macronus* and the editors would prefer to maintain this more frequently used spelling.—Eds.

pp. 1-27 (races of *gularis* and *flavicollis,* sub nom. *Mixornis*).

Rand and Rabor, 1960, Fieldiana: Zool. [Chicago], 35, pp. 429-430 (races of *striaticeps*).

MACRONOUS FLAVICOLLIS

Macronous flavicollis flavicollis (Bonaparte)

[*Mixornis*] *flavicollis* "Müll. Mus. Lugd." Bonaparte, 1850, Consp. Av., 1, p. 217 — Java.

Java.

Macronous flavicollis prillwitzi (Hartert)

Mixornis prillwitzi Hartert, 1901, Bull. Brit. Orn. Club, 12, p. 32 — Pulau Kangean.

Kangean.

MACRONOUS GULARIS

Macronous gularis rubicapillus (Tickell)

M[*otacilla*]. *Rubicapilla* Tickell, 1833, Journ. Asiat. Soc. Bengal, 2, p. 576 — "jungles of Borabhúm and Dholbhúm" [= Manbhum and Singhbhum Districts, Bihar].

Macronus gularis mayri Koelz, 1951, Journ. Zool. Soc. India, 3, p. 27 — Sukna, Darjeeling, West Bengal.

Macronus gularis assamicus Koelz, 1951, Journ. Zool. Soc. India, 3, p. 27 — Tura, Garo Hills, Assam.

Lowlands of India from eastern Nepal and eastern Madhya Pradesh, eastward through Bengal and Assam, to border of Burma.

Macronous gularis ticehursti (Stresemann)

Mixornis gularis ticehursti Stresemann, 1940, in Stresemann and Heinrich, Mitt. Zool. Mus. Berlin, 24, pp. 153 (in list), 200 — Dudaw Taung [lat. 21° 05′ N., long. 94° 19′ E.], Burma.

Western Burma from Upper Chindwin District southward into Arakan.

Macronous gularis sulphureus (Rippon)

Stachyridopsis sulphurea Rippon, 1900, Bull. Brit. Orn. Club, 11, p. 11 — Nammehet [lat. 20° 26′ N., long. 97° 28′ E.], Southern Shan State.

Mixornis gularis minor Gyldenstolpe, 1916, Kongl. Svensk

Vet.-Akad. Handl., 56, p. 60 — Sathani Pha Kho
[lat. 18° 15′ N., long. 99° 55′ E.], Thailand.

Eastern Burma from Kachin State, southward through
Shan States (excepting Kengtung) and Karenni, to south-
central Tenasserim; northern plateau of Thailand (except
areas occupied by *lutescens*) and western provinces southward to Kanchanaburi.

Macronous gularis lutescens (Delacour)

Mixornis rubricapilla lutescens Delacour, 1926, Bull.
Brit. Orn. Club, 47, p. 18 — Bao Ha [lat. 22° 10′ N.,
long. 104° 21′ E.], Tongking.

Southeastern Yunnan; Tongking; northern Annam; the
whole of Laos; Southern Shan State (Kengtung); northern
Thailand (provinces of the Mekong drainage); eastern
Thailand (provinces bordering the Mekong from Loei to
Ubon).

Macronous gularis kinneari (Delacour and Jabouille)

Mixornis kinneari Delacour and Jabouille, 1924, Bull. Brit.
Orn. Club, 45, p. 33 — Hai Lang [lat. 16° 42′ N., long.
107° 16′ E.], Annam.

Central Annam.

Macronous gularis versuricola (Oberholser)

Mixornis gularis versuricola Oberholser, 1922, Smiths.
Misc. Coll., 74 (2), p. 5 — Da Ban [lat. 12° 38′ N., long.
109° 06′ E.], Annam.

Southern Annam, northern Cochin China, and eastern
Cambodia.

Macronous gularis saraburiensis Deignan

Macronus gularis saraburiensis Deignan, 1956, Proc. Biol.
Soc. Washington, 69, p. 209 — Sathani Hin Lap [lat.
14° 40′ N., long. 101° 10′ E.], Thailand.

Western Cambodia and southwestern portion of eastern
plateau of Thailand.

Macronous gularis connectens (Kloss)

Mixornis rubricapilla connecteus [sic] Kloss, 1918, Ibis,
p. 207 — "about Lat. 10° N. [Malay Peninsula]."

Mixornis gularis deignani de Schauensee, 1946, Proc.

Acad. Nat. Sci. Philadelphia, 98, p. 67 — Khao Luang
[lat. 10° 40′ N., long. 99° 35′ E.], Thailand.
Coastal region along Gulf of Siam from Cambodia to
Isthmus of Kra; southern Tenasserim (Mergui District).

Macronous gularis inveteratus (Oberholser)

Mixornis gularis inveterata Oberholser, 1922, Smiths.
Misc. Coll., 74 (2), p. 5 — Ko Kut [lat. 11° 40′ N., long.
102° 35′ E.], Thailand.
Islets off coasts of southeastern Thailand and Cambodia.

Macronous gularis condorensis (Robinson)

Mixornis rubricapilla condorensis Robinson, 1920, Journ.
Nat. Hist. Soc. Siam, 4, p. 88 — Pulau Kondor, off the
mouths of the Mekong.
Pulau Kondor.

Macronous gularis archipelagicus (Oberholser)

Mixornis gularis archipelagica Oberholser, 1922, Smiths.
Misc. Coll., 74 (2), p. 4 — Domel Island, Mergui Ar-
chipelago, off southern Tenasserim.
Mergui Archipelago.

Macronous gularis chersonesophilus (Oberholser)

Mixornis gularis chersonesophila Oberholser, 1922.
Smiths. Misc. Coll., 74 (2), p. 3 — Trang Province,
Thailand.
Malay Peninsula from vicinity of Isthmus of Kra south-
ward to Perak and Trengganu.

Macronous gularis gularis (Horsfield) 2

Timalia gularis Horsfield, 1822, Zool. Res. in Java, no. 3,
pl. and text — Sumatra.
Mixornis pileata zaptera Oberholser, 1912, Smiths. Misc.
Coll., 60 (7), p. 9 — Pulau Tanah Masa, Batu Islands.
Mixornis pileata zarhabdota Oberholser, 1912, Smiths.
Misc. Coll., 60 (7), p. 9 — Pulau Tuangku, Banyak
Islands.
Malaya (except areas occupied by *chersonesophilus*);
Riouw and Lingga Archipelagos; Sumatra; Banyak and
Batu Islands.

Macronous gularis zopherus (Oberholser)

Mixornis pileata zophera Oberholser, 1917, Bull. U. S.
Nat. Mus., 98, pp. 5 (in list), 49 — Pulau Telaga, Anam-
ba Islands.
Anamba Islands.

Macronous gularis zaperissus (Oberholser)

Mixornis rubicapilla zaperissa Oberholser, 1932, Bull.
U. S. Nat. Mus., 159, pp. 6 (in list), 68 — Pulau Lin-
gung [= Pulau Lagong, North Natuna Islands].
Pulau Laut, Pulau Batang, and Pulau Lagong, North
Natuna Islands.

Macronous gularis everetti (Hartert)

Mixornis everetti Hartert, 1894, Novit. Zool., 1, p. 472 —
Pulau Bunguran, North Natuna Islands.
Pulau Bunguran, North Natuna Islands.

Macronous gularis ruficoma (Oberholser)

Mixornis bornensis ruficoma Oberholser, 1922, Smiths.
Misc. Coll., 74 (2), p. 6 — Tanjong Tedong, Bangka.
Bangka and Billiton.

Macronous gularis javanicus (Cabanis)

Myiothera gularis "Horsf." Temminck, 1827, Pl. col.,
livr. 74, pl. 442, fig. 1, and text — Java and Sumatra;
the figure and description apply to the form of Java.
Not *Timalia gularis* Horsfield, 1822 = *Macronous gularis
gularis* (Horsfield).
Mixornis javanica Cabanis, 1851, Mus. Hein., 1 (1850),
p. 77, footnote. New name for *Myiothera gularis* "Horsfield" Temminck, misapplied.
Western and central Java.

? Macronous gularis pontius (Oberholser)

Mixornis bornensis pontia Oberholser, 1922, Smiths. Misc.
Coll., 74 (2), p. 6 — Pulau Laut, off southeastern
Borneo.
Known only from type specimen; doubtfully distinct from
bornensis.

Macronous gularis bornensis (Bonaparte)

M[ixornis]. bornensis Bonaparte, 1850, Consp. Av., 1, p. 217 — Borneo; type from Banjermasin, *fide* Pucheran, 1853, Voyage au Pôle Sud, Zool., 3, p. 91.

T[imalia]. similis "Temm." Blyth, 1865, Ibis, p. 47 — Sumatra, error; the description applies to the form of Borneo.

Borneo, except area occupied by *montanus.*

Macronous gularis montanus (Sharpe)

Mixornis montana Sharpe, 1887, Ibis, p. 448 — Kina Balu, North Borneo.

Northeastern Borneo.

Macronous gularis cagayanensis (Guillemard)

Mixornis cagayanensis Guillemard, 1885, Proc. Zool. Soc. London, p. 419, pl. 25 — Cagayan Sulu, Sulu Sea.

Cagayan Sulu.

Macronous gularis argenteus (Chasen and Kloss)

Mixornis gularis argentea Chasen and Kloss, 1930, Bull. Raffles Mus., no. 4, p. 82 — Pulau Malawali, off northern Borneo.

Islets off northern Borneo (Pulau Malawali and Pulau Banggai).

Macronous gularis woodi (Sharpe)

Mixornis Woodi Sharpe, 1877, Trans. Linn. Soc. London, ser. 2, 1, p. 331 — Puerto Princesa, Palawan.

Pulau Balabac and Palawan.

MACRONOUS KELLEYI[1]

Macronous kelleyi (Delacour)

Mixornis kelleyi Delacour, 1932, Oiseau Rev. Franç. Orn., nouv. sér., 2, p. 425 — Pakse [lat. 15° 08′ N., long. 105° 48′ E.], Laos.

[1] This puzzling bird, considered by Delacour (*antea,* pp. 21-23) a geographical representative of *M. flavicollis,* would be treated by me as merely a color variant of *M. gularis* that appears more or less frequently in populations of *lutescens, kinneari,* and *versuricola,* were it not for the fact that the type series of 15 fresh-plumaged specimens were found sympatric with *M. g. lutescens.* In the circumstances, *M. kelleyi* must be given specific rank.

Eastern Thailand (?) ; southern Laos (Pakse) ; central
Annam (Khesanh and Bana) ; southern Annam (Dakto and
Kontoum).

MACRONOUS STRIATICEPS

Macronous striaticeps cumingi (Hachisuka)

Minodoria striaticeps cumingi Hachisuka, 1934, Tori, 8,
p. 221 — "Manila," Philippine Islands, error; corrected
to Samar, by Rand and Rabor, 1960, Fieldiana: Zool.
[Chicago], 35, p. 429.
Samar, Leyte, and Dinagat (?), Philippine Islands.

Macronous striaticeps boholensis Hachisuka

Macronus striaticeps boholensis Hachisuka, 1930, Suppl.
Publ. Orn. Soc. Japan, no. 14, p. 193 — Tagbilaran,
Bohol, Philippine Islands.
Bohol, Philippine Islands.

Macronous striaticeps mearnsi Deignan

Macronous mindanensis montanus Mearns, 1905, Proc.
Biol. Soc. Washington, 18, p. 4 — Todaya, Mount Apo,
Mindanao, Philippine Islands. Not *Mixornis montana*
Sharpe, 1887 = *Macronous gularis montanus* (Sharpe).
Macronus striaticeps mearnsi Deignan, 1951, Bull. Raffles
Mus., no. 23 (1950), p. 128. New name for *Macronous
mindanensis montanus* Mearns, preoccupied.
Mindanao (highlands), Philippine Islands.

Macronous striaticeps mindanensis Steere

Macronus Mindanensis Steere, 1890, List Birds Mammals
Steere Exped., p. 17 — Mindanao, Philippine Islands;
type from Ayala, *fide* Hachisuka, 1935, Birds Philippine Is., pt. 4, p. 407.
Mindanao (lowlands), Philippine Islands.

Macronous striaticeps striaticeps Sharpe

Macronus striaticeps Sharpe, 1877, Trans. Linn. Soc. Lon-
don, ser. 2, 1, p. 331 — Malamaui Islet, off Basilan,
Philippine Islands.
Basilan and Malamaui, Philippine Islands.

Macronous striaticeps kettlewelli Guillemard

Macronus kettlewelli Guillemard, 1885, Proc. Zool. Soc. London, p. 262, pl. 18, fig. 2 — Lukatlapas, Joló, Sulu Archipelago, Philippine Islands.

Sulu Archipelago (Joló, Tawi Tawi, Bongao Islet), Philippine Islands.

MACRONOUS PTILOSUS

Macronous ptilosus ptilosus Jardine and Selby

Macronous ptilosus Jardine and Selby, 1835, Illustr. Orn., pl. 150 — ". . . a collection . . . said to be brought from the islands of Java and Sumatra"; type locality inferentially restricted to "the Malay Peninsula and Sumatra," by Hartert, 1915, Bull. Brit. Orn. Club, 36, p. 36, and inferentially further restricted to Malacca, by Chasen, 1935, Handlist Malaysian Birds, p. 227.

Malay Peninsula from Siamese province of Surat Thani southward to Johore.

Macronous ptilosus trichorrhos (Temminck)

Timalia trichorrhos Temminck, 1836, Pl. col., livr. 100, pl. 594, fig. 1 — Sumatra and Borneo; restricted to "the lowlands of Sumatra," by Hartert, 1915, Bull. Brit. Orn. Club, 36, p. 36.

Macronus ptilosus batuensis Riley, 1937, Proc. Biol. Soc. Washington, 50, p. 61 — Pulau Tanah Bala, Batu Islands.

Sumatra and the Batu Islands (Pulau Tanah Bala).

Macronous ptilosus sordidus Chasen

Macronus ptilosus minor Riley, 1937 (Apr. 21), Proc. Biol. Soc. Washington, 50, p. 62 — Klabat Bay, Bangka. Not *Mixornis gularis minor* Gyldenstolpe, 1916 = *Macronous gularis sulphureus* (Rippon).

Macronus ptilosus sordidus Chasen, 1937 (Dec.), Treubia, 16, pp. 207 (in list), 228 — Billiton.

Bangka and Billiton.

Macronous ptilosus reclusus Hartert

Macronus ptilosus reclusus Hartert, 1915, Bull. Brit. Orn. Club, 36, p. 36 — Kina Balu, North Borneo.

Borneo.

Genus MICROMACRONUS Amadon

Micromacronus Amadon, 1962, Condor, 64, p. 3. Type, by original designation and monotypy, *Micromacronus leytensis* Amadon.

MICROMACRONUS LEYTENSIS

Micromacronus leytensis Amadon

Micromacronus leytensis Amadon, 1962, Condor, 64, p. 3 — Dagami, 1,500 ft., Barrio of Patok, eastern shoulder of Mount Lobu, Leyte.
Known only from type locality.

Genus TIMALIA Horsfield

Timalia Horsfield, 1821, Trans. Linn. Soc. London, 13, p. 150. Type, by monotypy, *Timalia pileata* Horsfield.

cf. Deignan, 1955, Bull. Brit. Orn. Club, 75, pp. 128-130.

TIMALIA PILEATA

Timalia pileata bengalensis Godwin-Austen

T[imalia]. Bengalensis Godwin-Austen, 1872 (June 25), Journ. Asiat. Soc. Bengal, 41, p. 143 — India; type from Khasi Hills, Assam, *fide* Kinnear, 1924, Bull. Brit. Orn. Club, 45, p. 9.
Timalia Jerdoni Walden, 1872 (July), Ann. Mag. Nat. Hist., ser. 4, 10, p. 61 — Khasi Hills, Assam.
Timalia pileata arundicola Koelz, 1953, Journ. Zool. Soc. India, 4, p. [153] — Karong, Manipur.
Submontane tracts and plains along Himalayas from Nepal eastward through Assam to northwestern Burma (Upper Chindwin District), and southeastward through East Bengal to northern Arakan.

Timalia pileata smithi Deignan

Timalia pileata smithi Deignan, 1955, Bull. Brit. Orn. Club, 75, p. 129 — Chiang Saen Kao [= King Chiang Saen (lat. 20° 15′ N., long. 100° 05′ E.)], Thailand.
Kachin and Shan States, Burma; southern Yunnan, southern Kweichow, and Kwangsi; Tongking; northern Annam; northern Laos; northern Thailand.

Timalia pileata intermedia Kinnear

Timalia pileata intermedia Kinnear, 1924, Bull. Brit. Orn. Club, 45, p. 9 — Pegu Division, Lower Burma. New name for *Timalia pileata jerdoni* "Walden" = Baker, 1922, Fauna Brit. India, Birds, ed. 2, 1, p. 227, not *Timalia Jerdoni* Walden, 1872.

Central and southern Burma southward to central Tenasserim (Amherst District) and southwestern Thailand (Kanchanaburi Province).

Timalia pileata patriciae Deignan

Timalia pileata patriciae Deignan, 1955, Bull. Brit. Orn. Club, 75, p. 129 — Ban Khlong Khlung [lat. 16° 10′ N., long. 99° 45′ E.], Thailand.

Western portion of central plains of Thailand from Kamphaeng Phet Province southward to Prachuap Khiri Khan.

Timalia pileata dictator Kinnear

Timalia pileata dictator Kinnear, 1930, Bull. Brit. Orn. Club, 50, p. 55 — Dran [lat. 11° 49′ N., long. 108° 38′ E.], Annam.

Eastern and southeastern Thailand; southern Laos; Cambodia; southern Annam; Cochin China.

Timalia pileata pileata Horsfield

Timalia pileata Horsfield, 1821, Trans. Linn. Soc. London, 13, p. 151 — Java.

Java.

GENUS **CHRYSOMMA** BLYTH

Chrysomma Blyth, 1843, Journ. Asiat. Soc. Bengal, 12, p. 181. Type, by original designation, *Timalia hypoleuca* Franklin.

Pyctoris "Hodgson" Blyth, 1844, Ann. Mag. Nat. Hist., 13, p. 116. Type, by original designation and monotypy, *Timalia hypoleuca* Franklin.

Pyctornis Moore, 1857, Proc. Zoc .. Soc. London, p. 94. *Nomen emendatum.*

Erythrops "Hodgs. MSS." = Sharpe, 1883, Cat. Birds Brit. Mus., 7, p. 510. In synonymy with *Pyctoris* Hodgson, *i.e.,* Blyth. Not *Erythrops* Rafinesque, 1820, Pisces, nor *Erythrops* Sars, 1869, Crustacea.

CHRYSOMMA SINENSE

Chrysomma sinense nasale (Legge)

Pyctorhis nasalis Legge, 1879, Ann. Mag. Nat. Hist., ser. 5, 3, p. 169 — Ceylon.

Ceylon.

Chrysomma sinense hypoleucum (Franklin)

Timalia hypoleuca Franklin, 1831, Proc. Comm. Zool. Soc. London, pt. 1, p. 118 — Ganges between Calcutta and Benares; restricted to the United Provinces [= Uttar Pradesh], by Ticehurst, 1922, Ibis, p. 543.

Chrysomma sinensis [sic] *nagaensis* [sic] Koelz, 1954, Contrib. Inst. Regional Explor., no. 1, p. 4 — Karong, Manipur.

Chrysomma sinensis [sic] *saurashtrensis* [sic] Koelz, 1954, Contrib. Inst. Regional Explor., no. 1, p. 4 — Sasan, Saurashtra.

Eastern West Pakistan and greater part of India south of Himalayas, southward to Nilgiri Hills and southeastward, through East Pakistan and Assam (south of the Brahmaputra), to western Burma.

Chrysomma sinense saturatius (Ticehurst)

Pyctorhis sinensis saturatior Ticehurst, 1922, Bull. Brit. Orn. Club, 42, p. 57 — Bhutan Duars.

Himalayas from the Duars of Sikkim eastward into Assam (north of the Brahmaputra).

Chrysomma sinense sinense (Gmelin)

[*Parus*] *sinensis* Gmelin, 1789, Syst. Nat., 1, pt. 2, p. 1012 — China; restricted to Kwangtung, by Stresemann and Heinrich, 1940, Mitt. Zool. Mus. Berlin, 24, p. 205.

Pyctorhis sinensis major La Touche, 1925, Birds Eastern China, 1, p. 72 — Kwangtung, Kwangsi, and Yunnan; lectotype from Mengtsz, Yunnan, *fide* Bangs, 1930, Bull. Mus. Comp. Zool., 70, p. 305.

The greater part of Burma (except area occupied by *hypoleucum*), southward to central Tenasserim; Thailand (northern plateau and western portion of central plains); Laos, Annam, and Tongking; Yunnan, Kwangsi, and Kwangtung.

Genus MOUPINIA David and Oustalet

Moupinia David and Oustalet, 1877, Oiseaux de la Chine, p. 219. Type, by original designation and monotypy, *Alcippe poecilotis* J. Verreaux.

MOUPINIA ALTIROSTRIS

Moupinia altirostris scindica (Harington)

Pyctorhis altirostris scindicus [sic] Harington, 1916, Journ. Bombay Nat. Hist. Soc., 23, p. 424 — Sukkur, West Pakistan; type from Mangrani, between Sukkur and Shikarpur, *fide* Ticehurst, 1922, Ibis, p. 543.

Known only from unique type.

Moupinia altirostris griseigularis (Hume)

Pyctorhis griseigularis Hume, 1877, Stray Feathers, 5, p. 116 — Bhutan Duars.

Grasslands at base of Himalayas from Bhutan Duars eastward to eastern Assam; grassy plains of Assam south of the Brahmaputra (Cachar and Sylhet Districts); northeastern Burma (Kachin State).

Moupinia altirostris altirostris (Jerdon)

Chrysomma altirostre Jerdon, 1862, Ibis, p. 22 — "Islands on the Burrampootra river, in Upper Burmah"; type from an island in the Irrawaddy off Thayetmyo, *fide* Sharpe, 1883, Cat. Birds Brit. Mus., 7, p. 513.

Grasslands that formerly covered the Irrawaddy-Sittang plains of south-central Burma; not reported since 1941 and now probably extinct.

MOUPINIA POECILOTIS

Moupinia poecilotis (Verreaux)

Alcippe pœcilotis J. Verreaux, 1870, Nouv. Arch. Mus. Hist. Nat. [Paris], 6, p. 35; *idem*, 1871, *ibid.*, 7, p. 37; *idem*, 1872, *ibid.*, 8, pl. 2, fig. 4 — "les montagnes du Thibet chinois"; types from Muping [= Paohing, Hsikang, *fide* Verreaux, 1871].

Pyctorhis gracilis Styan, 1899, Bull. Brit. Orn. Club, 8, p. 26 — Lungan [= Pingwu], northwestern Szechwan.

Moupinia poecilotis sordidior Rothschild, 1921, Novit. Zool., 28, p. 36 — Likiang Range, northwestern Yunnan.

Northwestern Szechwan, southeastern Hsikang, and northwestern Yunnan.

GENUS CHAMAEA GAMBEL

Chamæa Gambel, 1847, Proc. Acad. Nat. Sci. Philadelphia, 3, p. 154. Type, by original designation and monotypy, *Parus fasciatus* Gambel.

cf. Erickson, 1938, Univ. California Publ. Zool., 42, pp. 247-334 (biology).
Bowers, 1960, Condor, 62, pp. 91-120 (geographic variation).

CHAMAEA FASCIATA

Chamaea fasciata phaea Osgood

Chamæa fasciata phæa Osgood, 1899, Proc. Biol. Soc. Washington, 13, p. 42 — Newport, Lincoln County, Oregon.
Humid coastal belt of Oregon from the Columbia River southward to California border.

Chamaea fasciata rufula Ridgway

Chamæa fasciata rufula Ridgway, 1903, Proc. Biol. Soc. Washington, 16, p. 109 — Nicasio, Marin County, California.
Humid coastal belt of California from Oregon border southward to Marin County.

Chamaea fasciata intermedia Grinnell

Chamæa fasciata intermedia Grinnell, 1900, Condor, 2, p. 86 — Palo Alto, Santa Clara County, California.
Interior and southern areas of the San Francisco Bay region of California (eastern Sonoma and western Napa counties and from San Francisco southward to southern Santa Cruz County).

Chamaea fasciata fasciata (Gambel)

Parus fasciatus Gambel, 1845, Proc. Acad. Nat. Sci. Philadelphia, 2, p. 265 — California; restricted to Monterey, Monterey County, by Grinnell, 1932, Univ. California Publ. Zool., 38, p. 291.
Coastal area of California from Monterey County southward to central San Luis Obispo County.

Chamaea fasciata henshawi Ridgway

Chamaea fasciata henshawi Ridgway, 1882, Proc. U. S. Nat. Mus., 5, p. 13 — interior districts of California, including the western slope of the Sierra Nevada; inferentially restricted to Walker Basin, Kern County, by Ridgway, 1904, Bull. U. S. Nat. Mus., 50, pt. 3, p. 691.

Southwestern Oregon (eastern Josephine and western Jackson counties); interior of northern and central California; coastal area of southern California from Santa Barbara County southeastward to Mexican border.

Chamaea fasciata canicauda Grinnell and Swarth

Chamæa fasciata canicauda Grinnell and Swarth, 1926, Univ. California Publ. Zool., 30, p. 169 — La Grulla, Sierra San Pedro Mártir, Baja California.

Northwestern Baja California from the international boundary southward to lat. 30° N.

GENUS **TURDOIDES** CRETZSCHMAR

Turdoides Cretzschmar, 1827, in Rüppell's Atlas, Vög., hft. 4, p. 6, pl. 4. Type, by monotypy, *Turdoides leucocephalus* Cretzschmar.

Cratopus Jardine, 1831 (Feb.), Edinburgh Journ. Nat. Geogr. Sci., 3, p. 97. Type, by monotypy, *Cratopus bicolor* Jardine. Not *Cratopus* Schönherr, 1826, Coleoptera.

Argya Lesson, 1831 (Feb.), Traité Orn. livr. 6, p. 402. Type, by subsequent designation (G. R. Gray, 1855, Cat. Gen. Subgen. Birds, p. 44), "*Ixos squammiceps*, Rüpp." = *Malurus squamiceps* Cretzschmar.

Crateropus Swainson, 1831 (Aug.), Zool. Ill., ser. 2, 2, no. 17, pl. 80. Type, by monotypy, *Crateropus reinwardtii* Swainson.

Malacocircus Swainson, 1833, Zool. Ill., ser. 2, 3, no. 28, pl. 127. Type, by monotypy, *Malacocircus striatus* Swainson.

Malacocercus Strickland, 1841, Ann. Mag. Nat. Hist., 7, p. 27. *Nomen emendatum.*

Ischyropodus Reichenbach, 1850, Av. Syst. Nat., pl. 55. No species; generic details only. Type, by subsequent

designation (G. R. Gray, 1855, Cat. Gen. Subgen.
Birds, p. 44.), *Cratopus jardineii* A. Smith.
Acanthoptila Blyth, 1855, Journ. Asiat. Soc. Bengal, 24,
p. 478, footnote. Type, by original designation and
monotypy, *Timalia nipalensis* Hodgson.
Aethocichla Sharpe, 1876, in Layard, Birds South Africa,
new ed., pt. 3, p. 215. Type, by monotypy, *Crateropus
gymnogenys* Hartlaub.

cf. Vaurie, 1953, Amer. Mus. Novit., no. 1642, pp. 1-8
(*caudatus* and *altirostris*).

TURDOIDES NIPALENSIS

— **Turdoides nipalensis** (Hodgson) 2

Timalia nipalensis Hodgson, 1836, Asiat. Res., 19, p. 182
— Nepal.
Western and central Nepal.

TURDOIDES ALTIROSTRIS

Turdoides altirostris (Hartert)

Crateropus caudatus altirostris Hartert, 1909, Vög. pal.
Fauna, 5, p. 623, fig. 122b — Fao, southeastern Iraq
at the mouth of the Shatt al Arab.
Reed beds of the lower Tigris-Euphrates Valley in southeastern Iraq and southwestern Iran.

TURDOIDES CAUDATUS

Turdoides caudatus salvadorii (De Filippi)

Crateropus Salvadorii De Filippi, 1865, Note di un Viag-
gio in Persia, 1, p. 346 — Shiraz, Fars, Iran; the type
is labelled "Armadi, Karmán," *fide* Blanford, 1876,
Eastern Persia, etc., 2, p. 204, footnote.
Argya caudatus [sic] *theresæ* Meinertzhagen, 1930, Bull.
Brit. Orn. Club, 50, p. 55 — Baghdad, Iraq.
Southeastern Iraq and southwestern Iran (Luristan,
Fars).

Turdoides caudatus huttoni (Blyth)

Malacocercus Huttoni Blyth, 1847, Journ. Asiat. Soc. Bengal, 16, p. 476 — Kandahar, southeastern Afghanistan.
Eastern Iran (Baluchistan, Seistan), southern Afghanistan, and southern West Pakistan.

Turdoides caudatus eclipes (Hume)

Chatorhea eclipes Hume, 1877, Stray Feathers, 5, p. 337
— Peshawar, northern West Pakistan.
Northern West Pakistan from vicinity of Fort Sandeman
northeastward to Kashmir border.

Turdoides caudatus caudatus (Dumont)

Cossyphus caudatus Dumont, 1823, Dict. Sci. Nat., éd.
Levrault, 29, p. 268 — India.
Megalurus isabellinus Swainson, 1838, Animals in Menag-
eries, pt. 3, p. 291 — India.[1]
Indian Peninsula from foothills of Himalayas southward
to Kerala, westward to East Punjab, Rajasthan, and southeastern West Pakistan, and eastward to West Bengal; Lac-
cadive Islands; Rameswaram Island.

TURDOIDES EARLEI

Turdoides earlei sonivius (Koelz)

Argya earlei sonivia Koelz, 1954, Contrib. Inst. Regional
Explor., no. 1, p. 3 — Khinjar Lake, Sind.
West Pakistan (the Rann of Kutch and valley of the
Indus from its mouths to West Punjab).

Turdoides earlei earlei (Blyth)

M[alacocercus]. Earlei Blyth, 1844, Journ. Asiat. Soc.
Bengal, 13, p. 369 — near Calcutta, West Bengal.
Grasslands of northern India below the Himalayas from
western Nepal eastward to eastern Assam, and southeastward through Assam to Kachin State of Burma, plains of
central and southern Burma, and Arakan.

TURDOIDES GULARIS

Turdoides gularis (Blyth)

Chatarrhæa gularis Blyth, 1855, Journ. Asiat. Soc. Bengal, 24, p. 478 — Pegu Division, Lower Burma.
Plains of central and southern Burma (chiefly in dry
zone).

[1] Deignan (1946, Auk, 63, pp. 382-383) erroneously attributed this
name to the Indian subspecies of *Megalurus palustris* Horsfield, at the
same time restricting its type locality to Assam. Because *Turdoides
caudatus* does not occur in Assam, this action can have no possible
validity.

TURDOIDES LONGIROSTRIS

Turdoides longirostris (Moore)

> *Pyctorhis longirostris* "Hodgson" = Moore, 1854, in Horsfield and Moore, Cat. Birds Mus. East India Co., 1, p. 408 — Nepal.
>
> *Malacocercus (Layardia) rubiginosus* Godwin-Austen, 1874, Proc. Zool. Soc. London, p. 47 — Manipur. Not *Crateropus rubiginosus* Rüppell, 1845.
>
> *Argya longirostris arcana* Koelz, 1954, Contrib. Inst. Regional Explor., no. 1, p. 3 — Karong, Manipur.

Grasslands at foot of Himalayas from Nepal eastward to eastern Assam; grassy plateaus of hill tracts of Assam south of the Brahmaputra; northern Arakan.

TURDOIDES MALCOLMI

Turdoides malcolmi (Sykes)

> *Timalia Malcolmi* Sykes, 1832, Proc. Comm. Zool. Soc. London, pt. 2, p. 88 — the Deccan [= Poona, Bombay, *fide* Whistler and Kinnear, 1932, Journ. Bombay Nat. Hist. Soc., 35, p. 740].

Interior of peninsular India from East Punjab eastward to eastern Uttar Pradesh and southward to Mysore.

TURDOIDES SQUAMICEPS

Turdoides squamiceps squamiceps (Cretzschmar)

> *Malurus squamiceps* Cretzschmar, 1827, in Rüppell, Atlas, Vög., p. 19, pl. 12 — Aqaba, Transjordania.

Arabian Peninsula from the Dead Sea depression southward, along the Red Sea coast, to southwestern Saudi Arabia.

Turdoides squamiceps yemensis (Neumann)

> *Argya squamiceps yemensis* Neumann, 1904, Orn. Monatsb., 12, p. 29 — between Sheikh Othman and Lahej, Aden Protectorate.

Yemen and the Aden Protectorate.

Turdoides squamiceps muscatensis de Schauensee and Ripley

> *Turdoides squamiceps muscatensis* de Schauensee and Ripley, 1953, Proc. Acad. Nat. Sci. Philadelphia, 105, p. 85 — As Sib, near Muscat, Oman.

Arabian coast of the Gulf of Oman.

TURDOIDES FULVUS

Turdoides fulvus maroccanus Lynes

Turdoides (Crateropus) fulvus maroccanus Lynes, 1925, Mém. Soc. Sci. Nat. Maroc, 13, p. 49 — Taroudant, southwestern Morocco.

Southwestern Morocco (valley of the Oued Sous).

? Turdoides fulvus billypayni (Meinertzhagen)

Argya fulva billypayni Meinertzhagen, 1939, Bull. Brit. Orn. Club, 59, p. 69 — Ksar es Souk, southeastern Morocco.

Known only from type locality; doubtfully separable from *fulvus*.

Turdoides fulvus fulvus (Desfontaines)

Turdus fulvus Desfontaines, 1789, Hist. Acad. Roy. Sci., (1787), p. 498, pl. 11 — Gafsa, Tunisia.

Northern Algeria, Tunisia, and northwestern Libya (except coastal areas).

Turdoides fulvus buchanani (Hartert)

Crateropus fulvus buchanani Hartert, 1921, Novit. Zool., 28, p. 115 — Mount Baguezan, Aïr Mountains, Niger Territory.

Aïr Massif of the central Sahara.

Turdoides fulvus acaciae (Lichtenstein)

S[phenura]. Acaciae Lichtenstein, 1823, Verz. Doubl., p. 40 — Nubia.

Valley of the Nile from southern Egypt (First Cataract) southward to the central Sudan, and eastward to Red Sea coasts of the Sudan and Ethiopia.

TURDOIDES AYLMERI

Turdoides aylmeri aylmeri (Shelley)

Argya aylmeri Shelley, 1885, Ibis, p. 404, pl. 11, fig. 1 — British Somaliland.

British and Italian Somalilands and southeastern Ethiopia.

Turdoides aylmeri boranensis (Benson)

Argya aylmeri boranensis Benson, 1947, Bull. Brit. Orn. Club, 68, p. 10 — ten miles south of Yavello [lat. 4° 57′ N., long. 38° 12′ E.], Ethiopia.

South-central Ethiopia from the Arusi region southward to the Boran (Borama).

Turdoides aylmeri kenianus (Jackson)

Argya keniana Jackson, 1910, Bull. Brit. Orn. Club, 27, p. 7 — Emberre, east of Embu [lat. 0° 31′ S., long. 37° 26′ E.], Kenya.

Central Kenya.

Turdoides aylmeri loveridgei (Hartert)

Argya aylmeri loveridgei Hartert, 1923, Bull. Brit. Orn. Club, 43, p. 118 — Kampi-ya-Bibi, near Samburu Station, Mombasa District, Kenya.

Southeastern Kenya (Tsavo) and northeastern Tanganyika (Moshi).

Turdoides aylmeri mentalis (Reichenow)

Argya mentalis Reichenow, 1887, Journ. f. Orn., 35, p. 75 — Soboro, Tanganyika.

North-central Tanganyika.

TURDOIDES RUBIGINOSUS

Turdoides rubiginosus bowdleri Deignan, nom. nov.

Argya sharpii Ogilvie-Grant and Reid, 1901, Ibis, p. 662 — Shibeli River [ca. lat. 7° 12′ N., long. 42° 11′ E.], southeastern Ethiopia. Not *Crateropus sharpei* Reichenow, 1891.

Southeastern Ethiopia, from Diredawa southward to border of Italian Somaliland.

Turdoides rubiginosus rubiginosus (Rüppell)

Crateropus rubiginosus Rüppell, 1845, Syst. Uebers. Vög. Nord-ost.-Afr., p. 47, pl. 19 — Shoa, Ethiopia.

[*Argya, Malacocercus*] *Rufula* Heuglin, 1871, Orn. Nordost.-Afr., Nachtr., Index, p. 313. New name for *Argya rufescens* (Heuglin), 1869, *ibid.*, 1, p. 389 = *Crateropus rufescens* Heuglin, 1856.

Central and southwestern Ethiopia, southern Sudan (Equatoria), eastern Uganda, and western Kenya.

Turdoides rubiginosus heuglini (Sharpe)

Argya heuglini Sharpe, 1883, Cat. Birds Brit. Mus., 7, p. 391 — "Zanzibar," error, and Mombasa, Kenya.

Argya saturata Sharpe, 1895, Proc. Zool. Soc. London, p. 488 — "Zanzibar," error.

Coastal region of eastern Africa from Italian Somaliland (Juba region) southward, through eastern Kenya, to northeastern Tanganyika.

Turdoides rubiginosus schnitzeri Deignan, nom. nov.

Argya rubiginosa emini Reichenow, 1907, Orn. Monatsb., 15, p. 30 — Scamuye, Unyamwezi region, northwestern Tanganyika. Not *Crateropus plebeius emini* Neumann, 1904.

Northwestern Tanganyika.

TURDOIDES SUBRUFUS

Turdoides subrufus subrufus (Jerdon)

T[himalia]. subrufa Jerdon, 1839, Madras Journ. Lit. Sci., 10, p. 259 — The Wynaad, near Manantoddy, Madras.

Western Ghats of India from vicinity of Mahabaleshwar southward to northern Kerala (Malabar), and thence eastward through Madras to the Nilgiri and Shevaroy Hills.

Turdoides subrufus hyperythrus (Sharpe)

Argya hyperythra Sharpe, 1883, Cat. Birds Brit. Mus., 7, p. 390 — Madras; type locality restricted to Palghat, by Ripley, 1953, Postilla, Yale Univ., no. 17, p. 4.

Southwestern Madras and Kerala (except area occupied by *subrufus*).

TURDOIDES STRIATUS

Turdoides striatus rufescens (Blyth)

M[alacocercus]. rufescens Blyth, 1847, Journ. Asiat. Soc. Bengal, 16, p. 453 — Ceylon.

Humid forests of Ceylon.

Turdoides striatus malabaricus (Jerdon)

Malacocircus malabaricus Jerdon, 1845, Illus. Indian Ornith., pt. 2, text to pl. 19 — Forests of Malabar [= Travancore-Cochin (Kerala), *fide* Whistler, 1935, in Ali and Whistler, Journ. Bombay Nat. Hist. Soc., 38, p. 72].

Southwestern India from Kerala northward to Goa.

Turdoides striatus somervillei (Sykes)

Timalia Somervillei Sykes, 1832, Proc. Comm. Zool. Soc.
London, pt. 2, p. 88 — ghats of the Deccan.
M[alacocircus]. Sykesii Jerdon, 1863, Birds India, 2, pt. 1,
p. 63 — Bombay.

Western coast of India (Bombay State) from Goa northward to Surat Dangs.

Turdoides striatus sindianus (Ticehurst)

Crateropus terricolor sindianus Ticehurst, 1920, Bull.
Brit. Orn. Club, 40, p. 156 — Karachi, West Pakistan.

West Pakistan (valley of the Indus from its mouths to
vicinity of Peshawar) and northwestern India (East Punjab, western Uttar Pradesh, and less arid regions of
Rajasthan).

Turdoides striatus striatus (Dumont)

Cossyphus striatus Dumont, 1823 (Dec. 27), Dict. Sci.
Nat., éd. Levrault, 29, p. 268 — Bengal.
Malacocircus striata [sic] Swainson, 1833, Zool. Ill., ser.
2, 3, pl. 127 — "Ceylon," error; type from Bengal.
M[alacocercus]. terricolor "Hodgson" = Blyth, 1844,
Journ. Asiat. Soc. Bengal, 13, p. 367 — Nepal, ex
Hodgson.
M[alacocercus]. bengalensis Blyth, 1849, Cat. Birds Mus.
Asiat. Soc., p. 140 — Bengal, ex Edwards. Name based
on "Edwards, pl. 184, badly coloured."

Foot of the Himalayas from western Uttar Pradesh eastward to eastern Assam (valley of the Brahmaputra), southward to northern Madhya Pradesh, southeastern Bihar, and
the mouths of the Ganges, and thence southward along coast
to the mouths of the Godavari.

Turdoides striatus orientalis (Jerdon)

M[alacocircus]. orientalis Jerdon, 1845, Illus. Indian
Ornith., pt. 2, text to pl. 19 — Eastern Ghats of India;
restricted to Horsleykonda, west of Nellore, Andhra,
by Ripley, 1958, Postilla, Yale Univ., no. 35, p. 9.

All of central and southern India, except areas occupied
by *striatus, somervillei,* and *malabaricus.*

TURDOIDES AFFINIS

Turdoides affinis affinis (Jerdon)

[*Turdus*] *griseus* Gmelin, 1789, Syst. Nat., 1, pt. 2, p. 824 — Coromandel Coast of India. Not *Turdus griseus* Boddaert, 1783.

M[alacocircus]. affinis Jerdon, 1845, Illus. Indian Ornith., pt. 2, text to pl. 19 — Travancore [= Kerala].

[*Malacocircus*] *Elliotti* Jerdon, 1845, Illus. Indian Ornith., pt. 2, text to pl. 19 — Kanara; here restricted to Mangalore, South Kanara District, Madras.

Turdoides polioplocamus Oberholser, 1920, Proc. Biol. Soc. Washington, 33, p. 84. New name for *Turdus griseus* Gmelin, preoccupied.

The more arid lowlands of southern India from South Kanara on the west, southern Hyderabad, and the lower Godavari Valley on the east southward to Cape Comorin; Rameswaram Island (where intergrades with *taprobanus*).

Turdoides affinis taprobanus Ripley

Turdoides affinis taprobanus Ripley, 1958, Postilla, Yale Univ., no. 36, p. 10 — Alawna, Ceylon.[1]
Ceylon.

TURDOIDES MELANOPS

Turdoides melanops vepres Meinertzhagen

Turdoides melanops vepres Meinertzhagen, 1936, Bull. Brit. Orn. Club, 57, p. 69 — Nanyuki [lat. 0° 02' N., long. 37° 06' E.], Kenya.
Known only from type locality.

Turdoides melanops clamosus (van Someren)

Crateropus melanops clamosus van Someren, 1920, Bull. Brit. Orn. Club, 40, p. 95 — Naivasha [lat. 0° 44' S., long. 36° 26' E.], Kenya.
Kenya (Great Rift Valley).

Turdoides melanops sharpei (Reichenow)

Crateropus sharpei Reichenow, 1891, Journ. f. Orn., 39, p. 432 — Kakoma [lat. 5° 50' S., long. 32° 29' E.], Tanganyika.

---

[1] Replaces *Crateropus striatus* (Swainson) of Sharpe's Hand-list, not applicable, and *Turdoides striatus striatus* (Dumont) of Fauna Brit. India, ed. 2, 7, p. 36, neither applicable nor available.

Crateropus grisescens Reichenow, 1908, Orn. Monatsb.,
16, p. 47 — Nyawatura [lat. 1° 02' S., long. 30° 52' E.],
Tanganyika.
Western Kenya, southern Uganda, Ruanda-Urundi, and
northwestern Tanganyika.

Turdoides melanops melanops (Hartlaub)
Crateropus melanops Hartlaub, 1867, Proc. Zool. Soc.
London, pt. 3 (1866), p. 435, pl. 37 — Damaraland.
Southwestern Angola, northern South West Africa
(northern Damaraland), and the Bechuanaland Protec-
torate (northwestern Ngamiland).

TURDOIDES TENEBROSUS

Turdoides tenebrosus (Hartlaub)
Crateropus tenebrosus Hartlaub, 1883, Journ. f. Orn., 31,
p. 425 — Kudurma [lat. 4° 45' N., long. 29° 35' E.],
Sudan.
Crateropus tenebrosus claudei Bannerman, 1919, Bull.
Brit. Orn. Club, 39, p. 99 — Poko [lat. 3° 09' N., long.
26° 53' E.], Belgian Congo.
Northeastern Congo, southern Sudan, and southwestern
Ethiopia.

TURDOIDES REINWARDTII

Turdoides reinwardtii reinwardtii (Swainson)
Crateropus Reinwardii [sic] Swainson, 1831, Zool. Ill., ser.
2, 2, no. 17, pl. 80 — "Indian Islands," error; type locality corrected to Senegal, by Swainson, 1837, Birds
W. Africa, 1, p. 276.
Western Africa from Senegal southward through Portuguese and French Guinea to Sierra Leone.

Turdoides reinwardtii stictilaemus (Alexander)
Crateropus stictilæma [sic] Alexander, 1901, Bull. Brit.
Orn. Club, 12, p. 10 — Ghana.
Crateropus reinwardti houyi Neumann, 1915, Orn. Mo-
natsb., 23, p. 74 — Goré [lat. 7° 55' N., long. 16° 42' E.],
Ubangi-Shari.
Western Africa from Ghana eastward, through Nigeria
and the Cameroons, to western Ubangi-Shari.

TURDOIDES PLEBEJUS

Turdoides plebejus platycircus (Swainson)

Crateropus platycircus Swainson, 1837, Birds W. Africa,
1, p. 274 — western Africa; type commonly assumed to
have come from Gambia.
Crateropus plebejus permistus Neumann, 1906, Orn.
Monatsb., 14, p. 146 — Senegal.
Western Africa from Senegal southeastward to Ivory
Coast.

? Turdoides plebejus togoensis (Neumann)

Crateropus platycercus [sic] *togoensis* Neumann, 1904,
Journ. f. Orn., 52, p. 551 — Kete Krachi, Togo, Ghana.
Western Africa from Ghana eastward to western Nigeria;
doubtfully distinct from *platycircus*.

Turdoides plebejus uamensis (Reichenow)

Crateropus plebeius [sic] *gularis* Reichenow, 1910, Orn.
Monatsb., 18, p. 7 — Mba [lat. 7° 00′ N., long.
11° 50′ E.], British Cameroons. Not *Chatarrhæa gularis*
Blyth, 1855.
Crateropus uamensis Reichenow, 1921, Journ. f. Orn., 69,
p. 48 — Bozum [lat. 6° 18′ N., long. 16° 22′ E.], Ubangi-
Shari.
Crateropus plebeius [sic] *elberti* Reichenow, 1921, Journ.
f. Orn., 69, p. 461 — Uam district, Ubangi-Shari.
Grasslands of the Cameroons and western Ubangi-Shari.

Turdoides plebejus plebejus (Cretzschmar)

Ixos plebejus Cretzschmar, 1828, in Rüppell, Atlas Vög.,
p. 35, pl. 23 — Kordofan Province, Sudan.
Crateropus cordofanicus Butler, 1905, Ibis, p. 330, pl. 7
— Jebel Melbis, Kordofan, Sudan.
Crateropus plebejus anomalus Hartert, 1921, Novit. Zool.,
28, p. 116 — Farniso, near Kano [lat. 12° 00′ N., long.
8° 31′ E.], Nigeria.
Interior of Africa from northern Nigeria eastward,
through southern Chad, to the central Sudan (Kordofan).

Turdoides plebejus leucocephalus Cretzschmar

Turdoides leucocephala [sic] Cretzschmar, 1827, in Rüppell, Atlas Vög., p. 6, pl. 4 — Sennar [= Blue Nile Province, Sudan].

Crateropus leucocephalus abyssinicus Neumann, 1904,
Journ. f. Orn., 52, p. 550 — northern and central
Ethiopia.

Eastern Sudan (east of the Nile from Khartoum southward to Fung) and eastward through northern Ethiopia
to Red Sea coast.

Turdoides plebejus cinereus (Heuglin)

Crateropus cinereus Heuglin, 1856, Sitzungsb. K. Akad.
Wiss. Wien., Math.-Naturwiss. Cl., 19, p. 282 — banks
of the White Nile south of lat. 6° N.

Crateropus buxtoni Sharpe, 1891, Ibis, p. 445 — Turkwel
District, Kenya.

Crateropus jardinei hypobrunneus Reichenow, 1915,
Journ. f. Orn., 63, p. 129 — Amadi [lat. 3° 38′ N., long.
26° 46′ E.], Belgian Congo.

Interior of Africa from eastern Ubangi-Shari eastward,
through the southern Sudan, to southwestern Ethiopia, and
southward to the northeastern Congo, Uganda, and western
Kenya.

TURDOIDES JARDINEII

Turdoides jardineii hypostictus (Cabanis and Reichenow)

Crateropus hypostictus Cabanis and Reichenow, 1877,
Journ. f. Orn., 25, pp. 25, 103 — Loango Coast.

Southward from Cabinda, southern Moyen Congo, and
Kasai District of the Congo to Benguella Province, Angola.

Turdoides jardineii tanganjicae (Reichenow)

Crateropus Tanganjicae Reichenow, 1886, Journ. f. Orn.,
34, p. 115, pl. 3, fig. 1 — Mpala [lat. 6° 45′ S., long.
29° 31′ E.], Belgian Congo.

Crateropus carruthersi Ogilvie-Grant, 1907, Bull. Brit.
Orn. Club, 19, p. 106 — Upper Congo; type from Niem-
bo [lat. 4° 30′ S., long. 28° 15′ E.], Congo.

Southeastern Congo from northern end of Lake Tanganyika (where intergrades with *emini*) southward to southern Katanga Province (where intergrades with *kirkii*);
northwestern Northern Rhodesia.

Turdoides jardineii emini (Neumann)

Crateropus plebeius emini Neumann, 1904, Journ. f. Orn.,
52, p. 549 — Wala River, Tabora District, Tanganyika.

Uganda, Ruandi-Urundi, and Tanganyika (Tabora and Mkalama areas).

Turdoides jardineii kikuyuensis (Neumann)

Crateropus plebeius kikuyuensis Neumann, 1906, Orn. Monatsb., 14, p. 7 — Escarpment [lat. 1° 01' S., long. 36° 36' E.], Kenya.

Crateropus reichenowi Madarász, 1910, Arch. f. Zool., 1, p. 177 — Ngare-Dobash [= Mara River, Kenya or Tanganyika].

Southwestern Kenya and adjacent areas of northern Tanganyika.

Turdoides jardineii kirkii (Sharpe)

Crateropus kirkii Sharpe, 1876, in Layard, Birds South Africa, new ed., pt. 3, p. 213 — near rivers in the Zambesi country; type from Mazaro, on the Zambesi, *ca.* 75 miles above mouth of the Shire.

Coastal region of eastern Africa from Kenya (Lamu) southward to central Mozambique, and inland through the Zambesi drainage to Nyasaland and the eastern Rhodesias.

Turdoides jardineii tamalakanei de Schauensee

Crateropus affinis Bocage, 1869, Proc. Zool. Soc. London, p. 436 — Leullengues, Mossamedes, southwestern Angola. Not *Malacocircus affinis* Jerdon, 1845.

Turdoides jardinei tamalakanei de Schauensee, 1932, Proc. Acad. Nat. Sci. Philadelphia, 83, p. 469 — Maun [lat. 19° 55' S., long. 23° 30' E.], Bechuanaland.

Southwestern Northern Rhodesia, western Southern Rhodesia, northern Bechuanaland, and southern Angola.

Turdoides jardineii jardineii (Smith)

Cratopus Jardineii A. Smith, 1836, Rep. Exped. Centr. Africa, p. 45 — banks of rivers beyond Kurrichane, northwestern Transvaal.

Western Transvaal and neighboring districts of Bechuanaland.

Turdoides jardineii natalensis Roberts

Turdoides jardinei natalensis Roberts, 1932, Ann. Transvaal Mus., 15, p. 29 — Weenen, Natal.

? *Turdoides jardineii convergens* Clancey, 1958, Durban

Mus. Novit., 5, p. 123 — Manhiça, Sul do Save, southern Mozambique.
Southeastern Southern Rhodesia, southern Mozambique, eastern Transvaal, Swaziland, and Natal.

TURDOIDES SQUAMULATUS

Turdoides squamulatus jubaensis van Someren

Turdoides squamulata [sic] *jubaensis* van Someren, 1931, Journ. East Africa Uganda Nat. Hist. Soc., 37 (1930), p. 196 — Serenli [lat. 2° 30′ N., long. 42° 07′ E.], Italian Somaliland.
Southern Italian Somaliland (valley of the Juba from Ethiopian border southward to Serenli).

Turdoides squamulatus squamulatus (Shelley)

Crateropus squamulatus Shelley, 1884, Ibis, p. 45 — Mombasa [lat. 4° 04′ S., long. 39° 41′ E.], Kenya.
Coastal area of Kenya from Lamu southward to Tanganyika border.

TURDOIDES LEUCOPYGIUS

Turdoides leucopygius leucopygius (Rüppell)

Ixos leucopygius Rüppell, 1840, Neue Wirbelth., Vögel, p. 82, pl. 30, fig. 1 — central forest region along the coast of Abyssinia.
Coastal region of Ethiopia from Massaua to Adigrat.

Turdoides leucopygius limbatus (Rüppell)

Crateropus limbatus Rüppell, 1845, Syst. Uebers. Vög. Nord-ost.-Afr., p. 48 — Ali Amba, Shoa, Ethiopia.
Northwestern Ethiopia from valley of the Anseba southward to Shoa.

Turdoides leucopygius smithii (Sharpe)

Crateropus smithii Sharpe, 1895, Bull. Brit. Orn. Club, 4, p. 41 — "Somaliland"; type from Sheik Hussein, Arusi, Ethiopia, *fide* Sclater, 1930, Syst. Av. Aethiop., p. 354.
Southeastern Ethiopia and western British Somaliland.

Turdoides leucopygius lacuum (Neumann)

Crateropus smithi lacuum Neumann, 1903, Bull. Brit. Orn. Club, 14, p. 15 — Alelu, north of Lake Abassi [lat. 7° 04′ N., long. 38° 27′ E.], Ethiopia.

Turdoides leucopygia [sic] *clarkei* Macdonald, 1939, Bull.
Brit. Orn. Club, 60, p. 10 — "Gummaro," an estate and
stream a few miles west of the British Consulate at
Goré [lat. 8° 05′ N., long. 35° 30′ E.], Ethiopia.
Southwestern Ethiopia (except area occupied by *omoen-sis*).

— **Turdoides leucopygius omoensis** (Neumann)

Crateropus smithi omoensis Neumann, 1903, Bull. Brit.
Orn. Club, 14, p. 15 — Senti River (southern affluent
to the Omo), between Uba and Gofa, southwestern
Ethiopia.
Southwestern Ethiopia from the lower Omo valley east-
ward to Abaya Lakes and Boran region; southeastern Sudan
(Boma Hills).

— **Turdoides leucopygius ater** Friedmann

Turdoides melanops ater Friedmann, 1927, Proc. New
England Zool. Club, 10, p. 11 — Kamaniola [lat.
2° 46′ S., long. 29° 00′ E.], Belgian Congo.
Ruanda-Urundi, the southeastern Congo, northeastern
Northern Rhodesia (southward to the Muchinga Escarp-
ment), and southwestern Tanganyika.

— **Turdoides leucopygius hartlaubii** (Bocage)

Crateropus Hartlaubii Barboza du Bocage, 1868, Journ.
Sci. Math. Phys. Nat. Lisboa, 2, p. 48 — Biballa, Mossamedes, Angola.
Western Northern Rhodesia, northern Bechuanaland, and
southern Angola.

TURDOIDES HINDEI

Turdoides hindei (Sharpe)

Crateropus hindei Sharpe, 1900, Bull. Brit. Orn. Club, 11,
p. 29 — Athi River, Kenya.
Known only from eastern foothills of east Kenya highlands.

TURDOIDES HYPOLEUCUS

—**Turdoides hypoleucus hypoleucus** (Cabanis)

Crateropus hypoleucus Cabanis, 1878, Journ. f. Orn., 26,
p. 205 — Kitui [lat. 1° 22′ S., long. 38° 01′ E.], Kenya.

Known only from a restricted area of central Kenya southward from Mount Kenya (Athi River system).

Turdoides hypoleucus rufuensis (Neumann)

Crateropus hypoleucus rufuensis Neumann, 1906, Orn. Monatsb., 14, p. 148 — Useguha region, northeastern Tanganyika.

Turdoides hypoleuca [sic] *kilosa* Vincent, 1935, Bull. Brit. Orn. Club, 55, p. 176 — Kibedya, Kilosa [lat. 6° 49′ S., long. 37° 02′ E.], Tanganyika.

Northeastern Tanganyika.

TURDOIDES BICOLOR

Turdoides bicolor (Jardine)

Cratopus bicolor Jardine, 1831, Edinburgh Journ. Nat. Geogr. Sci., 3, p. 97, pl. 3 — South Africa.

South West Africa (Damaraland; Namaqualand), Bechuanaland (Kalahari Desert), and western Transvaal.

TURDOIDES GYMNOGENYS

Turdoides gymnogenys gymnogenys (Hartlaub)

Crateropus gymnogenys Hartlaub, 1865, Proc. Zool. Soc. London, p. 86 — Benguella Province, Angola.

Southwestern Angola.

Turdoides gymnogenys kaokensis (Roberts)

Aethocichla gymnogenys kaokensis Roberts, 1937, Ostrich, 8, p. 100 — Huab River, Kaokoveld, South West Africa.

Aethocichla gymnogenys tsumebensis Roberts, 1937, Ostrich, 8, p. 101 — Tsumeb, northeastern South West Africa.

Northern South West Africa.

GENUS **BABAX** DAVID

Babax David, 1875, Journ. trois. Voy. Expl. Chine, 1, p. 181. Type, by monotypy, *Pterorhinus lanceolatus* J. Verreaux.

Kaznakowia Bianchi, 1906, Bull. Acad. Imp. Sci., St. Pétersbourg, ser. 5, 23 (1905), p. 45. Type, by original designation, *Kaznakowia koslowi* Bianchi.

BABAX LANCEOLATUS

Babax lanceolatus lanceolatus (Verreaux)

Pterorhinus lanceolatus J. Verreaux, 1870, Nouv. Arch. Mus. Hist. Nat. [Paris], 6, p. 36; *idem*, 1871, *ibid.*, 7, p. 40, pl. 2, fig. 2 — "les montagnes du Thibet chinois"; type specimen from Muping [= Paohing], Hsikang, *fide* Oustalet, 1893, *ibid.*, sér. 3, 5, p. 193.

Babax Bonvaloti Oustalet, 1892, Ann. Sci. Nat., Zool., sér. 7, 12 (1891), p. 273 — "So," error; corrected to Tara [*ca.* lat. 30° 18′ N., long. 98° 30′ E.], Hsikang, by Oustalet, 1893, Nouv. Arch. Mus. Hist. Nat. [Paris], sér. 3, 5, p. 193.

Babax yunnanensis Rippon, 1905, Bull. Brit. Orn. Club, 15, p. 96 — hills east of Tengyueh [= Tengchung], Yunnan.

Southern Shensi; southern Kansu; western Szechwan; eastern half of Hsikang; northwestern Yunnan; northeastern Burma (Kachin and Northern Shan States).

Babax lanceolatus woodi Finn

Babax woodi Finn, 1902, in Wood and Finn, Journ. Asiat. Soc. Bengal, 71, p. 125, pl. 7 — Mount Victoria, Southern Chin Hills, Upper Burma.

Babax victoriæ Rippon, 1905, Bull. Brit. Orn. Club, 15, p. 97 — Mount Victoria, Southern Chin Hills, Upper Burma.

Babax lanceolatus oribata [sic] Koelz, 1954, Contrib. Inst. Regional Explor., no. 1, p. 3 — Blue Mountain, Lushai Hills, Assam.

Southeastern Assam (Lushai Hills) and western Burma (Chin Hills).

Babax lanceolatus latouchei Stresemann

Babax lanceolatus latouchei Stresemann, 1929, Orn. Monatsb., 37, p. 140 — Yao Shan, Kwangsi.

Hupeh; Fukien; Kwangsi.

BABAX WADDELLI

Babax waddelli waddelli Dresser

Babax waddelli Dresser, 1905 (June), Proc. Zool. Soc. London, p. 54, pl. 4 — near the Chaksam Ferry [lat. 29° 15′ N., long. 90° 32′ E.], Tsangpo Valley, Tibet.

Southeastern Tibet (Lhasa, Loti, Chushul Dzong, Chaksam).

? Babax waddelli lumsdeni Kinnear

Babax lanceolatus lumsdeni Kinnear, 1938, Bull. Brit. Orn. Club, 58, p. 76 — Le La [lat. 28° 25′ N., long. 92° 57′ E.], Tibet-Hsikang Boundary.

Known only from type locality; doubtfully distinct from *waddelli*.

Babax waddelli jomo Vaurie

Babax waddelli jomo Vaurie, 1955, Amer. Mus. Novit., no. 1753, p. 5 — below Tsechen, about four miles south of Gyangtse, southeastern Tibet.

Southeastern Tibet (vicinity of Gyangtse).

BABAX KOSLOWI

Babax koslowi (Bianchi)

Kaznakowia koslowi Bianchi, 1906, Bull. Acad. Imp. Sci. St. Pétersbourg, ser. 5, 23 (1905), p. 45 — Bar Chu and Dzer Chu, tributaries of the Mekong near Chamdo, Hsikang.

Valley of the Mekong in southern Tsinghai and northern Hsikang.

Genus GARRULAX Lesson

Garrulax Lesson, 1831, Traité Orn., livr. 8, p. 647. Type, by subsequent designation (Ripley, 1961, Synopsis Birds India Pakistan, p. 380), *Garrulax rufifrons* Lesson.

Ianthocincla Gould, 1835, Proc. Zool. Soc. London, pt. 3, p. 47. Type, by original designation, *Cinclosoma ocellatum* Vigors.

Garrulaxis "Less." Lafresnaye, 1838, Essai d'une nouvelle manière de grouper les genres et les espèces de l'Ordre des Passereaux, p. 25. *Nomen emendatum.*

Xanthocincla Jerdon, 1839, Madras Journ. Lit. Sci., 10, p. 255. *Nomen emendatum.*

Trochalopteron Blyth, 1843, Journ. Asiat. Soc. Bengal, 12, p. 952, footnote. Type, by subsequent designation (Baker, 1930, Fauna Brit. India, Birds, ed. 2, 7, p. 30), *Trochalopteron subunicolor* Blyth.

Trochalopterum Agassiz, 1846, Nomencl. Zool. Index Univ. *Nomen emendatum.*

Pterocyclus G. R. Gray, 1846, Gen. Birds, 1, p. [226], pl. 57. Type, by subsequent designation (G. R. Gray, 1855, Cat. Gen. Subgen. Birds, p. 45), *Cinclosoma erythrocephalum* Vigors.

Grammatoptila Reichenbach, 1850, Avium Syst. Nat., pl. 85. No species; generic details only. Type, by subsequent designation (G. R. Gray, 1855, Cat. Gen. Subgen. Birds, p. 45), *Garrulus striatus* Vigors.

Leucodioptron "Schiff." Bonaparte, 1854, Compt. Rend. Acad. Sci. Paris, 38, pp. 54, 55. Type, by original designation and monotypy, *Turdus canorus* Linnaeus.

Leucodiophron "Schiff. 1853" G. R. Gray, 1855, Cat. Gen. Subgen. Birds, p. 45. Type, by monotypy, *Turdus sinensis* Linnaeus, 1766 = *Turdus canorus* L., 1758.

Pterorhinus Swinhoe, 1868, Ibis, p. 60. Type, by original designation and monotypy, *Pterorhinus davidi* Swinhoe.

Leucodiopterum Swinhoe, 1868, Ibis, p. 61. *Nomen emendatum.*

Kittasoma "Bl. 1855" G. R. Gray, 1869, Hand-list Gen. Spec. Birds, pt. 1, p. 284. In synonymy with *Grammatoptila* Reichenbach.

Stactocichla Sharpe, 1883, Cat. Birds Brit. Mus., 7, p. 449. Type, by monotypy, *Garrulax merulinus* Blyth.

Melanocichla Sharpe, 1883, Cat. Birds Brit. Mus., 7, p. 451. Type, by monotypy, *Timalia lugubris* S. Müller.

Rhinocichla Sharpe, 1883, Cat. Birds Brit. Mus., 7, p. 452. Type, by subsequent designation (Anonymous, 1883, Ibis, p. 573), *Timalia mitrata* S. Müller.

Dryonastes Sharpe, 1883, Cat. Birds Brit. Mus., 7, p. 454. Type, by subsequent designation (Anonymous, 1883, Ibis, p. 573) *Ianthocincla ruficollis* Jardine and Selby.

Allocotops Sharpe, 1888, Ibis, p. 389. Type, by original designation and monotypy, *Allocotops calvus* Sharpe.

GARRULAX CINEREIFRONS

Garrulax cinereifrons Blyth

Garrulax cinereifrons "Kelaart" Blyth, 1851, Journ. Asiat. Soc. Bengal, 20, p. 176 — Ceylon.

Wet zone of southwestern Ceylon.

GARRULAX PALLIATUS

Garrulax palliatus palliatus (Bonaparte)

[*Janthocincla*] *palliata* "Müll." Bonaparte, 1850, Consp. Av., 1, p. 371 — Sumatra.
Highlands of western Sumatra.

Garrulax palliatus schistochlamys Sharpe

Garrulax schistochlamys Sharpe, 1888, Ibis, p. 479 — Kina Balu, North Borneo.
Highlands of northern Borneo.

GARRULAX RUFIFRONS

Garrulax rufifrons rufifrons Lesson

Garrulax rufifrons Lesson, 1831, Traité Orn., livr. 8, p. 648 — Java; here restricted to Mount Salak [lat. 6° 42′ S., long. 106° 44′ E.].
Mountains of western Java.

Garrulax rufifrons slamatensis Siebers

Garrulax rufifrons slamatensis Siebers, 1929, Treubia, 11, p. 150 — Kaligua, Mount Selamat [lat. 7° 13′ S., long. 109° 05′ E.], Java.
Known only from type locality.

GARRULAX PERSPICILLATUS

Garrulax perspicillatus (Gmelin)

[*Turdus*] *perspicillatus* Gmelin, 1789, Syst. Nat., 1, pt. 2, p. 830 — China; restricted to Amoy, Fukien, by Meinertzhagen, 1928, Ibis, p. 516.
Dryonastes perspicillatus shensiensis Riley, 1911, Proc. Biol. Soc. Washington, 24, p. 43 — fifteen miles south of Sian-fu, Shensi.
Dryonastes tsinlingensis Reichenow, 1917, Journ. f. Orn., 65, p. 391 — Chin Ling Mountains, Shensi.
Dryonastes perspicillatus annamensis Delacour, 1927, Bull. Brit. Orn. Club, 47, p. 157 — Hue [lat. 16° 29′ N., long. 107° 34′ E.], Annam. Not *Stactocichla merulina annamensis* Robinson and Kloss, 1919 = *Garrulax merulinus annamensis* (Robinson and Kloss).
Valley of the Yangtze from Szechwan eastward to Kiang-

su, and southward through coastal provinces to Kwangsi; eastern Tongking and northern half of Annam.

GARRULAX ALBOGULARIS

Garrulax albogularis whistleri Baker

Garrulax albogularis whistleri Baker, 1921, Bull. Brit. Orn. Club, 42, p. 29 — Simla, Himachal Pradesh.

Himalayas from northern West Pakistan southeastward to northwestern Uttar Pradesh.

Garrulax albogularis albogularis (Gould)

Ianthocincla albogularis Gould, 1836, Proc. Zool. Soc. London, pt. 3, p. 187 — Nepal.

Himalayas from western Nepal eastward to eastern Bhutan.

Garrulax albogularis eous Riley

Garrulax albogularis eous Riley, 1930 (June 5), Proc. Biol. Soc. Washington, 43, p. 79 — Fuchuan Shan, Mekong-Salween Divide, northwestern Yunnan.

Garrulax albogularis laetus Riley, 1930 (July 18), Proc. Biol. Soc. Washington, 43, p. 134 — Mount Omei, southwestern Szechwan.

Southwestern Szechwan, southeastern Hsikang, Yunnan, and northwestern Tongking.

Garrulax albogularis ruficeps Gould

Garrulax ruficeps Gould, 1863, Proc. Zool. Soc. London, p. 281 — Formosa.

Formosa.

GARRULAX LEUCOLOPHUS

Garrulax leucolophus leucolophus (Hardwicke)

Corvus leucolophus Hardwicke, 1815, Trans. Linn. Soc. London, 11, p. 207, pl. 15 — mountains above Hardwar, Saharanpur, Uttar Pradesh.

Himalayas from Himachal Pradesh eastward, through Nepal, Sikkim, Bhutan, and Assam (north of the Brahmaputra), to Mishmi Hills.

Garrulax leucolophus patkaicus Reichenow

> G[*arrulax*]. *patkaicus* Reichenow, 1913, Journ. f. Orn., 61, p. 557 — Patkai Range, northwestern Burma.
>
> *Garrulax leucolophus hardwickii* Ticehurst, 1926, Bull. Brit. Orn. Club, 46, p. 113 — Naga Hills, Assam.

Assam (south of the Brahmaputra), northern Burma, and western Yunnan; in Burma southward to Arakan and Northern Shan State.

Garrulax leucolophus belangeri Lesson

> *Garrulax Belangeri* Lesson, 1832, in Bélanger, Voyage aux Indes-Orientales, Zool., pt. 4, p. 258, Atlas, Oiseaux, pl. 4 — Pegu Division, Lower Burma.

Southern Burma (except areas occupied by *patkaicus* and *diardi*) southward to central Tenasserim; southwestern Thailand (valley of the Mae Klong).

Garrulax leucolophus diardi (Lesson)

> *Turdus Diardi* Lesson, 1831, Traité Orn., livr. 6, p. 408 — Cochinchina.
>
> [*Garrulax*] *leucogaster* Walden, 1867, Proc. Zool. Soc. London, (1866), p. 549 — "some part of Siam"; type specimen from Cambodia, *fide* Macdonald, 1946, in de Schauensee, Proc. Acad. Nat. Sci. Philadelphia, 98, p. 61.
>
> *Garrulax leucolophus peninsulae* de Schauensee, 1946, Proc. Acad. Nat. Sci. Philadelphia, 98, p. 60 — near Khao Luang [lat. 11° 40′ N., long. 99° 35′ E.], Thailand. Not *Trochalopterum peninsulæ* Sharpe, 1887 = *Garrulax erythrocephalus peninsulae* (Sharpe).
>
> *Garrulax leucolophus peninsularis* de Schauensee, 1946, Proc. Acad. Nat. Sci. Philadelphia, 98, p. 122. New name for *Garrulax leucolophus peninsulae* de Schauensee, preoccupied. Not *Melanocichla peninsularis* Sharpe, 1888 = *Garrulax lugubris lugubris* (S. Müller).

Southern Shan State, Burma (east of the Salween); all Thailand (except area occupied by *belangeri*) southward to peninsular province of Prachuap Khiri Khan; southeastern Yunnan; Tongking; Laos; Annam; Cochin China; Cambodia.

Garrulax leucolophus bicolor Hartlaub

G[arrulax]. bicolor "Müll." Hartlaub, 1844, Syst. Verz.
Naturh. Samml. Ges. Mus. (Bremen), Abth. 1, p. 44,
footnote — western Sumatra.

Garrulax leucolophus obscurus Junge, 1948, Zool. Meded.
Leiden, 29, p. 324 — Balek [lat. 4° 42′ N., long. 96° 44′
E.], Sumatra. Not Garrulax merulinus obscurus Dela-
cour and Jabouille, 1930.

Highlands of western Sumatra.

GARRULAX MONILEGER

Garrulax monileger monileger (Hodgson)

Cinc[losoma]. Monilegera [sic] Hodgson, 1836, Asiat.
Res., 19, p. 147 — Nepal.

Himalayas from Nepal eastward, through Bhutan and
Assam (both north and south of the Brahmaputra, except
area occupied by badius), to northeastern Burma, southward in Burma to Arakan and Northern Shan State.

Garrulax monileger badius Ripley

Garrulax moniliger badius Ripley, 1948, Proc. Biol. Soc.
Washington, 61, p. 102 — Tezu, Mishmi Hills, Assam.

Northeastern Assam (Mishmi Hills).

Garrulax monileger stuarti de Schauensee

Garrulax moniliger bakeri de Schauensee, 1935, Proc.
Acad. Nat. Sci. Philadelphia, 87, p. 409 — Nong Ho,
a pond near Muang Chiang Mai [lat. 18° 45′ N., long.
99° 00′ E.], Thailand. Not Trochalopteron phœniceum
bakeri Hartert, 1909 = Garrulax phoeniceus bakeri
(Hartert).

Garrulax moniliger stuarti de Schauensee, 1955, Auk,
72, p. 92. New name for Garrulax moniliger bakeri
de Schauensee, preoccupied.

Southeastern Burma from Southern Shan State (except
area occupied by schauenseei) southward through Karenni State to northern Tenasserim; northwestern Thailand
southward to northern Tak Province.

Garrulax monileger fuscatus Baker

Garrulax moniliger fuscata [sic] Baker, 1918, Bull. Brit.
Orn. Club, 38, p. 64 — Tavoy District, Tenasserim.

Central Tenasserim (Amherst and Tavoy Districts) and southwestern Thailand from southern Tak Province to Prachuap Khiri Khan.

Garrulax monileger mouhoti Sharpe

Garrulax mouhoti Sharpe, 1883, Cat. Birds Brit. Mus., 7, pp. 434 (in key), 444 — Cambodia.

Garrulax moniliger leucotis Baker, 1917, Bull. Brit. Orn. Club, 38, p. 8 — Ban Kabin Buri [lat. 14° 00′ N., long. 101° 45′ E.], Thailand.

Southeastern Thailand, Cambodia, Cochin China, and southern Annam.

Garrulax monileger pasquieri Delacour and Jabouille

Garrulax moniliger pasquieri Delacour and Jabouille, 1924, Bull. Brit. Orn. Club, 45, p. 32 — Khesanh, Quangtri Province, Annam.

Central Annam (Thuatien and Quangtri Provinces).

Garrulax monileger schauenseei Delacour and Greenway

Garrulax moniliger schauenseei Delacour and Greenway, 1939, Bull. Brit. Orn. Club, 59, p. 132 — Chiang Khwang [lat. 19° 19′ N., long. 103° 22′ E.], Laos.

Eastern Southern Shan State of Burma, eastern northern plateau of Thailand, and northern Laos.

Garrulax monileger tonkinensis Delacour

Garrulax moniliger tonkinensis Delacour, 1927, Bull. Brit. Orn. Club, 47, p. 158 — Bac Kan [lat. 22° 08′ N., long. 105° 50′ E., Tongking.

Northern Annam (Vinh and Thanhoa Provinces), Tongking, and Kwangsi.

Garrulax monileger melli Stresemann

Garrulax moniliger melli Stresemann, 1923, Journ. f. Orn., 71, p. 364 — Man-tsi-shan, Kwangtung.

Hill tracts of southeastern China from Kwangtung northward to Anhwei.

Garrulax monileger schmackeri Hartlaub

Garrulax Schmackeri Hartlaub, 1898, Abh. Nat. Ver. Bremen, 14 (1897), p. 349, pl. 4 — Hainan.

Hainan.

GARRULAX PECTORALIS

Garrulax pectoralis pectoralis (Gould)

Ianthocincla pectoralis Gould, 1836, Proc. Zool. Soc. London, pt. 3, p. 186 — Nepal.

Nepal.

Garrulax pectoralis melanotis Blyth

G[arrulax]. melanotis Blyth, 1843, Journ. Asiat. Soc. Bengal, 12, p. 949 — Arakan.

G[arrulax]. McClellandii Blyth, 1843, Journ. Asiat. Soc. Bengal, 12, p. 949 — Assam. Based on *Ianthocincla pectoralis* "Gould" McClelland, 1840, in Horsfield, Proc. Zool. Soc. London, pt. 7 (1839), p. 160, not *Ianthocincla pectoralis* Gould, 1836.

G[arrulax]. uropygialis "Caban." Bonaparte, 1850, Consp. Av., 1, p. 371 — Assam.

Garrulax waddelli Ogilvie-Grant, 1894, Bull. Brit. Orn. Club, 3, p. 29 — Rangit River, Sikkim.

Himalayas from Darjeeling District and Sikkim eastward, through Bhutan, Assam (both north and south of the Brahmaputra), and northern Burma, to western Yunnan, in Burma southward to Arakan and Northern Shan State.

Garrulax pectoralis subfusus Kinnear

Garrulax pectoralis meridionalis Robinson and Kloss, 1919, Bull. Brit. Orn. Club, 40, p. 11 — Hat Sanuk [ca. lat. 11° 47′ N., long. 99° 38′ E.], Thailand. Not *Trochalopterum meridionale* Blanford, 1880 = *Garrulax cacchinans meridionalis* (Blanford).

Garrulax pectoralis subfusa (sic) Kinnear, 1924, Bull. Brit. Orn. Club, 44, p. 103 — Mitan, Amherst, Tenasserim.

Garrulax pectoralis subsuffusa (sic) Kinnear, 1924, Bull. Brit. Orn. Club, 45, p. 28. *Lapsus* for *subfusus*.

Garrulax pectoralis confusa (sic) Baker, 1930, Fauna Brit. India, Birds, ed. 2, 7, p. 28. Lapsus for *subfusus*.

Southeastern Burma from Southern Shan State southward to central Tenasserim; western Thailand from Chiang Rai Province southward to Prachuap Khiri Khan; northwestern Laos.

Garrulax pectoralis robini Delacour

Garrulax pectoralis robini Delacour, 1927, Bull. Brit. Orn. Club, 47, p. 157 — Tam Dao [lat. 21° 27′ N., long. 105° 40′ E.], Tongking.
Tongking and adjacent northern Laos.

Garrulax pectoralis picticollis Swinhoe

Garrulax picticollis Swinhoe, 1872, Proc. Zool. Soc. London, p. 554 — Chekiang.
Hill tracts of southeastern China from Kwangtung northward to Anhwei.

Garrulax pectoralis semitorquatus Ogilvie-Grant

Garrulax semitorquata (sic) Ogilvie-Grant, 1900, Bull. Brit. Orn. Club, 10, p. 49 — Five-Finger Mountains, Hainan.
Hainan.

GARRULAX LUGUBRIS

Garrulax lugubris lugubris (Müller)

Timalia lugubris S. Müller, 1835, Tijdschr. Natuur. Gesch. Phys., 2, p. 344, pl. "5, fig. 2" [= pl. 9, fig. 2] — Sumatra.
Melanocichla peninsularis Sharpe, 1888, Proc. Zool. Soc. London, p. 274 — Gunong Batu Puteh, Perak.
Highlands of Malaya from northern Perak southward to southern Selangor and highlands of western Sumatra.

Garrulax lugubris calvus (Sharpe)

Allocotops calvus Sharpe, 1888, Ibis, p. 389 — Kina Balu, North Borneo.
Highlands of northeastern Borneo.

GARRULAX STRIATUS

Garrulax striatus striatus (Vigors)

Garrulus striatus Vigors, 1831, Proc. Comm. Zool. Soc. London, pt. 1, p. 7 — Himalayas; restricted to Naini Tal, Kumaun, Uttar Pradesh, by Baker, 1920, Journ. Bombay Nat. Hist. Soc., 27, p. 245.
Himalayas from northern East Punjab eastward to Kumaun.

Garrulax striatus vibex Ripley

> *Garrulax striatus vibex* Ripley, 1950, Proc. Biol. Soc. Washington, 63, p. 103 — Godavari, Central Valley, Nepal.

Himalayas of western and central Nepal.

Garrulax striatus sikkimensis (Ticehurst)

> *Grammatoptila striata sikkimensis* Ticehurst, 1924, Bull. Brit. Orn. Club, 44, p. 104 — Sikkim.

Himalayas from eastern Nepal eastward to eastern Bhutan.

Garrulax striatus cranbrooki (Kinnear)

> *Grammatoptila austeni* Oates, 1889, Fauna Brit. India, Birds, 1, p. 104 — Dafla and Naga Hills, Assam. Not *Trochalopteron Austeni* Godwin-Austen, 1870 = *Garrulax austeni austeni* (Godwin-Austen).
>
> *Grammatoptila striata cranbrooki* Kinnear, 1932, Bull. Brit. Orn. Club, 53, p. 79 — Adung Valley [at lat. 28° 10′ N., long. 97° 40′ E.], Burma.
>
> *Garrulax striatus brahmaputra* Hachisuka, 1953, Auk, 70, p. 92. New name for *Grammatoptila austeni* Oates, preoccupied.

Central Bhutan eastward, through hill tracts of Assam, to northeastern Burma (Kachin State) and western Burma (Chin Hills).

GARRULAX STREPITANS

Garrulax strepitans strepitans Blyth

> *Garrulax strepitans* "Tickell" Blyth, 1855, Journ. Asiat. Soc. Bengal, 24, p. 268 — "the mountainous interior of the Tenasserim provinces."

Northwestern Laos, eastern Southern Shan State of Burma (Kengtung), western Thailand from Chiang Rai Province southward to Kanchanaburi, and central Tenasserim (Amherst District).

Garrulax strepitans ferrarius Riley

> *Garrulax ferrarius* Riley, 1930, Proc. Biol. Soc. Washington, 43, p. 190 — Khao Kuap [lat. 12° 25′ N., long. 102° 50′ E.], Thailand.

Southeastern Thailand.

GARRULAX MILLETI

Garrulax milleti Robinson and Kloss

Garrulax milleti Robinson and Kloss, 1919, Ibis, p. 574, pl. 12 — Dalat [lat. 11° 55′ N., long. 108° 26′ E.], Annam.

Southern half of Annam.

GARRULAX MAESI

Garrulax maesi grahami (Riley)

Dryonastes grahami Riley, 1922, Proc. Biol. Soc. Washington, 35, p. 59 — Shin Kai Si, Mount Omei, Szechwan.

Southwestern Szechwan, southeastern Hsikang, and northeastern Yunnan.

Garrulax maesi maesi (Oustalet)

Dryonastes Maesi Oustalet, 1890, Bull. Soc. Zool. France, 15, p. 155 — Tongking; restricted to the Tam Dao Mountains [ca. lat. 21° 30′ N., long. 105° 34′ E.], by Delacour, 1927, Bull. Brit. Orn. Club, 47, p. 83.

Mountains of Kwangsi and Tongking.

Garrulax maesi varennei (Delacour)

Dryonastes varennei Delacour, 1926, Bull. Brit. Orn. Club, 47, p. 15 — Chiang Khwang [lat. 19° 19′ N., long. 103° 22′ E.], Laos.

Northeastern and central Laos (Chiang Khwang and Thakkek Provinces).

Garrulax maesi castanotis (Ogilvie-Grant)

Dryonastes castanotis Ogilvie-Grant, 1899, Ibis, p. 584 — Five-finger Mountains, Hainan.

Hainan.

GARRULAX CHINENSIS

Garrulax chinensis nuchalis Godwin-Austen

Garrulax nuchalis Godwin-Austen, 1876, Ann. Mag. Nat. Hist., ser. 4, 18, p. 411 — Lhota, Naga Hills, Assam.

Lower hills of northeastern Assam (south and east of the Brahmaputra) and of northern Burma (southeastward to Myitkyina District).

Garrulax chinensis lochmius Deignan

Garrulax chinensis lochmius Deignan, 1941, Zoologica [New York], 26, p. 241 — King Chiang Saen [lat. 20° 15′ N., long. 100° 05′ E.], Thailand.

Southeastern Burma from Northern Shan State southward to northern Tenasserim; northern Thailand and northwestern portion of eastern plateau; western portion of northern Laos; southwestern Yunnan.

Garrulax chinensis propinquus (Salvadori)

Dryonastes propinquus Salvadori, 1915, Ann. Mus. Civ. Genova, ser. 3, 6 (1914), p. 6 — near Thagata, southwest of Mulayit Taung, Amherst District, Tenasserim.

Central Tenasserim (Amherst and Tavoy Districts) and southwestern Thailand (Kanchanaburi and Rat Buri Provinces).

Garrulax chinensis germaini (Oustalet)

Dryonastes Germaini Oustalet, 1890, Bull. Soc. Zool. France, 15, p. 157 — Lower Cochinchina.

Cochin China and southernmost Annam (Phantiet and Phanrang Provinces).

Garrulax chinensis chinensis (Scopoli)

Lanius (chinensis) Scopoli, 1786, Del. Flor. Fauna Insubr., fasc. 2, p. 86 — China, ex Sonnerat; restricted to Canton, by Meinertzhagen, 1928, Ibis, p. 514.

Cr[ateropus]. leucogenys Blyth, 1842, Journ. Asiat. Soc. Bengal, 11, p. 180 — "Upper Bengal," error; corrected to China, by Blyth, 1843, Journ. Asiat. Soc. Bengal, 12, p. 949.

Garrulax chinensis, var. *lugens* Oustalet, 1879, Bull. Soc. Philom. [Paris], sér. 7, 3, p. 216 — Laos; here restricted to Chiang Khwang [lat. 19° 19′ N., long. 103° 22′ E.].

Dryonastes chinensis lowei La Touche, 1922, Bull. Brit. Orn. Club, 42, p. 52 — Hokow, southeastern Yunnan.

Northern half of Annam, eastern portion of northern Laos, Tongking, southeastern Yunnan, southern Kwangsi, and southwestern half of Kwangtung.

Garraulax chinensis monachus Swinhoe

   Garrulax monachus Swinhoe, 1870, Ibis, p. 248 — Hainan.

Hainan.

GARRULAX VASSALI

Garrulax vassali (Ogilvie-Grant)

   Dryonastes vassali Ogilvie-Grant, 1906, Bull. Brit. Orn. Club, 19, p. 13 — Bali Region [= Nhatrang Province], southern Annam.

Southern Laos (Saravane Province) and southern half of Annam.

GARRULAX GALBANUS

Garrulax galbanus galbanus Godwin-Austen

   Garrulax galbanus Godwin-Austen, 1874, Proc. Zool. Soc. London, p. 44, pl. 10 — Manipur.
   Garrulax galbanus galbanatus Koelz, 1954, Contrib. Inst. Regional Explor., no. 1, p. 2 — Blue Mountain, Lushai Hills, Assam.

Southeastern Assam (Manipur and Lushai Hills) and western Burma (Chin Hills).

Garrulax galbanus courtoisi Menegaux

   Garrulax Courtoisi Menegaux, 1923, Bull. Mus. Hist. Nat. Paris, 29, p. 287 — Wuyuan, northeastern Kiangsi.

Known only from type locality.

GARRULAX DELESSERTI

Garrulax delesserti delesserti (Jerdon)

   C[rateropus]. Delesserti Jerdon, 1839, Madras Journ. Lit. Sci., 10, p. 256 — Kotagiri, Nilgiri Hills.

Hill tracts of southwestern India from Travancore northward to North Kanara.

Garrulax delesserti gularis (McClelland)

   Ianthocincla gularis McClelland, 1840 (March), in Horsfield, Proc. Zool. Soc. London, (1839), p. 159 — Assam; restricted to Cachar, *apud* Baker, 1920, Journ. Bombay Nat. Hist. Soc., 27, p. 240; Sadiya suggested by Kinnear, 1937, in Ludlow and Kinnear, Ibis, p. 30.

Turdus (s.-g. *Crateropus*) *griseiceps* Delessert, 1840
(April), Rev. Zool. [Paris], 3, p. 101 — Bhutan.

Garrulax gularis auratus Delacour, 1926, Bull. Brit. Orn.
Club, 47, p. 15 — Chiang Khwang [lat. 19° 19′ N., long.
103° 22′ E.], Laos.

Garrulax gularis gratior Koelz, 1954, Contrib. Inst. Regional Explor., no. 1, p. 2 — Sangau, Lushai Hills, Assam.

Bhutan; hill tracts of Assam, both north and south of the
Brahmaputra; northern Burma; northern Laos.

GARRULAX VARIEGATUS

Garrulax variegatus similis (Hume)

Trochalopteron simile Hume, 1871, Ibis, p. 408 — no locality; type from Naoshera, Mirpur, Kashmir, *fide*
Sharpe, 1883, Cat. Birds Brit. Mus., 7, p. 360.

Himalayas from northernmost West Pakistan southeastward to western Kashmir.

Garrulax variegatus variegatus (Vigors)

Cinclosoma variegatum Vigors, 1831, Proc. Comm. Zool.
Soc. London, pt. 1, p. 56 — Himalayas; inferentially
restricted to Simla, Himachal Pradesh, by Hume, 1878,
Stray Feathers, 7, p. 457.

Himalayas from Chamba District, Himachal Pradesh,
southeastward into Nepal.

GARRULAX DAVIDI

? Garrulax davidi chinganicus (Meise)

Ianthocincla davidi chinganica Meise, 1934, Abh. Ber.
Mus. Dresden, 18, p. 41 — Arun River near Buchedu,
Khingan Range, northern Manchuria.

Khingan Range from Heilungkiang to Jehol; doubtfully
distinct from *davidi*.

Garrulax davidi davidi (Swinhoe)

Pterorhinus davidi Swinhoe, 1868, Ibis, p. 61 — Peking,
Hopeh.

Janthocincla davidi funebris Stresemann, 1927, Orn.
Monatsb., 35, p. 134 — Lanhuku, Tatung Range, northeastern Tsinghai.

Northern China from Hopeh westward through southern Inner Mongolia and Kansu to eastern Tsinghai.

? Garrulax davidi experrectus (Bangs and Peters)

Ianthocincla davidi experrecta Bangs and Peters, 1928, Bull. Mus. Comp. Zool., 68, p. 339 — Liyuanku, northern slopes of Richthofen Range, northern Kansu.

Northern slopes of Richthofen Range; doubtfully distinct from *davidi*.

Garrulax davidi concolor (Stresemann)

Janthocincla davidi concolor Stresemann, 1923, Journ. f. Orn., 71, p. 365 — Sungpan, northwesern Szechwan.

Known only from type locality.

GARRULAX SUKATSCHEWI

Garrulax sukatschewi (Berezowski and Bianchi)

TrochalopteronSukatschewi Berezowski and Bianchi, 1891, Aves expeditionis Potanini per provinciam Gan-su et confinia 1884-1887, p. 59, pl. 1, fig. 1 — Hsiku and Minchow districts, southern Kansu.

Southern Kansu.

GARRULAX CINERACEUS

Garrulax cineraceus cineraceus (Godwin-Austen)

Trochalopteron cineraceum Godwin-Austen, 1874, Proc. Zool. Soc. London, p. 45, pl. 11 — no locality; type from Thobal Valley, Manipur, *fide* Sharpe, 1883, Cat. Birds Brit. Mus., 7, p. 336.

Hill tracts of Assam south of the Brahmaputra and western Burma (Chin Hills).

Garrulax cineraceus strenuus Deignan

Trochalopteron Styani Oustalet, 1901, Nouv. Arch. Mus. Hist. Nat. [Paris], sér. 4, 3, p. 276 — Tatsienlu [Kangting], "Szechwan," and Tzeku, Yunnan; restricted to Tzeku, by Berlioz, 1930, Oiseau Rev. Franç. Orn., 11, p. 20. Not *Trochalopteron Styani* Oustalet, 1898 = *Garrulax cineraceus cinereiceps* (Styan), 1887.

Garrulax cineraceus strenuus Deignan, 1957, Proc. Biol. Soc. Washington, 70, p. 190 — Tsehchung Mountains, Mekong Valley, northwestern Yunnan.

Eastern Northern Shan State Burma, western Yunnan, and southeastern Hsikang.

Garrulax cineraceus cinereiceps (Styan)

Trochalopteron cinereiceps Styan, 1887, Ibis, p. 167, pl. 6 — "Yunnan"; type specimen agrees with birds of the Yangtze Valley, *fide* Sims, 1957, in Deignan, Proc. Biol. Soc. Washington, 70, p. 190.

Trochalopteron Styani Oustalet, 1898, Bull. Mus. Hist. Nat. Paris, 4, p. 226. New name for *Trochalopteron cinereiceps* Styan.

Trochalopterum ningpoense David and Oustalet, 1890, Naturaliste, sér. 2, 4, p. 187 — Ningpo, Chekiang.

Valley of the Yangtze from western Szechwan to Anhwei, and southward through coastal provinces from Chekiang to Kwangtung.

GARRULAX RUFOGULARIS

Garrulax rufogularis occidentalis (Hartert)

Ianthocincla rufogularis occidentalis Hartert, 1909, Vög. pal. Fauna, p. 635 — Dehra Dun, northwestern Uttar Pradesh.

Himalayas from northern West Pakistan eastward to northwestern Uttar Pradesh.

Garrulax rufogularis grosvenori Ripley

Garrulax rufogularis grosvenori Ripley, 1950, Proc. Biol. Soc. Washington, 63, p. 104 — Rekcha, Dailekh District, western Nepal.

Western Nepal.

Garrulax rufogularis rufogularis (Gould)

Ianthocincla rufogularis Gould, 1835, Proc. Zool. Soc. London, pt. 3, p. 48 — Himalayas; restricted to Sikkim, by Baker, 1920, Journ. Bombay Nat. Hist. Soc., 27, p. 241.

Cinc[losoma]. Rufimenta [sic] Hodgson, 1836, Asiat. Res., 19, p. 148 — Nepal; restricted to Kathmandu, by Ripley, 1950, Proc. Biol. Soc. Washington, 63, p. 104.

Himalayas from central Nepal eastward, through Bhutan, into Assam (north of the Brahmaputra).

Garrulax rufogularis assamensis (Hartert)

Ianthocincla rufogularis assamensis Hartert, 1909, Vög. pal. Fauna, p. 635 — Margherita, northeastern Assam.
Northeastern Assam.

Garrulax rufogularis rufitinctus (Koelz)

Ianthocincla rufogularis rufitincta Koelz, 1952, Journ. Zool. Soc. India, 4, p. 37 — Pynursla, Khasi Hills, Assam.
Hill tracts of Assam south of the Brahmaputra (Khasi Hills).

Garrulax rufogularis rufiberbis (Koelz)

Ianthocincla rufogularis rufiberbis Koelz, 1954, Contrib. Inst. Regional Explor., no. 1, p. 3 — between Langyang and Htawgaw, Kachin State, Upper Burma.
Northern Burma.

Garrulax rufogularis intensior Delacour and Jabouille

Garrulax rufogularis intensior Delacour and Jabouille, 1930, Oiseau Rev. Franç. Orn., 11, p. 398 — Chapa [lat. 22° 20′ N., long. 103° 50′ E.], Tongking.
Known only from type locality.

GARRULAX LUNULATUS

Garrulax lunulatus lunulatus (Verreaux)

Janthocincla lunulata J. Verreaux, 1870, Nouv. Arch. Mus. Hist. Nat. [Paris], 6, p. 36, pl. 3, fig. 2; *idem*, 1871, *ibid.*, 7, p. 41 — "les montagnes du Thibet chinois"; type from western Szechwan, *fide* Verreaux, 1871.
Southern Kansu, southern Shensi, and Szechwan.

Garrulax lunulatus bieti (Oustalet)

Ianthocincla Bieti Oustalet, 1897, Bull. Mus. Hist. Nat. Paris, 3, p. 163 — Tzeku, northwestern Yunnan.
Southeastern Hsikang and northwestern Yunnan.

GARRULAX MAXIMUS

Garrulax maximus (Verreaux)

Pterorhinus maximus J. Verreaux, 1870, Nouv. Arch. Mus. Hist. Nat. [Paris], 6, p. 36, pl. 3, fig. 1; *idem*,

1871, *ibid.*, 7, p. 38 — "les montagnes du Thibet chinois"; types from Muping [= Paohing], Hsikang, *fide* Verreaux, 1871.

Janthocincla maxima khamensis Serebrovski, 1927, Doklady Akad. Nauk. S.S.S.R., A, no. 20, p. 326 — River Bar-chu, basin of the Upper Mekong, Hsikang.

Southern Kansu, northwestern Szechwan, southeastern Hsikang, and northwestern Yunnan; southeastern Tibet (Pome district).

GARRULAX OCELLATUS

Garrulax ocellatus griseicauda Koelz

Garrulax ocellatus griseicauda Koelz, 1950, Amer. Mus. Novit., no. 1452, p. 7 — Wan, Garhwal, Uttar Pradesh.

Himalayas from northwestern Uttar Pradesh into western Nepal.

Garrulax ocellatus ocellatus (Vigors)

Cinclosoma ocellatum Vigors, 1831, Proc. Comm. Zool. Soc. London, pt. 1, p. 55 — Himalayas; restricted to Darjeeling, West Bengal, by Baker, 1922, Fauna Brit. India, Birds, ed. 2, 1, p. 156.

Himalayas from central Nepal eastward to Bhutan; southern Tibet.

Garrulax ocellatus maculipectus Hachisuka

Ianthocincla ocellata similis Rothschild, 1921, Novit. Zool., 28, p. 34 — Shweli-Salween Divide, Yunnan. Not *Trochalopteron simile* Hume, 1871 = *Garrulax variegatus similis* (Hume).

Garrulax ocellatus maculipectus Hachisuka, 1953, Auk, 70, p. 92. New name for *Ianthocincla ocellata similis*, preoccupied.

Northeastern Burma and northwestern Yunnan.

Garrulax ocellatus artemisiae (David)

Cinclosoma Artemisiæ David, 1871, Ann. Mag. Nat. Hist., ser. 4, 7, p. 256 — Muping [= Paohing], Hsikang.

Southwestern Szechwan and adjacent portion of Hsikang.

GARRULAX CAERULATUS

Garrulax caerulatus caerulatus (Hodgson)

Cinc[losoma]. *Cærulatus* (sic) Hodgson, 1836, Asiat.
Res., 19, p. 147 — Nepal.

Himalayas from central Nepal eastward, through Bhutan,
into Assam (north of the Brahmaputra).

Garrulax caerulatus subcaerulatus Hume

Garrulax subcærulatus Hume, 1878, Stray Feathers, 7,
p. 140 — near Shillong, Khasi Hills, Assam.

Hill tracts of Assam south of the Brahmputra (Khasi
Hills).

Garrulax caerulatus livingstoni Ripley

Garrulax caerulatus livingstoni Ripley, 1952, Journ. Bombay Nat. Hist. Soc., 50, p. 497 — Mount Japvo, Naga
Hills, Assam.

Dryonastes caerulatus biswasi Koelz, 1953, Journ. Zool.
Soc. India, 4, p. 153 — Kohima, Naga Hills, Assam.

Hill tracts of eastern Assam (Naga Hills and Manipur)
and adjacent northwestern Burma.

Garrulax caerulatus kaurensis (Rippon)

Dryonastes kaurensis Rippon, 1901, Bull. Brit. Orn. Club,
12, p. 13 — "Kauri-Kachin tract, to the east of Bhamo,
and bordering on the south of the Tapeng River," Kachin State, Burma.

Known only from type locality.

Garrulax caerulatus latifrons (Rothschild)

Ianthocincla caerulata latifrons Rothschild, 1926, Novit.
Zool., 33, p. 266 — Shweli-Salween Divide, Yunnan.

Northeastern Burma (Myitkyina District) and adjacent
portion of western Yunnan.

Garrulax caerulatus ricinus (Riley)

Dryonastes berthemyi ricinus Riley, 1930, Proc. Biol. Soc.
Washington, 43, p. 80 — Ndamucho, south of Lutien
[lat. 27° 12′ N., long. 99° 28′ E.], Yunnan.

Known only from type locality.

Garrulax caerulatus berthemyi (Oustalet)

Ianthocincla Berthemyi Oustalet, 1876, Bull. Soc. Philom. Paris, sér. 6, 13, p. 92 — western Fukien.

Mountains of northwestern Fukien.

Garrulax caerulatus poecilorhynchus Gould

Garrulax pœcilorhyncha (sic) Gould, 1863, Proc. Zool. Soc. London, p. 281 — Formosa.

Formosa.

GARRULAX MITRATUS

Garrulax mitratus major (Robinson and Kloss)

Rhinocichla mitrata major Robinson and Kloss, 1919, Bull. Brit. Orn. Club, 40, p. 16 — Gunong Ijau, Perak.

Highlands of Malaya from northern Perak southward to southern Selangor and Pahang.

Garrulax mitratus mitratus (Müller)

Timalia mitrata S. Müller, 1835, Tijdschr. Natuur. Gesch. Phys., 2, p. 345, pl. "5, fig. 3" [= pl. 9, fig. 3] — Sumatra.

Highlands of western Sumatra.

Garrulax mitratus damnatus (Harrisson and Hartley)

Rhinocichla mitrata damnata Harrisson and Hartley, 1934, Bull. Brit. Orn. Club, 54, p. 154 — Mount Dulit, Sarawak.

Mountains of eastern Sarawak (Mount Dulit, Mount Derian, and Kelabit Plateau).

Garrulax mitratus griswoldi (Peters)

Rhinocichla mitrata griswoldi Peters, 1940, Bull. Mus. Comp. Zool., 87, p. 204 — Mount Tibang, Borneo.

Highlands of central Borneo (Schwaner and Müller Ranges).

Garrulax mitratus treacheri (Sharpe)

Ianthocincla treacheri Sharpe, 1879, Proc. Zool. Soc. London, p. 248, pl. 23 — Kina Balu, Borneo.

Known only from type locality.

GARRULAX RUFICOLLIS

Garrulax ruficollis (Jardine and Selby)

Ianthocincla ruficollis Jardine and Selby, 1838, Ill. Orn., new ser. no. 4, pl. 21 and text — Himalayas.

Himalayas from eastern Nepal eastward, through Assam (both north and south of the Brahmaputra), to northeastern Burma.

GARRULAX MERULINUS

Garrulax merulinus merulinus Blyth

Garrulax merulinus Blyth, 1851, Journ. Asiat. Soc. Bengal, 20, p. 521 — Cherrapunji, Khasi Hills, Assam.

Stactocichla merulina toxostomina Koelz, 1952, Journ. Zool. Soc. India, 4, p. 38 — Karong, Manipur.

Stactocichla merulina minima Koelz, 1953, Contrib. Inst. Regional Explor., no. 1, p. 3 — Tasubum, Upper Chindwin, Burma.

Hill tracts of Assam (south of the Brahmaputra) eastward, through northern Burma, into western Yunnan (west of the Salween).

? Garrulax merulinus laoensis de Schauensee

Garrulax (Stactocichla) merulinus laoensis de Schauensee, 1938, Proc. Acad. Nat. Sci. Philadelphia, 90, p. 27 — Doi Pha Hom Pok [lat. 20° 05′ N., long. 99° 10′ E.], Thailand.

Known only from type locality; perhaps not different from merulinus.

Garrulax merulinus obscurus Delacour and Jabouille

Garrulax merulinus obscurus Delacour and Jabouille, 1930, Oiseau Rev. Franç. Orn., 11, p. 399 — Chapa [lat. 22° 20′ N., long. 103° 50′ E.], Tongking.

Garrulax merulinus taweishanicus T. H. Cheng, 1960, in T. H. Cheng and P. L. Cheng, Acta Zool. Sinica, 12, pp. 264, 276 — Ta-wei Shan, southeastern Yunnan.

Southeastern Yunnan, northwestern Tongking, and eastern northern Laos.

Garrulax merulinus annamensis (Robinson and Kloss)

Stactocichla merulina annamensis Robinson and Kloss, 1919, Ibis, p. 577 — Dran [lat. 11° 49′ N., long. 108° 38′ E.], Annam.

Southern Annam (Lang Bian Plateau).

GARRULAX CANORUS

Garrulax canorus canorus (Linnaeus)

[*Turdus*] *Canorus* Linnaeus, 1758, Syst. Nat., ed. 10, 1, p. 169 — Bengal and China; type locality restricted to Amoy, Fukien, by Meinertzhagen, 1928, Ibis, p. 512.

Trochalopterum canorum yunnanensis (sic) La Touche, 1922, Bull. Brit. Orn. Club, 42, p. 52 — Hokow, southeastern Yunnan. Not *Trochalopterum yunnanense* Rippon, 1906 = *Garrulax elliotii* (J. Verreaux).

Trochalopterum canorum namtiense La Touche, 1923, Ibis, p. 317. New name for *Trochalopterum canorum yunnanense* La Touche, preoccupied.

Trochalopteron touchena (sic) Mathews, 1934, Bull. Brit. Orn. Club, 55, p. 24. New name for *Trochalopterum canorum yunnanense* La Touche, preoccupied.

Yangtze Valley from Szechwan to Kiangsu, thence southward and westward through coastal provinces of China to southeastern Yunnan; Tongking, northern Annam, and eastern northern Laos.

Garrulax canorus owstoni (Rothschild)

Trochalopteron canorum owstoni Rothschild, 1903, Bull. Brit. Orn. Club, 14, p. 8 — Mount Wuchi, Hainan.

Hainan.

Garrulax canorus taewanus Swinhoe

Garrulax Taewanus Swinhoe, 1859, Journ. North-China Branch Roy. Asiat. Soc., 1, p. 228 — Formosa.

Formosa.

GARRULAX SANNIO

Garrulax sannio albosuperciliaris Godwin-Austen

Garrulax albo-superciliaris Godwin-Austen, 1874, Proc. Zool. Soc. London, p. 45 — near Kaibi, Manipur.

Hill tracts of eastern Assam (Naga Hills and Manipur).

Garrulax sannio comis Deignan

Garrulax sannio comis Deignan, 1952, Postilla, Yale Univ., no. 11, p. 3 — Likiang Plain, Yunnan.

Northeastern Burma from Kachin State southward to Southern Shan State; southeastern Hsikang; Yunnan;

northern Laos; northernmost Annam; Tongking (west of the Black River-Red River divide).

Garrulax sannio sannio Swinhoe

> Garrulax sannio Swinhoe, 1867, Ibis, p. 403 — Interior of Fukien.

Tongking (except area occupied by comis) ; Kwangsi; Kwangtung; Fukien; Kiangsi; northeastern Hunan.

Garrulax sannio oblectans Deignan

> Garrulax sannio oblectans Deignan, 1952, Postilla, Yale Univ., no. 11, p. 3 — near Ipin, Szechwan.

Southwestern Hupeh, northern Kweichow, and Szechwan.

GARRULAX CACHINNANS

Garrulax cachinnans jerdoni Blyth

> Garrulax (?) Jerdoni Blyth, 1851, Journ. Asiat. Soc. Bengal, 20, p. 522 — southern India; type from Balasore Peak, "a high hill at the edge of the Ghats separating Malabar from the Wynaad," fide Jerdon, 1863, Birds India, 2, p. 50.

Known only from type locality and near-by Brahmagiri Hills in Coorg.

Garrulax cachinnans cachinnans Jerdon

> C[rateropus]. cachinnaus (sic) Jerdon, 1839, Madras Journ. Lit. Sci., 10, p. 255, pl. 7 — Nilgiri Hills.
> Crat[eropus]. Delessertii Lafresnaye, 1840, Rev. Zool. [Paris], 3, p. 65 — Nilgiri Plateau. Not Crateropus delesserti Jerdon, 1839 = Garrulax delesserti delesserti (Jerdon).
> ? Trochalopterum cinnamomeum Davison, 1886, Ibis, p. 204 — unknown locality.

Nilgiri Hills, southern India.

Garrulax cachinnans fairbanki (Blanford)

> Trochalopteron Fairbanki Blanford, 1869, Journ. Asiat. Soc. Bengal, 38, p. 175, pl. 17a — Palni Hills, southern India.

Palni and Anaimalai Hills and the High Range of Kerala southward to, but not including, the Achankovil Gap, southern India.

Garrulax cachinnans meridionalis (Blanford)

Trochalopterum meridionale Blanford, 1880, Journ. Asiat.
Soc. Bengal, 49, p. 142 — southern Kerala; type speci-
men from the Tirunelveli boundary, *fide* Sharpe, 1883,
Cat. Birds Brit. Mus., 7, p. 375.

Hill tracts of southern Kerala southward from the Achankovil Gap.

GARRULAX LINEATUS

Garrulax lineatus bilkevitchi (Zarudny)

Trochalopteron (Janthocincla) lineatum bilkevitchi Zarudny, 1910, Orn. Monatsb., 18, p. 188 — mountains
near Kulyab, southeastern Tadzhikistan.

Ianthocincla lineatum [sic] *ziaratensis* Ticehurst, 1920,
Bull. Brit. Orn. Club, 41, p. 55 — Ziarat (42 miles ENE.
of Quetta), West Pakistan.

Tadzhikistan, eastern Afghanistan, and northwestern
West Pakistan.

Garrulax lineatus gilgit (Hartert)

Ianthocincla lineatum [sic] *gilgit* Hartert, 1909, Vög. pal.
Fauna, p. 636 — Gilgit, northwestern Kashmir.

Northeastern West Pakistan.

Garrulax lineatus lineatus (Vigors)

Cinclosoma lineatum Vigors, 1831, Proc. Comm. Zool.
Soc. London, pt. 1, p. 56 — Himalayas; restricted to
"N.W. Himalayas," by Hume, 1875, Stray Feathers, 3,
p. 396, and further restricted to "the district Simla-
Almora," by Ticehurst and Whistler, 1924, Ibis, p. 471.

Ianthocincla lineatum [sic] *grisescentior* Hartert, 1909,
Vög. pal. Fauna, p. 636 — Simla, Himachal Pradesh.

Himalayas from central Kashmir southeastward to northwestern Uttar Pradesh.

Garrulax lineatus setafer (Hodgson)

Cinc[losoma]. setafer Hodgson, 1836, Asiat. Res., 19, p.
148 — Nepal.

Nepal, Sikkim, and Darjeeling District of West Bengal.

Garrulax lineatus imbricatus Blyth

G[arrulax]. imbricatus Blyth, 1843, Journ. Asiat. Soc.
Bengal, 12, p. 951 — Bhutan.

Bhutan and southeastern Tibet.

GARRULAX VIRGATUS

Garrulax virgatus (Godwin-Austen)

Trochalopteron virgatum Godwin-Austen, 1874, Proc. Zool. Soc. London, p. 46 — Razami, Naga Hills, Assam.

Trochalopteron virgatum querulum Koelz, 1952, Journ. Zool. Soc. India, 4, p. 38 — Hmuntha, Lushai Hills, Assam.

Hill tracts of Assam (south of the Brahmaputra) and southwestern Burma (Chin Hills).

GARRULAX AUSTENI

Garrulax austeni austeni (Godwin-Austen)

Trochalopteron Austeni "Jerdon" Godwin-Austen, 1870, Journ. Asiat. Soc. Bengal, 39, pt. 2, p. 105 — Hengdang Peak, principal trigonometrical station of observation at head of the Jhiri River, Barail Range, Assam.

Hill tracts of Assam south of the Brahmaputra.

Garrulax austeni victoriae (Rippon)

Ianthocincla victoriæ Rippon, 1906, Bull. Brit. Orn. Club, 16, p. 47 — Mount Victoria, Southern Chin Hills, Upper Burma.

Garrulax austeni exvictoriæ Meinertzhagen, 1928, Ibis, p. 509. New name for *Ianthocincla victoriae* Rippon, when considered congeneric with *Babax victoriae* Rippon, 1905.

Known only from type locality.

GARRULAX SQUAMATUS

Garrulax squamatus (Gould)

Ianthocincla squamata Gould, 1835, Proc. Zool. Soc. London, pt. 3, p. 48 — Himalayas; restricted to Sikkim, by Baker, 1921, Journ. Bombay Nat. Hist. Soc., 27 (1920), p. 243.

Trochalopteron squamatum subsquamatum Koelz, 1952, Journ. Zool. Soc. India, 4, p. 38 — Pynursla, Khasi Hills, Assam.

Himalayas from Nepal eastward, through Sikkim, Bhutan, Assam, and northern Burma, to western Yunnan (west of the Salween), and southeastward through Assam to southwestern Burma (Chin Hills) ; reported from northwestern Tongking (Chapa).

GARRULAX SUBUNICOLOR

Garrulax subunicolor subunicolor (Blyth)

Tr[ochalopteron]. subunicolor Blyth, 1843, Journ. Asiat. Soc. Bengal, 12, p. 952, footnote — Nepal.

Himalayas from Nepal eastward, through Sikkim and Bhutan, to eastern Assam (Mishmi Hills).

Garrulax subunicolor griseatus (Rothschild)

Ianthocincla subunicolor griseata Rothschild, 1921, Novit. Zool., 28, p. 33 — Shweli-Salween Divide, Yunnan.

Northeastern Burma (Kachin State) and northwestern Yunnan.

Garrulax subunicolor fooksi Delacour and Jabouille

Garrulax subunicolor fooksi Delacour and Jabouille, 1930, Oiseau Rev. Franç. Orn., 11, p. 399 — Chapa [lat. 22° 20′ N., long. 103° 50′ E.], Tongking.

Mountains of northwestern Tongking.

GARRULAX ELLIOTII

Garrulax elliotii prjevalskii (Menzbier)

Trochalopteron prjevalskii Menzbier, 1887, Ibis, p. 300 — Kansu.

Ianthocincla elliotii perbona Bangs and Peters, 1928, Bull. Mus. Comp. Zool., 68, p. 340 — Liyuanku, northern slopes of Richthofen Range, northern Kansu.

Kansu and eastern Tsinghai.

Garrulax elliotii elliotii (Verreaux)

Trochalopteron elliotii J. Verreaux, 1870, Nouv. Arch. Mus. Hist. Nat. [Paris], 6, p. 36; *idem*, 1871, *ibid.*, 7, p. 44 — "les montagnes du Thibet chinois"; types from western Szechwan, *fide* Verreaux, 1871.

Trochalopteron Bonvaloti Oustalet, 1892, Ann. Sci. Nat., Zool., sér. 7, 12 (1891), p. 275 — "Tioungeu," Tibet, error; corrected to Chatu, near Lamda [lat. 31° 11′ N., long. 96° 59′ E.], Hsikang, by Oustalet, 1893, Nouv. Arch. Mus. Hist. Nat. [Paris], sér. 3, 5, p. 194.

Trochalopterum yunnanense Rippon, 1906, Bull. Brit. Orn. Club, 19, p. 32 — Yangtze River, western Yunnan.

Ianthocincla elliotti honoripeta Hartert, 1910, Vög. pal.
Fauna, **1**, p. xliv, footnote 1. New name for *Trochalopteron Bonvaloti* Oustalet, 1892, when considered congeneric with *Babax Bonvaloti* Oustalet.
Garrulax elliotii exyunnanensis Meinertzhagen, 1928, Ibis,
p. 510. New name for *Trochalopterum yunnanense* Rippon, 1906, when considered congeneric with *Babax
yunnanensis* Rippon.

Southern Shensi, western Hupeh, Szechwan, eastern
Hsikang, and northwestern Yunnan.

GARRULAX HENRICI

Garrulax henrici (Oustalet)

Trochalopteron Henrici Oustalet, 1892, Ann. Sci. Nat.,
Zool., sér. 7, **12** (1891), p. 274 — "So," Tibet, error;
types from Aio and Sutu, Hsikang, *fide* Oustalet, 1893,
Nouv. Arch. Mus. Hist. Nat. [Paris], sér. 3, **5**, p. 196.
Garrulax tibetanus Dresser, 1905 (Jan. 24), Abstr. Proc.
Zool. Soc. London, no. 13, p. 2 — Tibet; *idem*, 1905
(June), Proc. Zool. Soc. London, p. 54, pl. 5, fig. 2 —
near the Chaksam Ferry [lat. 29° 15′ N., long. 90° 32′
E.], Tsangpo Valley, Tibet.

Southeastern Tibet and southwestern Hsikang.

GARRULAX AFFINIS

Garrulax affinis affinis Blyth

G[arrulax]. affinis Blyth (ex Hodgson MS), 1843, Journ.
Asiat. Soc. Bengal, **12**, p. 950 — Nepal; restricted to
East Nepal, by Rand, 1953, Nat. Hist. Misc. [Chicago],
no. 116, p. 2, and corrected to Central Nepal, by Rand
and Fleming, 1956, Fieldiana Zool. [Chicago], **39**, p. 3.
Garrulax affinis flemingi Rand, 1953, Nat. Hist. Misc.
[Chicago], no. 116, p. 2 — Lete [*ca.* lat. 28° 40′ N.,
long. 83° 40′ W.], Nepal.

Western and central Nepal.

Garrulax affinis bethelae Rand and Fleming

Garrulax affinis bethelae Rand and Fleming, 1956, Fieldiana Zool. [Chicago], **39**, p. 2 — Thangii [= Thangu],
14,000 ft., Sikkim.

Himalayas from eastern Nepal to eastern Bhutan.

Garrulax affinis oustaleti (Hartert)

Ianthocincla affinis oustaleti Hartert, 1909, Vögel pal. Fauna, p. 633 — Tzeku, Yunnan.

Southwestern Hsikang, northeastern Assam (Mishmi Hills), northeastern Burma (Kachin State), and northwestern Yunnan (valleys of the Salween and Mekong).

Garrulax affinis muliensis Rand

Garrulax affinis muliensis Rand, 1953, Nat. Hist. Misc. [Chicago], no. 116, p. 5 — Muli [*ca.* lat. 28° 15′ N., long. 100° 50′ E.], Hsikang.

Northwestern Yunnan (valley of the Yangtze) and southeastern Hsikang.

Garrulax affinis blythii (Verreaux)

Trochalopteron Blythii J. Verreaux, 1870, Nouv. Arch. Mus. Hist. Nat. [Paris], 6, p. 37; *idem*, 1871, *ibid.*, 7, p. 45 — "les montagnes du Thibet chinois"; types from western Szechwan, *fide* Verreaux, 1871.[1]

Mountains of southwestern Szechwan and of adjacent portion of Hsikang (except area occupied by *muliensis*).

Garrulax affinis saturatus Delacour and Jabouille

Garrulax affinis saturatus Delacour and Jabouille, 1930, Oiseau Rev. Franç. Orn., 11, p. 400 — Fan Si Pan [lat. 22° 17′ N., long. 104° 47′ E.], Tongking.

Known only from type locality.

Garrulax affinis morrisonianus (Ogilvie-Grant)

Trochalopterum morrisonianum Ogilvie-Grant, 1906, Bull. Brit. Orn. Club, 16, p. 120 — Mount Morrison, Formosa.

Highlands of Formosa.

GARRULAX ERYTHROCEPHALUS

Garrulax erythrocephalus erythrocephalus (Vigors)

Cinclosoma erythrocephalum Vigors, 1832, Proc. Comm. Zool. Soc. London, pt. 1 (1831), p. 171 — Himalayas;

---

[1] Rand, 1953, has restricted the type locality to the vicinity of Muping [= Paohing], Hsikang, but Verreaux, 1871, says that "M. A. David, qui l'a découverte pour la première fois dans la Chine occidentale, nous apprend qu'elle est aussi représentée . . . plus loin à l'ouest, dans les montagnes boisées du Moupin et du Kokonoor." The type must, accordingly, have come from within the confines of present-day Szechwan.

restricted to "the district Simla-Almora," by Ticehurst
and Whistler, 1924, Ibis, p. 471, and further restricted
to Simla, Himachal Pradesh, by Vaurie, 1953, Bull.
Brit. Orn. Club, 73, p. 78.

Himalayas from Chamba District of Himachal Pradesh
southeastward to Kumaun District of Uttar Pradesh.

Garrulax erythrocephalus kali Vaurie

Garrulax erythrocephalus kali Vaurie, 1953, Bull. Brit.
Orn. Club, 73, p. 78 — Lete, Kali River valley, west-
central Nepal.

Western and central Nepal.

Garrulax erythrocephalus nigrimentum (Oates)

Trochalopterum nigrimentum Oates, 1889, Fauna Brit.
India, Birds, 1, pp. 88 (in key), 91, fig. 27 — "The Himalayas from Nepal to the Daphla hills in Assam";
inferentially restricted to Sikkim, by Kinnear, 1937,
in Ludlow and Kinnear, Ibis, p. 32.

Sikkim and Bhutan.

Garrulax erythrocephalus imprudens Ripley

Garrulax erythrocephalus imprudens Ripley, 1948, Proc.
Biol. Soc. Washington, 61, p. 102 — Tidding Saddle,
above Dreyi, Mishmi Hills, Assam.

Hill tracts of Assam north and east of the Brahmaputra.

Garrulax erythrocephalus chrysopterus (Gould)

Ianthocincla chrysoptera Gould, 1835, Proc. Zool. Soc.
London, pt. 3, p. 48 — Himalayas; type characteristic
of population of Khasi Hills, Assam, *fide* Oates, 1889,
Fauna Brit. India, Birds, 1, p. 90.

G[arrulax]. ruficapillus Blyth, 1851, Journ. Asiat. Soc.
Bengal, 20, p. 521 — Cherrapunji, Khasi Hills, Assam.

Hill tracts of Assam south of the Brahmaputra (Khasi
Hills).

? Garrulax erythrocephalus godwini (Harington)

Trochalopterum erythrocephalum godwini Harington,
1914, Bull. Brit. Orn. Club, 33, p. 92 — Hengdan Peak,
Barail Range, Assam.

Hill tracts of Assam south of the Brahmaputra (Barail
Range) ; doubtfully distinct from *erythrolaema*.

Garrulax erythrocephalus erythrolaema (Hume)

> Trochalopterum erythrolæma Hume, 1881, Stray Feathers, 10, p. 153 — near Machi, eastern Manipur.
>
> Trochalopterum holerythrops Rippon, 1904, Bull. Brit. Orn. Club, 14, p. 83 — Mount Victoria [lat. 21° 15' N., long. 93° 55' E.], Upper Burma.

Hill tracts of eastern Manipur and southwestern Burma (Chin Hills and the Arakan Yomas).

Garrulax erythrocephalus woodi (Baker)

> Trichalopterum (sic) erathrolæma (sic) woodi Baker, 1914 (Nov. 4), Bull. Brit. Orn. Club, 35, p. 17 — Loi Song [lat. 23° 51' N., long. 97° 49' E.], Upper Burma.
>
> Trochalopterum erythrocephalum woodi Harington, 1914 (Nov. 20), Journ. Bombay Nat. Hist. Soc., 23, p. 317 — Loi Song, Upper Burma.
>
> Ianthocincla forresti Rothschild, 1921, Novit. Zool., 28, p. 35 — Shweli-Salween Divide, Yunnan.

Northeastern Burma (Kachin and Northern Shan States) and adjacent portion of Yunnan (west of the Salween).

Garrulax erythrocephalus connectens (Delacour)

> Trochalopterum erythrocephalum connectans (sic) Delacour, 1929, Bull. Brit. Orn. Club, 49, p. 58 — Phu Ke, near Chiang Khwang [lat. 19° 19' N., long. 103° 22' E.], Laos.
>
> Garrulax erythrocephalus hendeei Bangs and Van Tyne, 1930, Field Mus. Nat. Hist. Publ., Zool. Ser., 18, p. 3 — Chapa [lat. 22° 20' N., long. 103° 50' E.], Tongking.

Mountains of northwestern Tongking and of eastern northern Laos.

Garrulax erythrocephalus subconnectens Deignan

> Garrulax erythrocephalus subconnectens Deignan, 1938, Proc. Biol. Soc. Washington, 51, p. 90 — Doi Phu Kha [lat. 19° 05' N., long. 101° 05' E.], Thailand.

Known only from type locality; specimens from eastern portion of Southern Shan State of Kengtung should perhaps be referred here.

Garrulax erythrocephalus schistaceus Deignan

> Garrulax erythrocephalus schistaceus Deignan, 1938, Proc. Biol. Soc. Washington, 51, p. 89 — Doi Luang Chiang Dao [lat. 19° 25' N., long. 98° 55' E.], Thailand.

Garrulax erythrocephalus shanus de Schauensee, 1946,
Proc. Acad. Nat. Sci. Philadelphia, 98, p. 110 — Kiu
Loi, *ca.* 20 miles west of Kengtung, Southern Shan
State, Upper Burma.

Western portion of Southern Shan State of Kengtung
and high mountains of northern portion of northwestern
Thailand (Doi Pha Hom Pok, Doi Luang Chiang Dao).

Garrulax erythrocephalus melanostigma Blyth

G[*arrulax*]. *melanostigma* Blyth, 1855, Journ. Asiat. Soc.
Bengal, 24, p. 268 — Mulayit Taung [lat. 16° 11′ N.,
long. 98° 32′ E.], Lower Burma.

Southern Shan State of Burma (west of the Salween),
high mountains of northwestern Thailand (except those occupied by *schistaceus*), and Amherst District of Tenasserim
(Mulayit Taung).

Garrulax erythrocephalus ramsayi (Ogilvie-Grant)

Trochalopterum ramsayi Ogilvie-Grant, 1904, Bull. Brit.
Orn. Club, 14, p. 92 — Karenni State, Upper Burma.

Karenni State of Burma and Tavoy District of Tenasserim (Nwalabo Taung).

Garrulax erythrocephalus peninsulae (Sharpe)

Trochalopterum peninsulæ Sharpe, 1887, Proc. Zool. Soc.
London, p. 436, pl. 37 — Larut Range, Perak.

High mountains of peninsular Thailand (Nakhon Si
Thammarat Province) and of Malaya from northern Perak
southward to southern Selangor and Pahang.

GARRULAX YERSINI[1]

Garrulax yersini (Robinson and Kloss)

Trochalopteron yersini Robinson and Kloss, 1919, Ibis, p.
575 — Lang Bian Peaks, southern Annam.

Southern Annam (Lang Bian Plateau).

GARRULAX FORMOSUS

Garrulax formosus formosus (Verreaux)

Trochalopteron formosum J. Verreaux, 1869, Nouv. Arch.
Mus. Hist. Nat. [Paris], 5, p. 35; *idem*, 1871, *ibid.*, 7,

[1] *Garrulax erythrocephalus* and *G. yersini* comprise a superspecies.

p. 43, pl. 2, fig. 1 — "le Thibet oriental"; type specimens from western Szechwan, *fide* Verreaux, 1871.

Trochalopterum milnii omeiensis (sic) Shaw, 1932, Bull. Fan Mem. Inst. Biol., 3, p. 220 — Omei-shan, Szechwan. Southwestern Szechwan and northeastern Yunnan.

Garrulax formosus greenwayi Delacour and Jabouille

Garrulax formosus greenwayi Delacour and Jabouille, 1930, Oiseau Rev. Franç. Orn., 11, p. 398 — Fan Si Pan [lat. 22° 17′ N., long. 104° 47′ E.], Tongking. Known only from type locality.

GARRULAX MILNEI

Garrulax milnei sharpei (Rippon)

Trochalopterum sharpei Rippon, 1901, Bull. Brit. Orn. Club, 12, p. 13 — "Kauri-Kachin tract, to the east of Bhamo, and bordering on the south of the Tapeng River," Kachin State, Burma.

Ianthocincla lustrabila (sic) Bangs and Phillips, 1914, Bull. Mus. Comp. Zool., 58, p. 285 — Loukouchai, southeastern Yunnan.

Trochalopterum milnei indochinensis Delacour, 1927, Bull. Brit. Orn. Club, 47, p. 158 — Tam Dao [lat. 21° 27′ N., long. 105° 40′ E.], Tongking.

Kachin State of Burma southward, through the Shan States, to northwestern Thailand; Yunnan; Tongking and eastern portion of northern Laos.

Garrulax milnei vitryi Delacour

Garrulax milni vitryi Delacour, 1932, Oiseau Rev. Franç. Orn., nouv. sér., 2, p. 424 — Phu Kong-Ntul [*ca.* lat. 15° 23′ N., long. 106° 25′ E.], Laos. Southern Laos (Boloven Plateau).

Garrulax milnei sinianus (Stresemann)

Trochalopteron milnei sinianum Stresemann, 1930, Orn. Monatsb., 38, p. 47 — Ku-chan, Yao Shan, Kwangsi. Known only from type locality.

Garrulax milnei milnei (David)

Trochalopteron Milnei David, 1874, Ann. Sci. Nat., Zool., sér. 5, 19, art. 9, p. 4 — western Fukien. Mountains of northwestern Fukien.

GENUS **LIOCICHLA** SWINHOE

Liocichla Swinhoe, 1877, Ibis, p. 473. Type, by monotypy, *Liocichla steerii* Swinhoe.

cf. Delacour, 1933, Bull. Brit. Orn. Club, 53, pp. 85-88.

LIOCICHLA PHOENICEA

Liocichla phoenicea phoenicea (Gould)

Ianthocincla phœnicea Gould, 1837, Icones Avium, pt. 1, pl. [3] — Himalayas; inferentially restricted to Nepal, *apud* Ripley, 1961, Synops. Birds India Pakistan, p. 396.

Himalayas from Nepal eastward, through Sikkim and Bhutan, to eastern Assam (Mishmi Hills).

Liocichla phoenicea bakeri (Hartert)

Trochalopteron phœniceum bakeri Hartert, 1908, Bull. Brit. Orn. Club, 23, p. 10 — Laisung, North Cachar, Assam.

Trochalopteron phoeniceum khasium Koelz, 1952, Journ. Zool. Soc. London, 4, p. 38 — Laitlyngkot, Khasi Hills, Assam.

Hill tracts of Assam (south of the Brahmaputra) and northwestern Burma.

Liocichla phoenicea ripponi (Oates)

Trochalopterum ripponi Oates, 1900, Bull. Brit. Orn. Club, 11, p. 10 — Shan States, Upper Burma; type from Fort Stedman, Southern Shan State, *fide* Berlioz, 1930, Oiseau Rev. Franç. Orn., 11, p. 93.

Eastern Burma from Kachin State southward to Southern Shan State (west of the Salween) and Karenni State; northwestern Thailand.

Liocichla phoenicea wellsi (La Touche)

Trochalopterum phœniceum wellsi La Touche, 1921, Bull. Brit. Orn. Club, 42, p. 15 — Mengtsz [lat. 23° 23′ N., long. 103° 27′ E.], Yunnan.

Southern Yunnan; northern Laos; Tongking.

LIOCICHLA STEERII

Liocichla (steerii) omeiensis Riley

Liocichla omeiensis Riley, 1926, Proc. Biol. Soc. Washington, 39, p. 57 — Mount Omei, Szechwan.

Known only from type locality.

Liocichla (steerii) steerii Swinhoe

Liocichla steerii Swinhoe, 1877, Ibis, p. 474, pl. 14 — mountains of southern Formosa.

Highlands of Formosa.

GENUS **LEIOTHRIX** SWAINSON

Leiothrix Swainson, 1832, in Swainson and Richardson, Fauna Bor.-Am., 2 (1831), pp. 233, 490. Type, by original designation, *Parus furcatus* Temminck = *Sylvia lutea* Scopoli.

Mesia Hodgson, 1837, India Review, 2 (1838), pp. 34, 88. Type, by monotypy, *Mesia argentauris* Hodgson.

LEIOTHRIX ARGENTAURIS

Leiothrix argentauris argentauris (Hodgson)

[*Mesia*] *Mesia Argentauris* Hodgson, 1837, India Review, 2, p. 88 — Nepal.

Himalayas from Garhwal eastward, through Nepal, Sikkim, and Bhutan, into Assam (north of the Brahmaputra).

Leiothrix argentauris aureigularis (Koelz)

Mesia argentauris aureigularis Koelz, 1953, Journ. Zool. Soc. India, 4, p. 153 — Tura Mountain, Garo Hills, Assam.

Hill tracts of Assam south of the Brahmaputra and southwestern Burma (Chin Hills).

Leiothrix argentauris vernayi (Mayr and Greenway)

Mesia argentauris vernayi Mayr and Greenway, 1938, Proc. New England Zool. Club, 17, p. 3 — Hai Bum, Patkai Hills, northwestern Burma.

Leiothrix argentauris gertrudis Ripley, 1948, Proc. Biol. Soc. Washington, 61, p. 103 — Dening, Mishmi Hills, Assam.

Northeastern Assam, northern Burma, and western Yunnan (west of the Salween).

Leiothrix argentauris galbana (Mayr and Greenway)

> *Mesia argentauris galbana* Mayr and Greenway, 1938,
> Proc. New England Zool. Club, 17, p. 3 — Doi Ang Ka
> [lat. 18° 35′ N., long. 98° 30′ E.], Thailand.

Shan and Karenni States, eastern Burma; hills of northern Thailand, on the west southward to the Amherst District of Tenasserim.

Leiothrix argentauris ricketti (La Touche)

> *Mesia argentauris ricketti* La Touche, 1923, Bull. Brit.
> Orn. Club, 43, p. 173 — Szemao, southern Yunnan.
> *Mesia argentauris rubrogularis* Kinnear, 1925, Bull. Brit.
> Orn. Club, 45, p. 75 — Ngai Tio [lat. 22° 40′ N., long.
> 103° 47′ E.], Tongking.

Southeastern Yunnan, Tongking, northern Laos, and northern Annam.

Leiothrix argentauris cunhaci (Robinson and Kloss)

> *Mesia argentauris cunhaci* Robinson and Kloss, 1919, Ibis,
> p. 591 — Dalat [lat. 11° 55′ N., long. 108° 26′ E.],
> Annam.

Southern Laos (Boloven Plateau) and southern Annam.

Leiothrix argentauris tahanensis (Yen)

> *Mesia argentauris tahanensis* "Robinson" Yen, 1934, Science Journal [College of Science, Sun Yatsen University, Canton] 6, pp. 378 (in key), 379 — Gunong Tahan
> [lat. 4° 38′ N., long. 102° 14′ E.], Pahang.[1]

Mountains of Malay Peninsula from Siamese province of Nakhon Si Thammarat southward to southern Selangor and Pahang.

Leiothrix argentauris rookmakeri (Junge)

> *Mesia argentauris rookmakeri* Junge, 1948, Zool. Meded.
> Leiden, 29, p. 325 — Pang Mok, north of Lake Takengon, Acheh district, northwestern Sumatra.

Known only from type locality.

[1] This name is sometimes cited from October, 1934, p. 63. Since it appears on the sixty-third page of Yen's paper, one may surmise that separately paged reprints were distributed two months prior to publication of this number of the Science Journal.

Leiothrix argentauris laurinae Salvadori

Leiothrix laurinae Salvadori, 1879, Ann. Mus. Civ. Genova, 14, p. 231 — Mount Singgalang [lat. 0° 24′ S., long. 100° 20′ E.], Sumatra.

Highlands of western Sumatra (except area occupied by *rookmakeri*).

LEIOTHRIX LUTEA

Leiothrix lutea kumaiensis Whistler

Leiothrix lutea kumaiensis Whistler, 1943, Bull. Brit. Orn. Club, 63, p. 62 — Dehra Dun, northwestern Uttar Pradesh.

Himalayas from East Punjab eastward to northwestern Uttar Pradesh.

Leiothrix lutea calipyga (Hodgson)

[*Mesia*] *Bahila Calipyga* Hodgson, 1837, India Review, 2, p. 88 — Nepal.

Himalayas from western Nepal eastward, through Sikkim and Bhutan, to eastern Assam.

? Leiothrix lutea luteola Koelz

Liothrix [sic] *lutea luteola* Koelz, 1952, Journ. Zool. Soc. India, 4, p. 39 — Mawryngkneng, Khasi Hills, Assam.

Hill tracts of Assam south of the Brahmaputra and southwestern Burma (Chin Hills and Arakan Yomas) ; doubtfully distinct from *calipyga*.

Leiothrix lutea yunnanensis Rothschild

Leiothrix luteus [sic] *yunnanensis* Rothschild, 1921, Novit. Zool., 28, p. 36 — Shweli-Salween Divide, western Yunnan.

Northeastern Burma, northwestern Yunnan, and southeastern Hsikang.

Leiothrix lutea kwangtungensis Stresemann

Leiothrix lutea kwangtungensis Stresemann, 1923, Journ. f. Orn., 71, p. 364 — Siuhang, Kwangtung.

Southeastern Yunnan, northeastern Tongking, Kwangsi, and Kwangtung.

Leiothrix lutea lutea (Scopoli)

Sylvia (*lutea*) Scopoli, 1786, Del. Flor. Fauna. Insubr., fasc. 2, p. 96 — China, ex Sonnerat; restricted to mountains of Anhwei, by Stresemann, 1923, Journ. f. Orn., 71, p. 364.

Hill tracts of Fukien and Chekiang and valley of the Yangtze from Anhwei westward to Szechwan.

? Leiothrix (lutea?) astleyi Delacour

Liothrix [sic] *astleyi* Delacour, 1921, Bull. Brit. Orn. Club, 41, p. 115 — "China."

Known only from aviary specimens; range and systematic status uncertain.

GENUS CUTIA HODGSON

Cutia Hodgson, 1837, Journ. Asiat. Soc. Bengal, 5 (1836), p. 772. Type, by original designation and monotypy, *Cutia nipalensis* Hodgson.

CUTIA NIPALENSIS

Cutia nipalensis nipalensis Hodgson

[*Cutia*] *Nipalensis* Hodgson, 1837, Journ. Asiat. Soc. Bengal, 5 (1836), p. 774 — Nepal.

Cutia nipalensis nagaensis Koelz, 1954, Contrib. Inst. Regional Explor., no. 1, p. 9 — Kohima, Naga Hills, Assam.

Himalayas from Nepal eastward to eastern Assam, both north and south of the Brahmaputra; western Burma (Chin Hills).

Cutia nipalensis melanchima Deignan

Cutia nipalensis melanchima Deignan, 1947, Journ. Washington Acad. Sci., 37, p. 105 — Khao Pha Cho [lat. 19° 00′ N., long. 99° 25′ E.], Thailand.

Eastern Burma from Kachin State southward through Shan States to Karenni; northwestern Thailand; northern Laos; northwestern Tongking.

Cutia nipalensis cervinicrissa Sharpe

Cutia cervinicrissa Sharpe, 1888, Proc. Zool. Soc. London, p. 276 — Gunong Batu Puteh, Perak.

Highlands of Malaya from southern Perak to southern Selangor.

Cutia nipalensis legalleni Robinson and Kloss

Cutia nipalensis legalleni Robinson and Kloss, 1919, Ibis, p. 588 — Lang Bian Peaks [*ca.* lat. 12° 02′ N., long. 108° 26′ E.], Annam.

Southern Annam (Lang Bian Plateau).

GENUS **PTERUTHIUS** SWAINSON

Pteruthius Swainson, 1832, in Swainson and Richardson, Fauna Bor.-Am., 2 (1831), p. 491. Type, by original designation and monotypy, *Lanius erythropterus* Gould, *i.e.*, Vigors = *Pteruthius flaviscapis validirostris* Koelz.

Allotrius Temminck, 1835, Pl. col., livr. 99, pl. 589. Type, by subsequent designation (G. R. Gray, 1855, Cat. Gen. Subgen. Birds, p. 54), *Allotrius aenobarbus* Temminck.

Ptererythrius vel *Pterythrius* Strickland, 1841, Ann. Mag. Nat. Hist., 7, p. 29. *Nomina emendata.*

Aenopogon "Agass." Fitzinger, 1863 (subgenus of "*Pterythrius* Cab.," *i.e.*, *Pteruthius* Swainson), Sitzungsb. K. Akad. Wiss. Wien., Math.-Naturwiss. Cl., 46, p. 196. Type, by monotypy, *Allotrius aenobarbus* Temminck.

Hilarocichla Oates, 1889, Fauna Brit. India, Birds, 1, p. 243. Type, by monotypy, *Pteruthius rufiventer* Blyth.

PTERUTHIUS RUFIVENTER

Pteruthius rufiventer rufiventer Blyth

Pt[*eruthius*]. *rufiventer* Blyth, 1842, Journ. Asiat. Soc. Bengal, 11, p. 183 — no locality; Darjeeling, West Bengal, *apud* Baker.

Himalayas from Nepal eastward, through Bhutan, Assam, and northern Burma, into Yunnan.

Pteruthius rufiventer delacouri Mayr

Pteruthius rufiventer delacouri Mayr, 1941, in Stanford and Mayr, Ibis, p. 96 — Loquiho, Fan Si Pan [lat. 22° 17′ N., long. 104° 47′ E.], Tongking.

Known only from type locality.

PTERUTHIUS FLAVISCAPIS

Pteruthius flaviscapis validirostris Koelz

Lanius erythropterus Vigors, 1831, Proc. Comm. Zool.

Soc. London, pt. 1 (1830-1831), p. 22 — Himalayas; restricted to Murree, Rawalpindi, West Pakistan, by Baker, 1922, Fauna British India, Birds, ed. 2, 1, p. 331. Not *Lanius erythropterus* Shaw, 1809.

Pteruthius erythropterus validirostris Koelz, 1951, Journ. Zool. Soc. India, 3, p. 28 — Kohima, Naga Hills, Assam.

Pteruthius erythropterus nocrecus Koelz, 1952, Journ. Zool. Soc. India, 4, p. 40 — Tura Mountain, Garo Hills, Assam.

Pteruthius erythropterus glauconotus Koelz, 1954, Contrib. Inst. Regional Explor., no. 1, p. 9 — Sangau, Lushai Hills, Assam.

Pteruthius validirostris ripleyi Biswas, 1960, Bull. Brit. Orn. Club, 80, p. 106. New name for *Lanius erythropterus* Vigors, preoccupied.

Himalayas from northern West Pakistan eastward, through Nepal and Bhutan, to eastern Assam, and southeastward through Assam to hill tracts of northwestern Burma (southward to Chin Hills).

Pteruthius flaviscapis ricketti Ogilvie-Grant

Pterythius [sic] *ricketti* Ogilvie-Grant, 1904, Bull. Brit. Orn. Club, 14, p. 92 — Fukien and southern Yunnan.

Pteruthius erythropterus yunnanensis Ticehurst, 1937, Bull. Brit. Orn. Club, 57, p. 147 — Shweli-Salween Divide, western Yunnan.

Northeastern Burma and northwestern Yunnan; southwestern Szechwan; Fukien; Kwangsi, southeastern Yunnan, Tongking, northern Annam, northern Laos, and eastern portion of northern Thailand.

Pteruthius flaviscapis annamensis Robinson and Kloss

Pterythius [sic] *æralatus annamensis* Robinson and Kloss, 1919, Ibis, p. 589 — Lang Bian Peaks, southern Annam.

Southern Annam (Lang Bian Plateau).

Pteruthius flaviscapis aeralatus Blyth

Pteruthius æralatus "Tickell" Blyth, 1855, Journ. Asiat. Soc. Bengal, 24, p. 267 — mountainous interior of the Tenasserim provinces, Lower Burma.

Shan and Karenni States of eastern Burma and hills of northwestern Thailand southward to Amherst District of

Tenasserim; specimens from mountains of western Cambodia have been referred here.

Pteruthius flaviscapis schauenseei Deignan

Pteruthius erythropterus schauenseei Deignan, 1946, Journ. Washington Acad. Sci., 36, p. 428 — Khao Luang [lat. 8° 30′ N., long. 99° 45′ E.], Thailand.

Mountains of Siamese portion of Malay Peninsula, south of Isthmus of Kra.

Pteruthius flaviscapis cameranoi Salvadori

Pteruthius cameranoi Salvadori, 1879, Ann. Mus. Civ. Genova, 14, p. 232 — Mount Singgalang [lat. 0° 24′ S., long. 100° 20′ E.], Sumatra.

Pteruthius flaviscapis leuser Chasen, 1939, Treubia, 17, p. 205 — Mount Leuser, Acheh district, northwestern Sumatra.

Highlands of Malaya from southern Perak southward to southern Selangor; highlands of western Sumatra.

Pteruthius flaviscapis flaviscapis (Temminck)

Allotrius flaviscapis Temminck, 1835, Pl. col., livr. 99, pl. 589, fig. 1 — Java and Sumatra; restricted to Java.

Highlands of Java.

Pteruthius flaviscapis robinsoni Chasen and Kloss

Pteruthius flaviscapis robinsoni Chasen and Kloss, 1931, Bull. Raffles Mus., no. 5, p. 86 — Kina Balu, North Borneo.

Highlands of northern Borneo.

PTERUTHIUS XANTHOCHLORUS

Pteruthius xanthochlorus occidentalis Harington

Pterythius [sic] *xanthochloris* [sic] *occidentalis* Harington, 1913, Bull. Brit. Orn. Club, 33, p. 82 — Dehra Dun, northwestern Uttar Pradesh.

Himalayas from East Punjab eastward to western Nepal.

Pteruthius xanthochlorus xanthochlorus Gray

Pteruthius xanthochlorus ["Hodgs."] J. E. Gray, 1846, Cat. Mamm. and Birds Nepal Thibet, p. 155 — Nepal.

Himalayas from central Nepal eastward, through Sikkim and Bhutan, into Assam (north of the Brahmaputra).

Pteruthius xanthochlorus hybridus Harington

> *Pterythius* [sic] *pallidus hybrida* [sic] Harington, 1913,
> Bull. Brit. Orn. Club, 33, p. 52 — Mount Victoria,
> Southern Chin Hills, Upper Burma.

Assam (Lushai and Naga Hills) ; Burma (Chin Hills).

Pteruthius xanthochlorus pallidus (David)

> *Allotrius xanthochloris* [sic] Hodgs., var. *pallidus* David,
> 1871, Nouv. Arch. Mus. Hist. Nat. [Paris], 7, p. 14 —
> "Kokonoor" = northwestern Szechwan.
> *Allotrius sophiæ* J. Verreaux, 1871, Nouv. Arch. Mus.
> Hist. Nat. [Paris], 7, p. 64 — "Kokonoor" = north-
> western Szechwan.
> *Pteruthius xanthochloris* [sic] *obscurus* Stresemann,
> 1925, Orn. Monatsb., 33, p. 59 — Fukien.

Northeastern Burma; southeastern Hiskang; western
Szechwan; northwestern and southeastern Yunnan; northwestern Fukien.

PTERUTHIUS MELANOTIS

Pteruthius melanotis melanotis Hodgson

> *Pt*[*eruthius*]. *melanotis* Hodgson, 1847, in Blyth, Journ.
> Asiat. Soc. Bengal, 16, p. 448 — the Terai, at base of
> southeastern Himalayas; type from Nepal, *fide* Gadow,
> 1883, Cat. Birds Brit. Mus., 8, p. 118.
> *Pteruthius melanotis melanops* Koelz, 1952, Journ. Zool.
> Soc. India, 4, p. 40 — Kohima, Naga Hills, Assam.

Himalayas from Nepal eastward, through Bhutan, Assam
(both north and south of the Brahmaputra), and northern
Burma, to northwestern Yunnan, in eastern Burma southward, through Shan States, to Karenni; southeastern Yunnan, northwestern Tongking, northern Annam, Laos, and
eastern portion of northern plateau of Thailand.

Pteruthius melanotis tahanensis Hartert

> *Pteruthius tahanensis* Hartert, 1902, Novit. Zool., 9, p.
> 576 — Gunong Tahan [lat. 4° 38′ N., long. 102° 14′ E.],
> Pahang.

Highlands of Malaya from southern Perak southward to
southern Selangor and northern Pahang.

PTERUTHIUS AENOBARBUS

Pteruthius aenobarbus aenobarbulus Koelz

Pteruthius aenobarbus aenobarbulus Koelz, 1954, Contrib. Inst. Regional Explor., no. 1, p. 9 — Nokrek, Garo Hills, Assam.

Known only from unique type.

Pteruthius aenobarbus intermedius (Hume)

Allotrius intermedius Hume, 1877, Stray Feathers, 5, pp. 112, 115 — central Tenasserim hills; type specimen from Mulayit Taung, Amherst District, *fide* Hume, 1878, in Hume and Davison, *ibid.*, 6, p. 370.

Pterythius [sic] *ænobarbus laotianus* Delacour, 1927, Bull. Brit. Orn. Club, 47, p. 162 — Chiang Khwang [lat. 19° 19′ N., long. 103° 22′ E.], Laos.

Eastern Burma from Kachin State southward to Southern Shan and Karenni States; hills of northwestern Thailand southward to Amherst District of Tenasserim; Laos and northwestern Tongking.

Pteruthius aenobarbus yaoshanensis Stresemann

Pteruthius aenobarbus yaoshanensis Stresemann, 1929, Orn. Monatsb., 37, p. 140 — Yao Shan, Kwangsi.

Known only from type locality.

Pteruthius aenobarbus indochinensis Delacour

Pterythius [sic] *ænobarbus indochinensis* Delacour, 1927, Bull. Brit. Orn. Club, 47, p. 163 — Jirinh [lat. 11° 31′ N., long. 108° 01′ E.], Annam.

Southern Annam (Lang Bian Plateau).

Pteruthius aenobarbus aenobarbus (Temminck)

Allotrius ænobarbus Temminck, 1835, Pl. col., livr. 99, pl. 589, fig. 2 — Java.

Highlands of western Java.

Genus GAMPSORHYNCHUS Blyth

Gampsorhynchus Blyth, 1844, Journ. Asiat. Soc. Bengal, 13, p. 370. Type, by monotypy, *Gampsorhynchus rufulus* Blyth.

GAMPSORHYNCHUS RUFULUS

Gampsorhynchus rufulus rufulus Blyth

G[ampsorhynchus]. rufulus Blyth, 1844, Journ. Asiat.
Soc. Bengal, 13, p. 371 — Darjeeling, West Bengal;
types from Arakan, Lower Burma, *fide* Finn, 1901, List
Birds Indian Mus., pt. 1, pp. 53-54.
Gampsorhynchus rufulus ahomensis Koelz, 1954, Contrib.
Inst. Regional Explor., no. 1, p. 4 — Nichuguard, Naga
Hills, Assam.

Lower hills from Sikkim eastward, through Bhutan and
Assam, to northeastern Burma, and southeastward through
Assam to southwestern Burma.

Gampsorhynchus rufulus torquatus Hume

Gampsorhynchus torquatus Hume, 1874, Proc. Asiat. Soc.
Bengal, no. 5, p. 107 — no locality; type specimen from
"the banks of the Younzaleen below the Pine forests
in the Salween district," Tenasserim, *fide* Hume, 1874,
Stray Feathers, 2, p. 446.
Gampsorhynchus rufulus luciæ Delacour, 1926, Bull. Brit.
Orn. Club, 47, p. 16 — Chiang Khwang [lat. 19° 19′ N.,
long. 103° 22′ E.], Laos.

Southeastern Burma from Karenni State southward to
central Tenasserim; southwestern Thailand (Kanchanaburi
Province); northern Thailand (except Chiang Rai Province) and northwestern portion of eastern plateau of Thailand; Tongking, central and southern Laos, southern Annam, and northern Cochin China.

Gampsorhynchus rufulus saturatior Sharpe

Gampsorhynchus saturatior Sharpe, 1888, Proc. Zool.
Soc. London, p. 273 — Gunong Batu Puteh, Perak.

Highlands of Malaya from southern Perak southward to
southern Selangor.

GENUS **ACTINODURA** GOULD

Actinodura Gould, 1836 (May), Proc. Zool. Soc. London,
pt. 4, p. 17. Type, by original designation and monotypy,
Actinodura egertoni Gould.
Sibia Hodgson, 1836 (*post* June), Asiat. Res., 19, p. 145.
Type, by monotypy, *Cinclosoma ? nipalense* Hodgson.

Ixops "Hodgson" Blyth, 1843, Journ. Asiat. Soc. Bengal, 12, p. 948. Type, by original designation and monotypy, *Cinclosoma* ? *nipalense* Hodgson.

Leiocincla Blyth, 1843, Journ. Asiat. Soc. Bengal, 12, p. 953. Type, by original designation and monotypy, *Leiocincla plumosa* Blyth = *Actinodura egertoni* Gould.

Actinura Agassiz, 1846, Nomencl. Zool. Index Univ., p. 7. *Nomen emendatum*

ACTINODURA EGERTONI

Actinodura egertoni egertoni Gould

Actinodura Egertoni Gould, 1836, Proc. Zool. Soc. London, pt. 4, p. 18 — Nepal.

Himalayas from Nepal eastward, through Sikkim and Bhutan, into Assam (north of the Brahmaputra).

Actinodura egertoni lewisi Ripley

Actinodura egertoni lewisi Ripley, 1948, Proc. Biol. Soc. Washington, 61, p. 105 — Dreyi, Mishmi Hills, Assam. Northeastern Assam (Mishmi Hills).

Actinodura egertoni khasiana (Godwin-Austen)

A[*ctinura*]. *khasiana* Godwin-Austen, 1876, Asiat. Soc. Bengal, 45, p. 76 — Khasi Hills, Assam.

Actinodura egertoni montivaga Koelz, 1954, Contrib. Inst. Regional Explor., no. 1, p. 7 — Kohima, Naga Hills, Assam.

Hill tracts of Assam south of the Brahmaputra.

Actinodura egertoni ripponi Ogilvie-Grant and La Touche

Actinodura ripponi Ogilvie-Grant and La Touche, 1907, Ibis, p. 186 — Mount Victoria [lat. 21° 15′ N., long. 93° 55′ E.], Upper Burma.

Hill tracts of southwestern Burma (Chin Hills and Arakan Yoma) ; birds from northern half of Burma referred here need reconsideration.

ACTINODURA RAMSAYI

Actinodura ramsayi yunnanensis Bangs and Phillips

Actinodura ramsayi yunnanensis Bangs and Phillips, 1914, Bull. Mus. Comp. Zool., 58, p. 288 — Loukouchai, southeastern Yunnan.

Actinodura ramsayi minor Kinnear, 1925, Bull. Brit. Orn. Club, 45, p. 74 — Ngai Tio [lat. 22° 40′ N., long. 103° 47′ E.], Tongking.

Actinodura ramsayi kinneari Delacour, 1927, Bull. Brit. Orn. Club, 47, p. 162 — Tam Dao [lat. 21° 27′ N., long. 105° 38′ E.], Tongking.

Mountains of southeastern Yunnan and Tongking.

Actinodura ramsayi radcliffei Harington

Actinodura radcliffei Harington, 1910, Bull. Brit. Orn. Club, 27, p. 9 — "Ruby Mines District" [= Katha District, Sagaing Division, Upper Burma].

Northern Shan State, eastern portion of Southern Shan State (?), and northern Laos.

Actinodura ramsayi ramsayi (Walden)

Actinura Ramsayi Walden, 1875, Ann. Mag. Nat. Hist., ser. 4, 15, p. 402 — Karenni State, Upper Burma; type from Kyebogyi [lat. 19° 21′ N., long. 97° 14′ E.], *fide* Ramsay, 1881, Orn. Works Arthur, Ninth Marquis Tweeddale, p. 415, footnote.

Western portion of Southern Shan State and Karenni State of Burma; northwestern Thailand.

ACTINODURA NIPALENSIS

Actinodura nipalensis nipalensis (Hodgson)

Cinclosoma ? Nipalensis Hodgson, 1836, Asiat. Res., 19, p. 145 — Nepal; restricted to Kathmandu Valley, by Ripley, 1950, Proc. Biol. Soc. Washington, 63, p. 104.

Western and central Nepal.

Actinodura nipalensis vinctura Ripley

Actinodura nipalensis vinctura Ripley, 1950, Proc. Biol. Soc. Washington, 63, p. 104 — Mangalbare, Dhankuta District, eastern Nepal.

Eastern Nepal, Sikkim, and Bhutan.

ACTINODURA WALDENI

Actinodura waldeni daflaensis (Godwin-Austen)

Actinura daflaensis Godwin-Austen, 1875, Ann. Mag. Nat. Hist., ser. 4, 16, p. 340 — Dafla Hills, Assam.

Hill tracts of Assam north of the Brahmaputra (Miri and Dafla Hills).

Actinodura waldeni waldeni Godwin-Austen

Actinodura waldeni Godwin-Austen, 1874, Proc. Zool. Soc. London, p. 46, pl. 12 — Mount Japvo, Naga Hills, Assam.

Hill tracts of southeastern Assam (Naga Hills and Manipur); northwestern Burma.

Actinodura waldeni poliotis (Rippon)

Ixops poliotis Rippon, 1905, Bull. Brit. Orn. Club, 15, p. 97 — Mount Victoria [lat. 21° 15′ N., long. 93° 55′ E.], Upper Burma.

Known only from type locality.

Actinodura waldeni saturatior (Rothschild)

Ixops poliotis saturatior Rothschild, 1921, Novit. Zool., 28, p. 38 — Shweli-Salween Divide, western Yunnan.

Actinodura nipalensis wardi Kinnear, 1932, Bull. Brit. Orn. Club, 53, p. 79 — Adung Valley [lat. 28° 10′ E., long. 97° 40′ E.], Upper Burma.

Northeastern Burma (Kachin State) and northwestern Yunnan.

ACTINODURA SOULIEI

Actinodura souliei souliei Oustalet

Actinodura Souliei Oustalet, 1897, Bull. Mus. Hist. Nat. Paris, 3, p. 164 — Tzeku, northwestern Yunnan.

Northwestern Yunnan.

Actinodura souliei griseinucha Delacour and Jabouille

Actinodura souliei griseinucha Delacour and Jabouille, 1930, Oiseau Rev. Franç. Orn., 11, p. 403 — Fan Si Pan [lat. 22° 17′ N., long. 104° 47′ E.], Tongking.

Northwestern Tongking.

ACTINODURA MORRISONIANA

Actinodura morrisoniana Ogilvie-Grant

Actinodura morrisoniana Ogilvie-Grant, 1906, Bull. Brit. Orn. Club, 16, p. 119 — Mount Morrison, Formosa.

Highlands of Formosa.

Genus MINLA Hodgson

Minla Hodgson, 1837 (Apr.), India Review, 2, p. 32. Type, by original designation, *Minla ignotincta* Hodgson.
Siva Hodgson, 1837 (May), India Review, 2, p. 88. (Subgenus of *Mesia* Hodgson.) Type, by subsequent designation (G. R. Gray, 1841, List. Gen. Birds, ed. 2, p. 45), *Mesia* (*Siva*) *cyanouroptera* Hodgson.

MINLA CYANOUROPTERA

Minla cyanouroptera cyanouroptera (Hodgson)

Siva Cyanouroptera Hodgson, 1838, India Review, 2 (1837), p. 88 — Nepal.
Leiothrix lepida McClelland, 1840, Proc. Zool. Soc. London, pt. 7, p. 162 — Assam.
Siva cyanouroptera thalia Koelz, 1954, Contrib. Inst. Regional Explor., no. 1, p. 8 — Maoflang, Khasi Hills, Assam.
Siva cyanouroptera rama Koelz, 1954, Contrib. Inst. Regional Explor., no. 1, p. 8 — near Nokrek, Garo Hills, Assam.

Himalayas from northwestern Uttar Pradesh, through Nepal and Sikkim, to Bhutan and hill tracts of eastern Assam (both north and south of the Brahmaputra).

Minla cyanouroptera aglae (Deignan)

Siva cyanouroptera aglaë Deignan, 1942, Notulae Naturae [Philadelphia], no. 100, p. 1 — Mount Victoria, Southern Chin Hills, Upper Burma.

Hill tracts of southeastern Assam and western Burma.

Minla cyanouroptera sordida (Hume)

Siva sordida Hume, 1877, Stray Feathers, 5, p. 104 — Mulayit Taung [lat. 16° 11′ N., long. 98° 32′ E.], Tenasserim.
Siva cyanuroptera oatesi Harington, 1913, Bull. Brit. Orn. Club, 33, p. 62 — Byingye Taung [lat. 20° 01′ N., long. 96° 26′ E.], Upper Burma.

Southern Shan State (west of the Salween), Karenni, northwestern Thailand, and central Tenasserim (Mulayit Taung).

Minla cyanouroptera **wingatei** (Ogilvie-Grant)

> *Siva wingatei* Ogilvie-Grant, 1900, Bull. Brit. Orn. Club,
> 10, p. 38 — near Kunming, Yunnan.
> *Siva cyanouroptera yaoshanica* Hachisuka, 1941, Proc.
> Biol. Soc. Washington, 54, p. 50 — "Most probably
> Yaoshan, N. E. of Nanning," Kwangsi.

Northeastern Burma (Kachin and Northern Shan States
and Southern Shan State east of the Salween), northeastern
portion of northern plateau of Thailand, northern Laos,
northern Annam, Tongking, Kwangsi, and the whole of
Yunnan.

Minla cyanouroptera **croizati** Deignan

> *Minla cyanouroptera croizati* Deignan, 1958, Proc. Biol.
> Soc. Washington, 71, p. 162 — Ipin [Suifu], Szechwan.

Known only from type locality.

Minla cyanouroptera **rufodorsalis** (Engelbach)

> *Siva cyanouroptera rufodorsalis* Engelbach, 1946, Oiseau
> Rev. Franç. Orn., nouv. sér., 16, p. 61 — mountains of
> the Cardomoms, southwestern Cambodia.

Mountains of southeastern Thailand (?) and southwestern
Cambodia.

Minla cyanouroptera **orientalis** (Robinson and Kloss)

> *Siva sordida orientalis* Robinson and Kloss, 1919, Ibis, p.
> 587 — Lang Bian Peaks [*ca.* lat. 12° 02′ N., long. 108°
> 26′ E.], Annam.

Southern Annam (Lang Bian Plateau).

Minla cyanouroptera **sordidior** (Sharpe)

> *Siva sordidior* Sharpe, 1888, Proc. Zool. Soc. London, p.
> 276 — Batang Padang Mountains, Perak.

Mountains of peninsular Thailand (Surat Thani and
Nakhon Si Thammarat Provinces) and highlands of Malaya
from northern Perak to southern Selangor.

MINLA STRIGULA

Minla strigula **simlaensis** (Meinertzhagen)

> *Siva strigula simlaensis* Meinertzhagen, 1926, Bull. Brit.
> Orn. Club, 46, p. 128 — Simla, Himachal Pradesh.

Himalayas from East Punjab eastward to western Nepal.

Minla strigula strigula (Hodgson)

[*Mesia*] *Siva Strigula* Hodgson, 1837, India Review, 2, p. 89 — Nepal.

Himalayas from central Nepal eastward, through Bhutan and Assam (north of the Brahmaputra), as far as the Abor Hills.

? Minla strigula cinereigenae (Ripley)

Siva strigula cinereigenae Ripley, 1952, Journ. Bombay Nat. Hist. Soc., 50, p. 500 — Mount Japvo, Barail Hills, Assam.

Known only from type locality; doubtfully distinct from *yunnanensis*.

Minla strigula yunnanensis (Rothschild)

Siva strigula yunnanensis Rothschild, 1921, Novit. Zool., 28, p. 40 — Likiang Range, northwestern Yunnan.

Siva strigula victoriæ Meinertzhagen, 1926, Bull. Brit. Orn. Club, 46, p. 128 — Mount Victoria, Southern Chin Hills, Upper Burma.

Hill tracts of eastern Assam and western Burma eastward, through northern Burma and Yunnan, to northwestern Tongking and eastern portion of northern Laos.

Minla strigula castanicauda (Hume)

Siva castanicauda Hume, 1877, Stray Feathers, 5, p. 100 — Mulayit Taung [lat. 16° 11′ N., long. 98° 32′ E.], Tenasserim.

Hills of Karenni, northern Tenasserim, and northwestern Thailand.

Minla strigula malayana (Hartert)

Siva strigula malayana Hartert, 1902, Novit. Zool., 9, p. 567 — Gunong Tahan, Pahang.

Siva strigula omissa Rothschild, 1921, Novit. Zool., 28, p. 40 — Gunong Kerbau, Perak.

Highlands of Malaya from northern Perak southward to southern Selangor and Pahang.

MINLA IGNOTINCTA

Minla ignotincta ignotincta Hodgson

Minla Ignotincta Hodgson, 1837, India Review, 2, p. 32 — Nepal.

Himalayas from Nepal eastward, through Assam and northern Burma, to northwestern Yunnan, and southward through hill tracts of Assam and Burma to Chin Hills and Karenni.

Minla ignotincta mariae La Touche

Minla ignotincta mariæ La Touche, 1921, Bull. Brit. Orn. Club, 42, p. 30 — Milati, southeastern Yunnan.

Southeastern Yunnan and northwestern Tongking.

Minla ignotincta sini Stresemann

Minla ignotincta sini Stresemann, 1929, Journ. f. Orn., 77, pp. 325 (in list), 333 — Yao Shan, Kwangsi.

Known only from type locality.

Minla ignotincta jerdoni Verreaux

Minla jerdoni J. Verreaux, 1870, Nouv. Arch. Mus. Hist. Nat. [Paris], 6, p. 38; *idem*, 1871, *ibid.*, 7, p. 52; *idem*, 1872, *ibid.*, 8, pl. 2, fig. 1 — "les montagnes du Thibet chinois"; type from Chengtu, Szechwan, *fide* Verreaux, 1871.

Southwestern Szechwan.

GENUS **ALCIPPE** BLYTH

Alcippe Blyth, 1844, Journ. Asiat. Soc. Bengal, 13, pp. 370, 384. Type, by original designation, *Thimalia poioicephala* Jerdon.

Schœniparus Anonymous = Hume, 1874, Stray Feathers, 2, p. 449. Type, by subsequent designation (Sharpe, 1883, Cat. Birds Brit. Mus., 7, p. 606), *Minla rufogularis* Mandelli.

Fulvetta David and Oustalet, 1877, Oiseaux Chine, p. 220. Type, by subsequent designation (Sharpe, 1883, Cat. Birds Brit. Mus., 7, p. 628), *Siva cinereiceps* J. Verreaux.

Sittiparus Oates, 1889, Fauna Brit. India, Birds, 1, pp. 131 (in key), 171. Type, by original designation, *Minla cinerea* Blyth. Not *Sittiparus* Selys-Longchamps, 1884, Aves.

Lioparus Oates, 1889, Fauna Brit. India, Birds, 1, pp. 131 (in key), 174. Type, by original designation, *Lioparus chrysaeus* Oates = *Proparus chrysotis* Blyth.

Pseudominla Oates, 1894, Ibis, p. 480. New name for *Sittiparus* Oates, preoccupied.
Semiparus Hellmayr, 1901, Journ. f. Orn., 49, p. 171. New
name for *Sittiparus* Oates, preoccupied.
Proparoides Bianchi, 1902, Bull. Brit. Orn. Club, 12, p. 55.
New name for *Sittiparus* Oates, preoccupied.
Alcippornis Oberholser, 1922, Smiths. Misc. Coll., 74 (2),
p. 1. New name for "*Alcippe* Auct. nec Blyth"; type,
by original designation, *Alcippe cinerea* Blyth, *nec*
Eyton = *Hyloterpe brunneicauda* Salvadori = *Alcippe
brunneicauda brunneicauda* (Salvadori).
Pseudoalcippe Bannerman, 1923, Bull. Brit. Orn. Club, 44,
p. 26. Type, by original designation, *Turdinus atriceps*
Sharpe.

ALCIPPE CHRYSOTIS

Alcippe chrysotis chrysotis (Blyth)

Pr[*oparus*]. *chrysotis* "Hodgson" Blyth, 1845, Journ.
Asiat. Soc. Bengal, 13 (1844), p. 938 — no locality =
Nepal, ex Hodgson.
Lioparus chrysæus Oates, 1889, Fauna Brit. India, Birds,
1, p. 174 — "Nepal and Sikhim; the Daphla and Eastern
Nága hills; Manipur." Based upon *P*[*roparus*]. ?
chrysæus Hodgson, 1844, a *nomen nudum*.
Himalayas from Nepal eastward, through Sikkim and
Bhutan, to eastern Assam (north of the Brahmaputra).

Alcippe chrysotis albilineata (Koelz)

Lioparus chrysotis albilineatus Koelz, 1954, Contrib. Inst.
Regional Explor., no. 1, p. 7 — Karong, Manipur.
Hill tracts of Assam south of the Brahmaputra.

Alcippe chrysotis forresti (Rothschild)

Fulvetta chrysotis forresti Rothschild, 1926, Bull. Brit.
Orn. Club, 46, p. 64 — Shweli-Salween Divide, Yunnan.
Northeastern Burma and northwestern Yunnan.

Alcippe chrysotis amoena (Mayr)

Fulvetta chrysotis amœna Mayr, 1941, in Stanford and
Mayr, Ibis, p. 81 — Fan Si Pan [lat. 22° 17′ N., long.
104° 47′ E.], Tongking.
Northwestern Tongking.

Alcippe chrysotis swinhoii (Verreaux)

Proparus swinhoii J. Verreaux, 1870, Nouv. Arch. Mus. Hist. Nat. [Paris], 6, p. 38; *idem*, 1871, *ibid.*, 7, p. 51; *idem*, 1872, *ibid.*, 8, pl. 2, fig. 2 — "les montagnes du Thibet chinois"; types from western Szechwan and Muping [= Paohing], *fide* Verreaux, 1871.

Southeastern Hsikang and southwestern Szechwan.

ALCIPPE VARIEGATICEPS

Alcippe variegaticeps Yen

Alcippe variegaticeps Yen, 1932, Bull. Mus. Hist. Nat. Paris, sér. 2, 4, p. 383 — Yaoshan, Kwangsi.

Known only from type locality.

ALCIPPE CINEREA

Alcippe cinerea (Blyth)

Minla cinerea Blyth, 1847, Journ. Asiat. Soc. Bengal, 16, p. 449 — Darjeeling, West Bengal.

Alcippe Delacouri Yen, 1935, Science Journ. [College of Science, Sun Yatsen Univer. Canton], 6, p. 706. New name for *Minla cinerea* Blyth, considered preoccupied by "*Alcippe cinerea*" Blyth, 1844.

Himalayas from Nepal through Assam, both north and south of the Brahmaputra; northeastern Burma; northern Laos.

ALCIPPE CASTANECEPS

Alcippe castaneceps castaneceps (Hodgson)

[*Minla*] *Castaneceps* Hodgson, 1837, India Review, 2, p. 33 — Nepal.

Minla brunneicauda Sharpe, 1883, Cat. Birds Brit. Mus., 7, pp. 606 (in key), 609 — Khasi Hills, Assam. Not *Hyloterpe brunneicauda* Salvadori, 1879 = *Alcippe brunneicauda brunneicauda* (Salvadori).

Pseudominla castaneiceps garoensis Koelz, 1951, Journ. Zool. Soc. India, 3, p. 29 — Tura Mountain, Garo Hills, Assam.

Alcippe castaneceps wagstaffei Wynne, 1954, North Western Naturalist [Arbroath, Scotland], new ser., 2, pp. 297, 306. New name for *Minla brunneicauda* Sharpe, preoccupied, with insufficient indication.

Alcippe castaneceps wagstaffei Wynne, 1955, North Western Naturalist [Arbroath, Scotland], new ser., 3, p. 120. New name for *Minla brunneicauda* Sharpe, preoccupied.

Himalayas from Nepal eastward to northwestern Yunnan (west of the Salween), southward through hill tracts of Assam to Chin Hills of western Burma, and through hills of eastern Burma to northwestern Thailand and central Tenasserim.

Alcippe castaneceps exul Delacour

Alcippe castaneiceps exul Delacour, 1932, Oiseau Rev. Franç. Orn., nouv. sér., 2, p. 427 — Phu Kong-Ntul [*ca.* lat. 15° 23′ N., long. 106° 25′ E.], Laos.

Eastern portion of northern plateau of Thailand, the whole of Laos, and northwestern Tongking.

Alcippe castaneceps soror (Sharpe)

Minla soror Sharpe, 1887, Proc. Zool. Soc. London, p. 439, pl. 38, fig. 1 — Larut Range, Perak.

Highlands of Malaya from northern Perak southward to southern Selangor and Pahang.

Alcippe castaneceps klossi Delacour and Jabouille

Pseudominla atriceps Robinson and Kloss, 1919, Ibis, p. 583 — Lang Bian Peaks, southern Annam. Not *Turdinus atriceps* Sharpe, 1902 = *Alcippe atriceps* (Sharpe).

Alcippe castaneiceps klossi Delacour and Jabouille, 1931, Oiseaux Indochine Franç., 3, p. 308. New name for *Pseudominla atriceps* Robinson and Kloss, considered preoccupied by *Alcippe atriceps* (Jerdon), 1844, a secondary homonym created by Blyth, ex *Brachypteryx atriceps* Jerdon, 1839.

Southern Annam (Lang Bian Plateau).

ALCIPPE VINIPECTUS

Alcippe vinipectus kangrae (Ticehurst and Whistler)

Fulvetta vinipecta [sic] *kangræ* Ticehurst and Whistler, 1924, Bull. Brit. Orn. Club, 44, p. 71 — Palampur, East Punjab.

Himalayas from northern East Punjab eastward to northwestern Uttar Pradesh.

Alcippe vinipectus vinipectus (Hodgson)

[*Mesia*] *Siva Vinipectus* Hodgson, 1837, India Review, 2, p. 89 — Nepal.

Nepal, except area occupied by *chumbiensis*.

Alcippe vinipectus chumbiensis (Kinnear)

Fulvetta vinipectus chumbiensis Kinnear, 1939, Ibis, p. 751 — Yatung, Chumbi Valley, southeastern Tibet.

Easternmost Nepal, Sikkim, Chumbi Valley of Tibet, and Bhutan.

Alcippe vinipectus austeni (Ogilvie-Grant)

Proparus austeni Ogilvie-Grant, 1895, Bull. Brit. Orn. Club, 5, p. iii — Manipur and the Naga Hills, Assam.

Hill tracts of Assam south of the Brahmaputra.

Alcippe vinipectus ripponi (Harington)

Proparus ripponi Harington, 1913, Bull. Brit. Orn. Club, 33, p. 59 — Mount Victoria, Southern Chin Hills, Burma.

Known only from highest peaks of Chin Hills.

Alcippe vinipectus perstriata (Mayr)

Fulvetta vinipectus perstriata Mayr, 1941, Ibis, p. 79 — Chawngmawhka, Burma-Yunnan border.

Kachin State of Burma and adjacent regions of Yunnan eastward to the Salween.

Alcippe vinipectus valentinae Delacour and Jabouille

Alcippe vinipectus valentinae Delacour and Jabouille, 1930, Oiseau Rev. Franç. Orn., 11, p. 401 — Fan Si Pan [lat. 22° 17′ N., long. 104° 47′ E.], Tongking.

Known only from type locality.

Alcippe vinipectus bieti Oustalet

Alcippe (*Proparus*) *Bieti* Oustalet, 1891, Ann. Sci. Nat., Zool., sér. 7, 12, p. 284, pl. 9, fig. 2 — Tatsienlu [= Kangting, Hsikang].

Northwestern Yunnan (except area occupied by *perstriata*) and southeastern Hsikang.

ALCIPPE STRIATICOLLIS

Alcippe striaticollis (Verreaux)

Siva striaticollis J. Verreaux, 1870, Nouv. Arch. Mus. Hist. Nat. [Paris], 6, p. 38; *idem*, 1871, *ibid.*, 7, p. 50 — "les montagnes du Thibet chinois"; type from Muping [= Paohing], *fide* Verreaux, 1871.

Southwestern Kansu, northwestern Szechwan, and southeastern Hsikang.

ALCIPPE RUFICAPILLA

Alcippe ruficapilla ruficapilla (Verreaux)

Siva ruficapilla J. Verreaux, 1870, Nouv. Arch. Mus. Hist. Nat. [Paris], 6, p. 37; *idem*, 1871, *ibid.*, 7, p. 49; *idem*, 1872, *ibid.*, 8, pl. 5, fig. 2 — "les montagnes du Thibet chinois"; types from western Szechwan, *fide* Verreaux, 1871.

Southern Shensi and Szechwan.

Alcippe ruficapilla sordidior (Rippon)

Proparus sordidior Rippon, 1903, Bull. Brit. Orn. Club, 13, p. 60 — Gyi-dzin-shan, east of Talifu, Yunnan.

Northwestern Yunnan.

Alcippe ruficapilla danisi Delacour and Greenway

Alcippe (Fulvetta) ruficapilla danisi Delacour and Greenway, 1941, Proc. New England Zool. Club, 18, p. 47 — Phu Kobo [lat. 19° 16′ N., long. 103° 25′ E.], Laos.

Southeastern Yunnan, northwestern Tongking (?), and eastern portion of northern Laos.

ALCIPPE CINEREICEPS

Alcippe cinereiceps ludlowi (Kinnear)

Fulvetta ludlowi Kinnear, 1935, Bull. Brit. Orn. Club, 55, p. 134 — Sakden, eastern Bhutan.

Eastern Bhutan and southeastern Tibet.

Alcippe cinereiceps manipurensis (Ogilvie-Grant)

Proparus manipurensis Ogilvie-Grant, 1906, Bull. Brit. Orn. Club, 16, p. 123 — Owenkulno Peak, Manipur.

Proparus striaticollis yunnanensis Rothschild, 1922, Bull. Brit. Orn. Club, 43, p. 11 — Mekong-Salween Divide, northwestern Yunnan. Not *Alcippe fratercula yunnan-ensis* Harington, 1913.

Fulvetta insperata Riley, 1930, Proc. Biol. Soc. Washington, 43, p. 123 — Ndamucho, south of Lutien [lat. 27° 12′ N., long. 99° 28′ E.], Yunnan.

Fulvetta ruficapilla menghwaensis Chong, 1937, Sinensis, 8, p. 381 — Siaotsun, near Menghwa, western Yunnan.

Hill tracts of eastern Assam (Naga and Manipur Hills) and western Burma (Chin Hills) ; northeastern Burma and northwestern Yunnan.

Alcippe cinereiceps tonkinensis Delacour and Jabouille

Alcippe ruficapillus (sic) *tonkinensis* Delacour and Jabouille, 1930, Oiseau Rev. Franç. Orn., 11, p. 402 — Fan Si Pan [lat. 22° 17′ N., long. 104° 47′ E.], 2,800 metres, Tongking.

Southeastern Yunnan (?), northwestern Tongking, and eastern portion of northern Laos.

Alcippe cinereiceps guttaticollis (La Touche)

Proparus guttaticollis La Touche, 1897, Bull. Brit. Orn. Club, 6, p. 50 — Kuatun, Fukien.

Highlands of Fukien and northern Kwangtung.

Alcippe cinereiceps formosana (Ogilvie-Grant)

Proparus formosanus Ogilvie-Grant, 1906, Bull. Brit. Orn. Club, 16, p. 120 — Mount Morrison, Formosa.

Highlands of Formosa.

Alcippe cinereiceps fucata (Styan)

Proparus fucatus Styan, 1899, Bull. Brit. Orn. Club, 8, p. 26; *idem*, 1899, Ibis, p. 306, pl. 4, fig. 1 — Ichang, Hupeh.

Alcippe cinereiceps Berliozi Yen, 1932, Bull. Mus. Nat. Hist. Paris, sér. 2, 4, p. 381 — Ching-tung-shan, southern Hunan.

Hupeh and Hunan.

Alcippe cinereiceps cinereiceps (Verreaux)

Siva cinereiceps J. Verreaux, 1870, Nouv. Arch. Mus. Hist. Nat. [Paris], 6, p. 37; *idem*, 1871, *ibid.*, 7, p. 48; *idem*, 1872, *ibid.*, 8, pl. 5, fig. 3 — "les montagnes du

Thibet chinois"; type from Muping [= Paohing], *fide* Verreaux, 1871.
Western Hupeh, Szechwan, and southeastern Hsikang.

Alcippe cinereiceps fessa (Bangs and Peters)

Fulvetta cinereiceps fessa Bangs and Peters, 1928, Bull. Mus. Comp. Zool., 68, p. 342 — near Choni, southwestern Kansu.
Southwestern Kansu.

ALCIPPE RUFOGULARIS

Alcippe rufogularis rufogularis (Mandelli)

Minla Rufogularis Mandelli, 1873, Stray Feathers, 1, p. 416 — Bhutan Duars.
Himalayas from Bhutan Duars eastward through Assam (north of the Brahmaputra) to the Abor Hills.

Alcippe rufogularis collaris Walden

Alcippe collaris Walden, 1874, Ann. Mag. Nat. Hist., ser. 4, 14, p. 156 — Sadiya, northeastern Assam.
Hill tracts of eastern Assam (from the Mishmis southward to Manipur) and of East Pakistan.

Alcippe rufogularis major (Baker)

Schœniparus rufigularis major Baker, 1920, Bull. Brit. Orn. Club, 41, p. 11 — Ban Pak Mat [lat. 18° 52′ N., long. 101° 51′ E.], Laos.
Northeastern Burma (Kachin State), Northern and Southern Shan States (?), northern and central Laos (valley of the Mekong), and adjacent areas of northern and eastern Thailand.

Alcippe rufogularis stevensi (Kinnear)

Schœniparus rufogularis stevensi Kinnear, 1924, Bull. Brit. Orn. Club, 45, p. 10 — Bao Ha [lat. 22° 10′ N., long. 104° 21′ E.], Tongking.
Schœniparus rufogularis blanchardi Delacour and Jabouille, 1928, Bull. Brit. Orn. Club, 48, p. 132 — Phuqui, northern Annam.
Northwestern Tongking, eastern portion of northern Laos, and northern Annam.

Alcippe rufogularis kelleyi (Bangs and Van Tyne)

Schoeniparus rufogularis kelleyi Bangs and Van Tyne, 1930, Field Mus. Nat. Hist. Publ., Zool. Ser., 18, p. 4 — Phuoc Mon, Quangtri Province, Annam.

Central Annam.

Alcippe rufogularis khmerensis (de Schauensee)

Schoeniparus rufogularis khmerensis de Schauensee, 1938, Proc. Acad. Nat. Sci. Philadelphia, 90, p. 27 — Krat [= Ban Bang Phra (lat. 12° 15′ N., long. 102° 30′ E.)], Thailand.

Southeastern Thailand and southwestern Cambodia.

ALCIPPE BRUNNEA

Alcippe brunnea mandellii (Godwin-Austen)

Minla Mandellii Godwin-Austen, 1876, Ann. Mag. Nat. Hist., ser. 4, 17, p. 33 — Naga Hills, Assam.

Schoeniparus dubius certus Koelz, 1952, Journ. Zool. Soc. India, 4, p. 39 — Shillong Peak, Khasi Hills, Assam.

Hill tracts of Assam (south of the Brahmaputra) and western Burma (Chin Hills).

Alcippe brunnea intermedia (Rippon)

Schœniparus intermedius Rippon, 1900, Bull. Brit. Orn. Club, 11, p. 11 — Loi Mai [lat. 20° 25′ N., long. 97° 26′ E.], Southern Shan State, Upper Burma.

Kachin and Shan States of Burma and adjacent regions of Yunnan eastward to the Salween.

Alcippe brunnea dubia (Hume)

Proparus dubius Anonymous = Hume, 1874, Stray Feathers, 2, p. 447 — "outskirts of the Pine forests above the Salween," Salween District, Tenasserim, Lower Burma.

Hills of Karenni State and Tenasserim southward to Amherst District.

Alcippe brunnea genestieri Oustalet

Alcippe Genestieri Oustalet, 1897, Bull. Mus. Hist. Nat. Paris, 3, p. 210 — Tzeku, northwestern Yunnan.

Schœniparus variegatus Styan, 1899, Bull. Brit. Orn. Club, 8, p. 26; *idem*, 1899, Ibis, p. 306, pl. 4, fig. 2 — Suiyang, Kweichow.

Southeastern Hsikang, Yunnan (except areas occupied by *intermedia*), western Kweichow, northwestern Tongking, and eastern portion of northern Laos.

Alcippe brunnea superciliaris (David)

Ixulus superciliaris David, 1874, Ann. Sci. Nat., Zool., sér. 5, 19, art. 9, p. 4 — Fukien.

Hills of southeastern China from Kwangtung, northward through Fukien and Kiangsi, to Anhwei.

Alcippe brunnea brunnea Gould

Alcippe brunnea Gould, 1863, Proc. Zool. Soc. London, p. 280 — Formosa.

Alcippe obscurior Ogilvie-Grant, 1906, Bull. Brit. Orn. Club, 16, p. 121 — Mount Morrison, Formosa.

Formosa.

Alcippe brunnea arguta (Hartert)

Proparus brunnea (sic) *argutus* Hartert, 1910, Novit. Zool., 17, p. 231 — Mount Wuchi, Hainan.

Hainan.

Alcippe brunnea olivacea Styan

Alcippe olivacea Styan, 1896, Ibis, p. 312 — Ichang, Hupeh.

Schoeniparus brunneus weigoldi Stresemann, 1923, Journ. f. Orn., 71, p. 366 — mountains near Kwanhsien, Szechwan.

Western Hupeh and Szechwan (except northwestern highlands).

ALCIPPE BRUNNEICAUDA

Alcippe brunneicauda brunneicauda (Salvadori)

Hyloterpe brunneicauda Salvadori, 1879, Ann. Mus. Civ. Genova, 14, p. 210 — "Ayer Manchor," a waterfall at 10 kilometres from Padang Panjang [lat. 0° 29′ S., long. 100° 22′ E.] on the road to the Padang Highlands, Sumatra.

Alcippe cinerea hypocneca Oberholser, 1912, Smiths. Misc. Coll., 60 (7), p. 8 — Pulau Pini, Batu Islands.

Alcippornis brunneicauda epipolia Oberholser, 1932, Bull. U. S. Nat. Mus., 159, p. 63 — Pulau Bunguran, North Natuna Islands.

Malay Peninsula from Siamese provinces of Nakhon Si Thammart and Krabi southward to Johore; Sumatra; Batu and North Natuna Islands; northwestern Borneo (limits of range uncertain).

Alcippe brunneicauda eriphaea (Oberholser)

Alcippornis brunneicauda eriphaea Oberholser, 1922, Smiths. Misc. Coll., 74 (2), p. 2 — Gunong Liang Kubung [ca. lat. 0° 39' N., long. 113° 17' E.], Borneo.

Borneo (except area occupied by *brunneicauda*).

ALCIPPE POIOICEPHALA

Alcippe poioicephala poioicephala (Jerdon)

Thimalia poioicephala Jerdon, 1844, Madras Journ. Lit. Sci., 13, p. 169 — Coonoor Ghat, Nilgiri Hills.

Alcippe phæocephala Sharpe, 1883, Cat. Birds Brit. Mus., 7, pp. 618 (in key), 622. *Nomen emendatum.*

Western Ghats of India from Kerala northward to South Kanara, Coorg, and southern Mysore; Palni Hills.

Alcippe poioicephala brucei Hume

Alcippe phæocephala brucei Hume, 1870, Journ. Asiat. Soc. Bengal, 39, p. 122 — Mahabaleshwar, Satara North, Bombay.

Peninsular India (except area occupied by *poioicephala*) southward of a line drawn from the Surat Dangs to northern Orissa.

Alcippe poioicephala fusca Godwin-Austen

Alcippe fusca Godwin-Austen, 1876, Journ. Asiat. Soc. Bengal, 45, p. 197 — Naga Hills, Assam.

Assam (south of the Brahmaputra) and northwestern Burma.

Alcippe poioicephala phayrei Blyth

Alcippe Phayrei Blyth, 1845, Journ. Asiat. Soc. Bengal, 14, p. 601 — Arakan.

Southwestern Burma (Chin Hills and Arakan).

Alcippe poioicephala haringtoniae Hartert

Alcippe haringtoniæ Hartert, 1909, Bull. Brit. Orn. Club, 25, p. 10 — Bhamo [lat. 24° 15' N., long. 97° 14' E.],

Upper Burma.

Northeastern Burma (Myitkyina District) southward
through Shan States to northwestern Thailand.

Alcippe poioicephala alearis (Bangs and Van Tyne)

Alcippornis poiocephala alearis Bangs and Van Tyne,
1930, Field Mus. Nat. Hist. Publ., Zool. Ser., 18, p. 4 —
Muang Mun [lat. 21° 42′ N., long. 103° 21′ E.], Tongking.

Eastern portion of northern plateau of Thailand and
northwestern portion of eastern plateau; northern Laos,
northwestern Tongking, and northern Annam.

Alcippe poioicephala karenni Robinson and Kloss

Alcippe magnirostris Walden, 1875, in Blyth, Journ. Asiat.
Soc. Bengal, 43, extra no., p. 115 — Karenni State, Upper Burma. Not *Alcippe magnirostris* Moore, 1854.

Alcippe phæocephala karenni Robinson and Kloss, 1923,
Journ. and Proc. Asiat. Soc. Bengal, new ser., 18, p. 563.
New name for *Alcippe magnirostris* Walden, preoc-
cupied.

Alcippe poioicephala blythi Collin and Hartert, 1927,
Novit. Zool., 34, p. 50. New name for *Alcippe magnirostris* Walden, preoccupied.

Southeastern Burma from Karenni State southward to
central Tenasserim; southwestern Thailand from southern
Tak Province southward to Prachuap Khiri Khan Province.

Alcippe poioicephala davisoni Harington

Alcippe phæocephala davisoni Harington, 1915, Journ.
Bombay Nat. Hist. Soc., 23, pp. 447 (in key), 453 —
"Tavoy, Mergui and to the south . . ."; here restricted
to Mergui District, Tenasserim.

Mergui Archipelago; Malay Peninsula from southern
Tenasserim and Isthmus of Kra southward to Siamese
Province of Trang.

ALCIPPE PYRRHOPTERA

Alcippe pyrrhoptera (Bonaparte)

N[apothera]. pyrrhoptera "Boie, Mus. Lugd." Bonaparte,
1850, Consp. Av., 1, p. 358 — Java.

A[lcippe]. solitaria Cabanis, 1851, Mus. Hein., 1 (1850),
p. 87 — "Sumatra," error = Java.

A [lcippe]. dumetoria Cabanis, 1851, Mus. Hein., 1 (1850),
   p. 88 — Java.
Mixornis erythronota Reichenow, 1895, Journ. f. Orn., 43,
   p. 356 — Java.
Western and central Java.

ALCIPPE PERACENSIS

Alcippe peracensis grotei Delacour

Alcippe nipalensis major Delacour, 1926, Bull. Brit. Orn.
   Club, 47, p. 18 — Col des Nuages [lat. 16° 11′ N., long.
   108° 08′ E.], Annam. Not *Schœniparus rufigularis major* Baker, 1920 = *Alcippe rufogularis major* (Baker).
Alcippe nipalensis grotei Delacour, 1936, Orn. Monatsb.,
   44, p. 24. New name for *Alcippe nipalensis major*, preoccupied.
Northern and central Annam and adjacent regions of
southern Laos.

Alcippe peracensis annamensis Robinson and Kloss

Alcippe nepalensis annamensis Robinson and Kloss, 1919,
   Ibis, p. 582 — Dalat [lat. 11° 55′ N., long. 108° 26′ E.],
   Annam.
Southern Laos (Boloven Plateau), southern Annam, and
adjacent regions of Cochin China.

Alcippe peracensis eremita Riley

Alcippe nipalensis eremita Riley, 1936, Proc. Biol. Soc.
   Washington, 49, p. 25 — Khao Saming [lat. 12° 21′ N.,
   long. 102° 27′ E.], Thailand.
Southeastern Thailand.

Alcippe peracensis peracensis Sharpe

Alcippe peracensis Sharpe, 1887, Proc. Zool. Soc. London,
   p. 439 — Larut Range, Perak.
Highlands of Malaya from northern Perak southward to
southern Selangor and Pahang.

ALCIPPE MORRISONIA

Alcippe morrisonia yunnanensis Harington

Alcippe fratercula yunnanensis Harington, 1913, Bull.
   Brit. Orn. Club, 33, p. 63 — Gyi-dzin-shan, east of Talifu, Yunnan.

Southeastern Hsikang, northwestern Yunnan, and north-eastern Burma (Kachin State).

Alcippe morrisonia fraterculus Rippon

Alcippe fratercula (sic) Rippon, 1900, Bull. Brit. Orn. Club, 11, p. 11 — Southern Shan State, Upper Burma. *Alcippe nipalensis laotianus* Delacour, 1926, Bull. Brit. Orn. Club, 47, p. 19 — Chiang Khwang [lat. 19° 19′ N., long. 103° 22′ E.], Laos.

Southwestern Yunnan; southeastern Burma from Northern Shan State southward, through Karenni, to central Tenasserim; northern Thailand; northern Laos.

Alcippe morrisonia schaefferi La Touche

Alcippe nipalensis schaefferi La Touche, 1923, Bull. Brit. Orn. Club, 43, p. 81 — Milati, southeastern Yunnan.

Southeastern Yunnan, Tongking, and northern Annam.

Alcippe morrisonia rufescentior (Hartert)

Proparus nipalensis rufescentior Hartert, 1910, Novit. Zool., 17, p. 231 — Mount Wuchi, Hainan.

Hainan.

Alcippe morrisonia morrisonia Swinhoe

Alcippe morrisonia Swinhoe, 1863, Ibis, p. 296 — Formosa. *Alcippe morrisoniana* Sharpe, 1883, Cat. Birds Brit. Mus., 7, p. 621. *Nomen emendatum.*

Formosa.

Alcippe morrisonia hueti David

Alcippe Hueti David, 1874, Ann. Sci. Nat., Zool., sér. 5, 19, art. 9, p. 4 — Fukien.

Hill tracts of southeastern China from Kwangtung northward to Anhwei.

Alcippe morrisonia davidi Styan

Alcippe cinerea David, 1871, Nouv. Arch. Mus. Hist. Nat. [Paris], 7, p. 14 — western Szechwan. Not *Minla cinerea* Blyth, 1847 = *Alcippe cinerea* (Blyth).
Alcippe davidi Styan, 1896, Ibis, p. 310. New name for *Alcippe cinerea* David, considered preoccupied by "*Alcippe cinerea*" Blyth, 1844.

Western Hupeh and Szechwan.

ALCIPPE NIPALENSIS

Alcippe nipalensis nipalensis (Hodgson)

[*Mesia*] *Siva Nipalensis* Hodgson, 1837, India Review, 2 (1838), p. 89 — Nepal.

Alcippe nipalensis turensis Koelz, 1952, Journ. Zool. Soc. India, 4, p. 39 — Tura Mountain, Garo Hills, Assam.

Himalayas from Nepal eastward, through Sikkim and Bhutan, into eastern Assam north of the Brahmaputra (Abor Hills), and to Garo Hills south of the river.

Alcippe nipalensis commoda Ripley

Alcippe nipalensis commoda Ripley, 1948, Proc. Biol. Soc. Washington, 61, p. 104 — Dening, Mishmi Hills, Assam.

Alcippe nipalensis khasiensis Koelz, 1954, Contrib. Inst. Regional Explor., no. 1, p. 6 — Cherrapunji, Khasi Hills, Assam.

Assam, both south and east of the Brahmaputra (except areas occupied by *nipalensis*), and northern Burma.

Alcippe nipalensis stanfordi Ticehurst

Alcippe nepalensis stanfordi Ticehurst, 1930, Bull. Brit. Orn. Club, 50, p. 84 — Taungup-Prome Cart Road, Arakan Yoma, southwestern Burma.

Hill tracts of southwestern Burma (Chin Hills, Arakan Yoma); possibly also hills of southeastern Assam and East Pakistan.

ALCIPPE ABYSSINICA

Alcippe abyssinica monachus (Reichenow)

Turdinus monachus Reichenow, 1892 (Feb. 19), Sitzungsb. Allg. Deutschen Orn. Ges. Berlin, p. 4 — Buea [lat. 4° 10′ N., long. 9° 12′ E.], British Cameroons.

Turdinus monachus Reichenow, 1892 (Apr.), Journ. f. Orn., 40, pp. 193, 220 — Buea, British Cameroons.

Known only from Mount Cameroon, British Cameroons.

Alcippe abyssinica claudi (Alexander)

Lioptilus claudei [sic] Alexander, 1903, Bull. Brit. Orn. Club, 13, p. 34 — Mount Santa Isabel, 10,800 ft., Fernando Po.[1]

[1] The spelling of the specific name was corrected and an additional type specimen (from Bakaki) was designated by Alexander, 1903, Ibis, p. 382.

Island of Fernando Po.

Alcippe abyssinica ansorgei (Rothschild)

Lioptilus abyssinicus ansorgei Rothschild, 1918, Bull. Brit. Orn. Club, 38, p. 78 — Mucuio [lat. 13° 28′ S., long. 14° 44′ E.], Angola.

Bailundu highlands of west-central Angola, the southeastern Congo, and western Tanganyika.

Alcippe abyssinica stierlingi (Reichenow)

Turdinus stierlingi Reichenow, 1898, Orn. Monatsb., 6, p. 82 — Iringa [lat. 7° 47′ S., long. 35° 42′ E.], Tanganyika.

Alcippe stictigula Shelley, 1903, Bull. Brit. Orn. Club, 13, p. 61 — Mwenembe, Nyasaland.

Lioptilus stierlingi uluguru Hartert, 1922, Bull. Brit. Orn. Club, 42, p. 50 — Uluguru Mountains, Tanganyika.

Highlands of northern Nyasaland and southwestern and central Tanganyika.

Alcippe abyssinica abyssinica (Rüppell)

Drymophila abyssinica Rüppell, 1840, Neue Wirbelt., Vögel, 13, p. 108, pl. 40, fig. 2 — Simen Mountains, Begemdir Province, northern Ethiopia.

Alcippe kilimensis Shelley, 1889, Proc. Zool. Soc. London, p. 364 — Mount Kilimanjaro, Tanganyika.

Alcippe abyssinica micra Grote, 1928, Orn. Monatsb., 36, p. 77 — Mlalo [lat. 4° 34′ S., long. 38° 19′ E.], Tanganyika.

Pseudoalcippe abyssinicus [sic] *chyulu* van Someren, 1939, Journ. East Africa Uganda Nat. Hist. Soc., 14, p. 59 — Chyulu Hills, Camp 3 of Coryndon Museum Expedition, Kenya.

Highlands of northeastern Tanganyika, western Kenya, and western Ethiopia.

ALCIPPE ATRICEPS

Alcippe atriceps (Sharpe)

Turdinus atriceps Sharpe, 1902, Bull. Brit. Orn. Club, 13, p. 10 — Ruwenzori Mountains, Uganda.

Pseudoalcippe atriceps kivuensis Schouteden, 1937, Rev.

Zool. Bot. Africa, 30, p. 165 — Kivu Volcanoes, Belgian Congo-Ruanda boundary.

British Cameroons (Banso and Genderu Mountains); highlands of the northeastern Congo, western Uganda, and Ruanda-Urundi.

GENUS LIOPTILUS BONAPARTE

Lioptilus "Caban. 1850" = Bonaparte, 1851, Consp. Gen. Av., 1 (1850), p. 332. Type, by monotypy, *Turdus nigricapillus* Vieillot.

Lioptilornis Oberholser, 1921, Proc. Biol. Soc. Washington, 34, p. 136. New name for *Lioptilus* Cabanis, *i.e.*, Bonaparte, considered preoccupied by *Leioptila* Blyth.

Kupeornis Serle, 1949, Bull. Brit. Orn. Club, 69, p. 50. Type, by original designation and monotypy, *Kupeornis gilberti* Serle.

LIOPTILUS NIGRICAPILLUS

— **Lioptilus nigricapillus** (Vieillot)

Turdus nigri capillus [sic] Vieillot, 1818, Nouv. Dict. Hist. Nat., nouv. éd., 20, p. 256 — the south of Africa; restricted to the forests of Bruintjes Hoogte, near Somerset East, Cape Province, ex Levaillant.

Forested areas of eastern Cape Province, Natal, and northern Transvaal (Zoutpansberg).

LIOPTILUS GILBERTI

Lioptilus gilberti (Serle)

Kupeornis gilberti Serle, 1949, Bull. Brit. Orn. Club, 69, p. 50 — Kupé Mountain [lat. 4° 50′ N., long. 9° 40′ E.], British Cameroons.

Known only from type locality.

LIOPTILUS RUFOCINCTUS

Lioptilus rufocinctus Rothschild

Lioptilus rufocinctus Rothschild, 1908, Bull. Brit. Orn. Club, 23, p. 6 — Rugege Forest, southwestern Ruanda.

Ruanda and the eastern Congo from valley of the Ruzizi southward to Lake Tanganyika (north of Albertville).

LIOPTILUS CHAPINI

Lioptilus chapini chapini (Schouteden)

Kupeornis chapini Schouteden, 1949, Rev. Zool. Bot. Africa., 42, p. 344 — Mongbwalu [lat. 1° 57′ N., long. 30° 02′ E.], Belgian Congo.

Forests of the northeastern Congo from Lake Albert southward to Lake Edward.

Lioptilus chapini nyombensis Prigogine

Lioptilus chapini nyombensis Prigogine, 1960, Rev. Zool. Bot. Africa., 61, p. 16 — Butokolo [lat. 2° 42′ S., long. 28° 16′ E.], Belgian Congo.

Known only from type locality.

GENUS **PAROPHASMA** REICHENOW

Parophasma Reichenow, 1905, Vög. Afr., 3, 743. Type, by original designation and monotypy, *Parisoma galinieri* Guérin-Méneville.

PAROPHASMA GALINIERI

Parophasma galinieri (Guérin-Méneville)

Parisoma Galinieri Guérin-Méneville, 1843, Rev. Zool. [Paris], 6, p. 162 — Ethiopia.

Parisoma frontalis Rüppell, 1845, Syst. Uebers. Vög. Nord-ost.-Afr., pp. 43, 59 — Shoa, Ethiopia.

Crateropus melodus von Heuglin, 1862, Journ. f. Orn., 10, p. 299 — Semien and Wogara, Ethiopia.

Highlands of central and southern Ethiopia.

GENUS **PHYLLANTHUS** LESSON

Phyllanthus Lesson, 1844, Écho du Monde Savant, 11, col. 1165. Type, by monotypy, *Phyllanthus capucinus* Lesson = *Crateropus atripennis* Swainson.

PHYLLANTHUS ATRIPENNIS

Phyllanthus atripennis atripennis (Swainson)

Crateropus atripennis Swainson, 1837, Birds W. Africa, 1, p. 278 — western Africa; restricted to Senegal, *apud* Sclater.

Phyllanthus Capucinus Lesson, 1844, Écho du Monde
Savant, 11, col. 1165 — "Asia," error; here corrected
to Senegal.
G[arrulax]. poliocephalus Blyth, 1865, Ibis, p. 46 — no
locality; Senegal here designated.
Forests of western Africa from Senegal southeastward
to Liberia.

Phyllanthus atripennis rubiginosus (Blyth)

Garrulax rubiginosus Blyth, 1865, Ibis, p. 46 — no locality;
West Africa designated by Blyth, 1870, Ibis, p. 171;
here restricted to Accra, Ghana.
Crateropus haynesi Sharpe, 1871, Ibis, p. 415 — Accra,
Ghana.
Forests of western Africa from Ivory Coast eastward to
southern Nigeria.

Phyllanthus atripennis bohndorffi (Sharpe)

Crateropus Bohndorffi Sharpe, 1884, Journ. Linn. Soc.
London, Zool., 17, p. 422 — Sassa [lat. 5° 05′ N., long.
25° 30′ E.], Belgian Congo.
Phyllanthus czarnikowi Ogilvie-Grant, 1907, Bull. Brit.
Orn. Club, 19, p. 40 — Mawambi [lat. 1° 03′ N., long.
28° 36′ E.], Belgian Congo.
Forests of the northeastern Congo.

Genus CROCIAS Temminck

Laniellus Swainson, 1832, in Swainson and Richardson,
Fauna Bor.-Am., 2 (1831), pp. 125 (footnote), 127, 481.
No valid species; generic details only. Type, by subsequent designation (Gadow, 1883, Cat. Birds Brit. Mus.,
8, pp. 228 [in key], 230), *Laniellus leucogrammicus*
Gadow. Not *Laniellus* Blyth, 1833.
Crocias Temminck, 1836, Pl. col., livr. 100, pl. 592. Type,
by monotypy, *Crocias guttatus* Temminck = *Lanius albonotatus* Lesson.

CROCIAS LANGBIANIS

Crocias langbianis Gyldenstolpe

Crocias langbianis Gyldenstolpe, 1939, Ark. f. Zool., 31B,
p. 2 — Dalat [lat. 11° 55′ N., long. 108° 26′ E.], Annam.
Southern Annam (Lang Bian Plateau).

CROCIAS ALBONOTATUS

Crocias albonotatus (Lesson)

> *Lanius albonotatus* Lesson, 1832, in Bélanger, Voyage aux
> Indes-Orientales, Zool., pt. 4, p. 249 — Java.
> *Crocias guttatus* Temminck, 1836, Pl. col., livr. 100, pl.
> 592 — Java.
> *Laniellus leucogrammicus* Gadow, 1883, Cat. Birds Brit.
> Mus., 8, p. 230 — Java.

Highlands of western and central Java.

Genus **HETEROPHASIA** Blyth

> *Heterophasia* Blyth, 1842, Journ. Asiat. Soc. Bengal, 11,
> p. 186. Type, by monotypy, *Heterophasia cuculopsis*
> Blyth = *Sibia picaoides* Hodgson.
> *Leioptila* Blyth, 1847, Journ. Asiat. Soc. Bengal, 16, p. 449.
> Type, by monotypy, *Leioptila annectens* Blyth.
> *Malacias* Cabanis, 1851, Mus. Hein., 1 (1850), p. 113.
> Type, by monotypy, *Cinclosoma capistratum* Vigors.

cf. Kinnear, 1939, Ibis, pp. 751-752 (races of *capistrata*).

HETEROPHASIA ANNECTENS

Heterophasia annectens annectens (Blyth)

> *L[eioptila]. annectans* [sic] Blyth, 1847, Journ. Asiat.
> Soc. Bengal, 16, p. 450 — Darjeeling, West Bengal.

Himalayas from Sikkim eastward, through Bhutan and
Assam (both north and south of the Brahmaputra), to hill
tracts of northern and western Burma; western Yunnan
(west of the Salween).

Heterophasia annectens mixta Deignan

> *Heterophasia annectens mixta* Deignan, 1948, Proc. Biol.
> Soc. Washington, 61, p. 15 — Doi Pha Hom Pok [lat.
> 20° 05′ N., long. 99° 10′ E.], Thailand.

Southern Shan State (east of the Salween), northernmost
Thailand (Doi Pha Hom Pok), northern Laos, and northwestern Tongking.

Heterophasia annectens saturata (Walden)

> *L[eioptila]. saturata* Walden, 1875, in Ramsay, Ibis, p.
> 352, footnote — Karenni State, Upper Burma. Chestnut-
> backed phase!

Leioptila Davisoni Hume, 1877, Stray Feathers, 5, p. 110 — Mulayit Taung [lat. 16° 11' N., long. 98° 32' E.], Tenasserim. Black-backed phase!

Southern Shan State (west of the Salween), Karenni State, and hills of northwestern Thailand southward to Amherst District of Tenasserim.

Heterophasia annectens eximia (Riley)

Leioptila annectens eximia Riley, 1940, Proc. Biol. Soc. Washington, 53, p. 48 — Forests of Cam-ly [= Dalat (lat. 11° 55' N., long. 108° 26' E.)], Annam.

Known only from type locality.

HETEROPHASIA CAPISTRATA

Heterophasia capistrata capistrata (Vigors)

Cinclosoma capistratum Vigors, 1831, Proc. Comm. Zool. Soc. London, pt. 1 (1830-1831), p. 56 — Himalayas; restricted to Simla, Himachal Pradesh, by Ripley, 1961, Synop. Birds India Pakistan, p. 416.

M[alacias]. capistrata [sic] *pallida* [sic] Hartert, 1891, Kat. Mus. Senckenb., p. 21, footnote 47 — northwestern India; restricted to Simla, by Baker, 1921, Journ. Bombay Nat. Hist. Soc., 27, p. 460.

Himalayas from northern West Pakistan eastward to Garhwal.

Heterophasia capistrata nigriceps (Hodgson)

[*Sibia*] *Nigriceps* Hodgson, 1839, Journ. Asiat. Soc. Bengal, 8, p. 38 — Nepal; restricted to central Nepal, by Ripley, 1950, Journ. Bombay Nat. Hist. Soc., 49, p. 399.

Himalayas from Kumaun eastward to central Nepal.

Heterophasia capistrata bayleyi (Kinnear)

Leioptila capistrata bayleyi Kinnear, 1939, Ibis, p. 752 — Taktoo, near Sakden, eastern Bhutan.

Himalayas from eastern Nepal eastward, through Sikkim and Bhutan, into Assam (north of the Brahmaputra).

HETEROPHASIA GRACILIS

Heterophasia gracilis (McClelland)

Hypsipetes gracilis McClelland, 1840, Proc. Zool. Soc. London, pt. 7, p. 159 — Assam; restricted to the Naga Hills, by Koelz, 1954, Contrib. Inst. Regional Explor., no. 1, p. 7.

Leioptila gracilis dorsalis Stresemann, 1940, in Stresemann and Heinrich, Mitt. Zool. Mus. Berlin, 24, pp. 153 (in list), 217 — Mount Victoria [lat. 21° 15′ N., long. 93° 55′ E.], Upper Burma.

Leioptila gracilis ardosiaca Koelz, 1954, Contrib. Inst. Regional Explor., no. 1, p. 7 — Maoflang, Khasi Hills, Assam.

Hill tracts of Assam (south of the Brahmaputra) and western Burma (Chin Hills); northeastern Burma and western Yunnan (west of the Salween).

HETEROPHASIA MELANOLEUCA

Heterophasia melanoleuca desgodinsi (Oustalet)

Sibia Desgodinsi Oustalet, 1877, Bull. Soc. Philom. Paris, sér. 7, 1, p. 140 — Yer-ka-lo, on the Mekong at *ca.* lat. 29° 03′ 30″ N., Hsikang.

Heterophasia capistrata tecta Deignan, 1948, Proc. Biol. Soc. Washington, 61, p. 15 — Nguluko, near Likiang, northwestern Yunnan.

Northeastern Burma, northwestern Yunnan, southeastern Hsikang, and southwestern Szechwan.

Heterophasia melanoleuca tonkinensis (Yen)

Leioptila Desgodinsi tonkinensis Yen, 1934, Science Journ. (College of Science, Sun Yatsen Univer., Canton), 6, p. 324 — Fan Si Pan [lat. 22° 17′ N., long. 104° 47′ E.], Tongking.

Northwestern Tongking.

Heterophasia melanoleuca engelbachi (Delacour)

Leioptila desgodinsi engelbachi Delacour, 1930, Oiseau Rev. Franç. Orn., 11, p. 653 — Phu-king-toul, Boloven Plateau, southern Laos.

Southern Laos (Boloven Plateau).

Heterophasia melanoleuca robinsoni (Rothschild)

Lioptila robinsoni Rothschild, 1921, Novit. Zool., 28, p. 38 — Dalat [lat. 11° 55′ N., long. 108° 26′ E.], Annam.

Southern Annam (Lang Bian Plateau).

Heterophasia melanoleuca melanoleuca (Blyth)

Sibia melanoleuca "Tickell" Blyth, 1859, Journ. Asiat. Soc.
Bengal, 28, p. 413 — Tenasserim; type from Mulayit
Taung [lat. 16° 11′ N., long. 98° 32′ E.], *fide* Tickell,
1859, *ibid.*, 28, p. 451.

Leioptila melanoleuca radcliffei Baker, 1922, Fauna Brit.
India, Birds, ed. 2, 1, pp. 296 (in key), 300 — "N. E.
Central Burma . . . and . . . Taunghoo."

Leioptila melanoleuca laeta de Schauensee, 1929, Proc.
Acad. Nat. Sci. Philadelphia, 81, p. 470 — Doi Suthep
[lat. 18° 50′ N., long. 98° 55′ E.], Thailand.

Shan States (except area occupied by *castanoptera*)
southward, through hills of northwestern Thailand, to Amherst District of Tenasserim.

Heterophasia melanoleuca castanoptera (Salvadori)

Malacias castanopterus Salvadori, 1889, Ann. Mus. Civ.
Genova, ser. 2, 7, pp. 363, 411 — Karenni State, Upper
Burma.

Southern Shan State (west of the Salween) and Karenni
State.

HETEROPHASIA AURICULARIS

Heterophasia auricularis (Swinhoe)

Kittacincla auricularis Swinhoe, 1864, Ibis, p. 361 —
Formosa.

Highlands of Formosa.

HETEROPHASIA PULCHELLA

Heterophasia pulchella (Godwin-Austen)

Sibia pulchella Godwin-Austen, 1874, Ann. Mag. Nat.
Hist., ser. 4, 13, p. 160 — Barail Range, Assam.

Lioptila pulchella coeruleotincta Rothschild, 1921, Novit.
Zool., 28, p. 38 — Shweli-Salween Divide, western
Yunnan.

Leioptila pulchella nigroaurita Kinnear, 1944, in Ludlow,
Ibis, 86, p. 83 — Lhalung, Pachakshiri district, southeastern Tibet.

Southeastern Tibet; hill tracts of Assam, both north and
south of the Brahmaputra; northeastern Burma; western
Yunnan (west of the Salween).

HETEROPHASIA PICAOIDES

— **Heterophasia picaoides picaoides** (Hodgson)

[*Sibia*] *Pieaoïdes* [sic] Hodgson, 1839, Journ. Asiat. Soc. Bengal, 8, p. 38 — Nepal.

Himalayas from Nepal eastward, through Sikkim, Bhutan, and Assam (north of the Brahmaputra), to northeastern Burma.

Heterophasia picaoides cana (Riley)

Sibia picaoides cana Riley, 1929, Proc. Biol. Soc. Washington, 42, p. 166 — Doi Ang Ka [lat. 18° 35′ N., long. 98° 30′ E.], Thailand.

Heterophasia picaoides burmanica Ticehurst, 1934, Bull. Brit. Orn. Club., 55, p. 19 — Ta-ok Plateau [lat. 16° 20′ N., long. 98° 28′ E.], Tenasserim.

Northern Shan State southward, through Karenni and northern Thailand, to Amherst District of Tenasserim; northern Laos, northern Annam, and northwestern Tongking.

Heterophasia picaoides wrayi (Ogilvie-Grant)

Sibia wrayi Ogilvie-Grant, 1910, Bull. Brit. Orn. Club, 25, p. 98 — mountains of central portion of Malay Peninsula.

Highlands of Malaya from northern Perak southward to southern Selangor and northern Pahang.

Heterophasia picaoides simillima Salvadori

Heterophasia simillima Salvadori, 1879, Ann. Mus. Civ. Genova, 14, p. 232 — Mount Singgalang [lat. 0° 24′ S., long. 100° 20′ E.], Sumatra.

Highlands of western Sumatra.

GENUS **YUHINA** HODGSON

Yuhina Hodgson, 1836, Asiat. Res., 19, p. 165, pl. 9, fig. Type, by original designation, *Yuhina gularis* Hodgson.

Ixulus Hodgson, 1844, in J. E. Gray, Zool. Misc., no. 3, p. 82. Type, by monotypy, *Yuhina? flavicollis* Hodgson.

Erpornis Hodgson, 1844, in Blyth, Journ. Asiat. Soc. Bengal, 13, p. 379, footnote. Type, by original designation and monotypy, *Erpornis zantholeuca* Blyth.

Herpornis Agassiz, 1846, Nomencl. Zool. Index Univ., p. 516. *Nomen emendatum.*

Staphida "Swinhoe" Gould, 1871 (March), Birds Asia, 4, pt. 23, pl. 8. Type, by subsequent designation (G. R. Gray, 1871, Hand-list Gen. Spec. Birds, pt. 3, p. 173), *Siva torqueola* Swinhoe.

Staphida Swinhoe, 1871 (Oct.), Proc. Zool. Soc. London, p. 373. Type, by subsequent designation (Baker, 1930, Fauna Brit. India, Birds, ed. 2, 7, p. 63.), *Siva torqueola* Swinhoe. (Subgenus of *Ixulus* Hodgson.)

Staphidea Blyth, 1875, Journ. Asiat. Soc. Bengal, 44, extra no., p. 110. *Lapsus.*

Staphidia Anonymous, 1878, Ibis, p. 198. *Lapsus.*

YUHINA CASTANICEPS

Yuhina castaniceps rufigenis (Hume)

Ixulus rufigenis Hume, 1877, Stray Feathers, 5, p. 108 — Himalayas.

Darjeeling District of West Bengal and Sikkim.

Yuhina castaniceps plumbeiceps (Godwin-Austen)

Staphida plumbeiceps Godwin-Austen, 1877, Ann. Mag. Nat. Hist., ser. 4, 20, p. 519 — near Sadiya and Brahmakhand, northeastern Assam.

Staphidia castaneiceps conjuncta Mayr, 1941, Ibis, p. 86 — Chipwi-Laukkaung Road, Myitkyina District, Kachin State, Upper Burma.

Hill tracts of Assam (north and east of the Brahmaputra), northern Burma, and western Yunnan (west of the Salween).

Yuhina castaniceps castaniceps (Moore) 4

Ixulus castaniceps Moore, 1854, in Horsfield and Moore, Cat. Birds Mus. East India Co., 1, p. 411 — "Afghanistan," error; corrected to Khasi Hills, Assam, by Sharpe, 1883, Cat. Birds Brit. Mus., 7, p. 616.

Hill tracts of Assam (south of the Brahmaputra) and southwestern Burma (Chin Hills and Arakan Yomas).

Yuhina castaniceps striata (Blyth) 3

Ixulus striatus Blyth, 1859, Journ. Asiat. Soc. Bengal, 28, p. 413 — mountainous interior of the Tenasserim provinces, Lower Burma; type from "Near Teethoungplee, 3,000 feet," Amherst District, *fide* Tickell, 1859, *ibid.*, p. 452.

Southern Shan and Karenni States of eastern Burma; northwestern Thailand and Amherst District of Tenasserim.

Yuhina castaniceps torqueola (Swinhoe)

Siva torqueola Swinhoe, 1870, Ann. Mag. Nat. Hist., ser. 4, 5, p. 174 — Tingchow Mountains, Fukien.

Eastern portion of northern plateau of Thailand, northern Laos, northern half of Annam, Tongking, Kwangsi, Kwangtung, and Fukien.

Yuhina castaniceps everetti (Sharpe)

Staphidia everetti Sharpe, 1887, Ibis, p. 447 — Kina Balu, North Borneo.

Highlands of northern Borneo.

YUHINA BAKERI

Yuhina bakeri Rothschild

Siva occipitalis Blyth, 1845, Journ. Asiat. Soc. Bengal, 13 (1844), p. 937 — Darjeeling, West Bengal. Not *Yuhina occipitalis* Hodgson, 1836.

Yuhina bakeri Rothschild, 1926, Novit. Zool., 33, p. 276. New name for *Siva occipitalis* Blyth = *Yuhina occipitalis* (Blyth), preoccupied.

Yuhina occipitalis atrovinacea Koelz, 1954, Contrib. Inst. Regional Explor., no. 1, p. 8 — Laikul, Cachar, Assam.

Himalayas from Garhwal eastward to northeastern Burma and southward through hill tracts of Assam to Manipur and Tripura.

YUHINA FLAVICOLLIS

Yuhina flavicollis albicollis (Ticehurst and Whistler)

Ixulus flavicollis albicollis Ticehurst and Whistler, 1924, Bull. Brit. Orn. Club, 44, p. 71 — Dharmsala, Kangra, East Punjab.

Himalayas from northern East Punjab eastward into western Nepal.

Yuhina flavicollis flavicollis Hodgson

Yuhina? Flavicollis Hodgson, 1836, Asiat. Res., 19, p. 167 — Nepal; restricted to central Nepal, by Ripley, 1961, Synops. Birds India Pakistan, p. 408.

Himalyas from central Nepal eastward, through Bhutan, to eastern Assam (Abor Hills).

Yuhina flavicollis rouxi (Oustalet)

Ixulus Rouxi Oustalet, 1896, Bull. Mus. Hist. Nat. Paris, 2, p. 186 — banks of the Lisiang-kiang, or Black River, Yunnan.

Ixulus flavicollis harterti Harington, 1913, Bull. Brit. Orn. Club, 33, p. 62 — Sinlum, near Bhamo, Kachin State, Burma.

Ixulus flavicollus baileyi Baker, 1914, Bull. Brit. Orn. Club, 35, p. 17 — Mishmi Hills, Assam.

Hill tracts of Assam south of the Brahmaputra and of northeastern Assam (Mishmi Hills), northern half of Burma, Yunnan, northwestern Tongking, and adjacent portion of northern Laos.

Yuhina flavicollis constantiae Ripley

Yuhina flavicollis constantiae Ripley, 1953, Oiseau Rev. Franç. Orn., 23, p. 91 — Phu Kobo [lat. 19° 16′ N., long. 103° 25′ E.], Laos.

Mountains of northern Laos (Chiang Khwang Province).

Yuhina flavicollis rogersi Deignan

Yuhina flavicollis rogersi Deignan, 1937, Proc. Biol. Soc. Washington, 50, p. 217 — Doi Phu Kha [lat. 19° 05′ N., long. 101° 05′ E.], Thailand.

Known only from type locality.

Yuhina flavicollis clarki (Oates)

Ixulus clarki Oates, 1894, Bull. Brit. Orn. Club, 3, p. 41; *idem*, 1894, Ibis, p. 481, pl. 13, fig. 1 — Byin Gye Taung [lat. 20° 01′ N., long. 96° 26′ E.], Upper Burma.

Mountains of Southern Shan and Karenni States of Burma.

Yuhina flavicollis humilis (Hume)

Ixulus humilis Hume, 1877, Stray Feathers, 5, p. 106 — Mulayit Taung [lat. 16° 11′ N., long. 98° 32′ E.], Tenasserim.

Known only from type locality.

YUHINA GULARIS

Yuhina gularis vivax Koelz

Yuhina gularis vivax Koelz, 1954, Contrib. Inst. Regional Explor., no. 1, p. 8 — above Luni, Tehri, northwestern Uttar Pradesh.

Garhwal and Kumaun.

Yuhina gularis gularis Hodgson

Yuhina Gularis Hodgson, 1836, Asiat. Res., 19, p. 166 — Nepal.

Yuhina yangpiensis Sharpe, 1902, Bull. Brit. Orn. Club, 13, p. 12 — Yangpi, western Yunnan.

Yuhina gularis griseotincta Rothschild, 1921, Novit. Zool., 28, p. 42 — Shweli-Salween Divide, western Yunnan.

Yuhina gularis sordidior Kinnear, 1925, Bull. Brit. Orn. Club, 45, p. 74 — Ngai Tio, Loakay Province, Tongking.

Himalayas from Nepal eastward through Assam and Burma to northwestern Yunnan (except area occupied by *omeiensis*), southward through Assam to Chin Hills of western Burma and to Northern Shan State; northwestern Tongking.

Yuhina gularis omeiensis Riley

Yuhina gularis omeiensis Riley, 1930, Proc. Biol. Soc. Washington, 43, p. 134 — Mount Omei, Szechwan.

Northwestern Yunnan (Likiang Range), southwestern Szechwan, and southeastern Hsikang.

YUHINA DIADEMATA

Yuhina diademata Verreaux

Yuhina diademata J. Verreaux, 1869, Nouv. Arch. Mus. Hist. Nat. [Paris], 5, p. 35; *idem*, 1871, *ibid.*, 7, p. 53; *idem*, 1872, *ibid.*, 8, pl. 3, fig. 3 — "le Thibet oriental"; types from Muping [= Paohing], *fide* Verreaux, 1871.

Yuhina ampelina Rippon, 1900, Bull. Brit. Orn. Club, 11, p. 12 — Warar Bum, 30 miles east of Bhamo, Kachin State, Upper Burma.

Yuhina diademata obscura Delacour and Jabouille, 1930, Oiseau Rev. Franç. Orn., 11, p. 403 — Fan Si Pan [lat. 22° 17′ N., long. 104° 47′ E.], Tongking.

Yuhina diademata Delacouri Yen, 1934, Science Journ. [College of Science, Sun Yatsen Univ., Canton], 6, pp.

358 (in key), 362.[1] New name for *Yuhina diademata obscura* Delacour and Jabouille, considered preoccupied by *Yuhina occipitalis obscura* (*lapsus* for *obscurior*) Rothschild, 1921.

Northeastern Burma (Kachin State), southeastern Hsikang, southwestern Szechwan, Yunnan, and northwestern Tongking.

YUHINA OCCIPITALIS

Yuhina occipitalis occipitalis Hodgson

Yuhina Occipitales (sic) Hodgson, 1836, Asiat. Res., 19, p. 166 — Nepal.

Himalayas from Nepal eastward, through Sikkim, Bhutan, and southeastern Tibet, into northern Assam (Dafla Hills).

Yuhina occipitalis obscurior Rothschild

Yuhina occipitalis obscurior Rothschild, 1921, Novit. Zool., 28, p. 42 — Likiang Range, Yunnan.

Northeastern Burma (Kachin State) and northwestern Yunnan.

YUHINA BRUNNEICEPS

Yuhina brunneiceps Ogilvie-Grant

Yuhina brunneiceps Ogilvie-Grant, 1906, Bull. Brit. Orn. Club, 16, p. 121 — Mount Morrison, Formosa.

Highlands of Formosa.

YUHINA NIGRIMENTA

Yuhina nigrimenta nigrimenta Blyth

Y[*uhina*]. *nigrimenta* Blyth, 1845, Journ. Asiat. Soc. Bengal, 14, p. 562 — Nepal, *ex* Hodgson.

Yuhina nigrimentum titania Koelz, 1954, Contrib. Inst. Regional Explor., no. 1, p. 9 — Karong, Manipur.

Himalayas from Garhwal eastward to eastern Assam, both north and south of the Brahmaputra.

Yuhina nigrimenta intermedia Rothschild

Yuhina nigrimentum intermedia Rothschild, 1922, Bull. Brit. Orn. Club, 43, p. 11 — Mekong-Salween Divide, Yunnan.

[1] This name is sometimes cited from October 1934, p. 46; see footnote, p. 382.

Yuhina nigrimentum quarta Riley, 1930, Proc. Biol. Soc. Washington, 43, p. 134 — Tseo-jia-keo, Szechwan.

Northeastern Burma, southeastern Hsikang, southwestern Szechwan, Yunnan, northwestern Tongking, northern Annam, and eastern portion of northern Laos; southern Annam (?).

Yuhina nigrimenta pallida La Touche

Yuhina pallida La Touche, 1897, Bull. Brit. Orn. Club, 6, p. 50 — Kuatun, Fukien.

Highlands of Fukien.

YUHINA ZANTHOLEUCA

Yuhina zantholeuca zantholeuca (Blyth)

Erp[ornis]. zantholeuca Blyth, 1844, Journ. Asiat. Soc. Bengal, 13, p. 380 — Nepal.

Himalayas eastward through Assam, both north and south of the Brahmaputra, and northern Burma to northwestern Yunnan (east of the Salween); southward through Burma to Tavoy District of Tenasserim; western Thailand southward to Prachuap Khiri Khan Province.

Yuhina zantholeuca tyrannula (Swinhoe)

Herpornis tyrannulus Swinhoe, 1870, Ibis, p. 347, pl. 10 — central Hainan.

Eastern portion of northern plateau of Thailand and northwestern portion of eastern plateau; northern Laos, northern Annam, Tongking, and southeastern Yunnan; Hainan.

Yuhina zantholeuca griseiloris (Stresemann)

Herpornis xantholeuca griseiloris Stresemann, 1923, Journ. f. Orn., 71, p. 364 — Siuhang, Kwangtung.

Kwangtung and Fukien; Formosa.

Yuhina zantholeuca sordida (Robinson and Kloss)

Herpornis xantholeuca sordida Robinson and Kloss, 1919, Ibis, p. 588 — Da Ban [lat. 12° 38′ N., long. 109° 06′ E.], Annam.

Southern Annam, Cochin China, eastern Cambodia, southern Laos, and easternmost portion of eastern plateau of Thailand (Ubon Province).

Yuhina zantholeuca canescens (Delacour and Jabouille)

Erpornis xantholeuca canescens Delacour and Jabouille, 1928, Bull. Brit. Orn. Club, 48, p. 132 — Le Boc Kor [lat. 10° 37′ N., long. 104° 03′ E.], Cambodia.

Western Cambodia, southeastern Thailand, and the southwestern portion of eastern plateau of Thailand.

Yuhina zantholeuca interposita (Hartert)

Herpornis xantholeuca interposita Hartert, 1917, Bull. Brit. Orn. Club, 38, p. 20 — Temengor [lat. 5° 19′ N., long. 101° 24′ E.], Perak.

Malay Peninsula from southern Tenasserim (Mergui District) and Isthmus of Kra southward to Johore.

Yuhina zantholeuca saani (Chasen)

Erpornis zantholeuca saäni Chasen, 1939, Treubia, 17, p. 206 — near Pendeng [lat. 4° 08′ N., long. 97° 36′ E.], Sumatra.

Northwestern Sumatra.

Yuhina zantholeuca brunnescens (Sharpe)

Herpornis brunnescens Sharpe, 1876, Ibis, p. 41 — Sarawak.

Borneo.

GENERA SEDIS INCERTAE[1]

HERBERT G. DEIGNAN

GENUS MALIA SCHLEGEL

Malia Schlegel, 1880, Notes Leyden Mus., 2, p. 165. Type, by monotypy, *Malia grata* Schlegel.

MALIA GRATA

Malia grata recondita Meyer and Wiglesworth

Malia recondita A. B. Meyer and Wiglesworth, 1894, Abh. Ber. Dresden, 5, p. 1 — mountains between Menado

---

[1] MS read by J. Delacour, B. P. Hall, K. C. Parkes, A. L. Rand, and S. D. Ripley.

[lat. 1° 30′ N., long. 124° 50′ E.] and Mongondok [lat. 0° 48′ N., long. 124° 11′ E.], Celebes.
Mountains of the northern peninsula, Celebes.

Malia grata stresemanni Meise

Malia grata stresemanni Meise, 1931, Orn. Monatsb., **39**, p. 47 — Rano Rano [lat. 1° 30′ S., long. 120° 19′ E.], Celebes.
Central Celebes and the southeastern peninsula.

Malia grata grata Schlegel

Malia grata Schlegel, 1880, Notes Leyden Mus., **2**, p. 165 — Makassar District, Celebes.
Mount Lombobatang, southwestern peninsula, Celebes.

Genus MYZORNIS Blyth

Myzornis "Hodgson" Blyth, 1843, Journ. Asiat. Soc. Bengal, **12**, p. 984. Type, by original designation and monotypy, *Myzornis pyrrhoura* Blyth.

MYZORNIS PYRRHOURA

Myzornis pyrrhoura Blyth

Myzornis pyrrhoura Blyth (Hodgson MS), 1843, Journ. Asiat. Soc. Bengal, **12**, p. 984 — Nepal.
Himalayas from Nepal eastward, through Sikkim, to Bhutan and southeastern Tibet; northeastern Burma and northwestern Yunnan.

Genus HORIZORHINUS Oberholser

Cuphopterus Hartlaub, 1866 (Oct.), in Dohrn, Proc. Zool. Soc. London, p. 326, pl. 34. Type, by monotypy, *Cuphopterus dohrni* Hartlaub. Not *Cuphopterus* Morawitz, 1866 (Jan.), Hymenoptera.
Cuphornis[1] Giebel, 1872, Thesaurus Ornithologiae, **1**, p. 839. Type, by monotypy, *Cuphopterus dohrni* Hartlaub. *Nomen oblitum.*
Horizorhinus Oberholser, 1899, Proc. Acad. Nat. Sci. Philadelphia, p. 216. New name for *Cuphopterus* Hartlaub, preoccupied.

---

[1] The author would prefer to follow strict priority and employ this generic name. Eds.

HORIZORHINUS DOHRNI

Horizorhinus dohrni (Hartlaub)

Cuphopterus dohrni Hartlaub, 1866, in Dohrn, Proc. Zool. Soc. London, p. 326, pl. 34 — Príncipe Island, Gulf of Guinea.

Príncipe Island.

GENUS OXYLABES SHARPE

Oxylabes Sharpe, 1870, Proc. Zool. Soc. London, p. 386, fig. 1. Type, by original designation and monotypy, ? Ellisia madagascariensis Hartlaub = Motacilla madagascariensis Gmelin.

Nesobates Sharpe, 1902, Bull. Brit. Orn. Club, 54, p. 54. New name for Oxylabes Sharpe, considered preoccupied by Oxylabis Foerster, 1856, Hymenoptera.

OXYLABES MADAGASCARIENSIS

Oxylabes madagascariensis (Gmelin)

[Motacilla] madagascariensis Gmelin, 1789, Syst. Nat., 1, pt. 2, p. 952 — Madagascar; restricted to East Madagascar, by Salomonsen, 1934, Ann. Mag. Nat. Hist., ser. 10, 14, p. 68.

Forests of the humid eastern district, Madagascar.

GENUS MYSTACORNIS SHARPE

Mystacornis Sharpe, 1870, Proc. Zool. Soc. London, p. 392, fig. 2. Type, by original designation and monotypy, Bernieria crossleyi Grandidier.

MYSTACORNIS CROSSLEYI

Mystacornis crossleyi (Grandidier)

Bernieria Crossleyi Grandidier, 1870, Rev. Mag. Zool. [Paris], sér. 2, 22, p. 50 — Madagascar; inferentially restricted to Lake Alaotra, apud Salomonsen, 1934, Ann. Mag. Nat. Hist., ser. 10, 14, p. 73.

Forests of the humid eastern district, Madagascar.

SUBFAMILY **PANURINAE**[1, 2]

HERBERT G. DEIGNAN

GENUS **PANURUS** KOCH

Panurus Koch, 1816, Syst. baier. Zool., 1, p. 201, pl. 5, B, fig. 43. Type, by monotypy, *Parus biarmicus* Linnaeus.

PANURUS BIARMICUS

Panurus biarmicus biarmicus (Linnaeus)

[*Parus*] *biarmicus* Linnaeus, 1758, Syst. Nat., ed. 10, 1, p. 190 — Europe; restricted to Holstein, by Hartert, 1907, Vög. pal. Fauna, 1, p. 403.

Panurus biarmicus occidentalis Tschusi, 1904, Orn. Jahrb., 15, p. 228 — Venetia.

England (Norfolk) ; southern Europe from the Netherlands (formerly from Mecklenburg and Holstein) southward and eastward to eastern Spain, southern France, Italy, and Greece.

Panurus biarmicus russicus (Brehm)

Mystacinus Russicus C. L. Brehm, 1831, Handb. Naturg. Vög. Deutschl., p. 472 — Russia.

Panurus biarmicus alexandrovi Zarudny and Bilkevich, 1916, Mess. Orn., 7, p. 241 — Atrek River, Potemkin Peninsula, Bender-i-gyaz, and mouth of the Gyurgen River, southeastern shores of the Caspian Sea.

Panurus biarmicus turkestanicus Zarudny and Bilkevich, 1916, Mess. Orn., 7, p. 241 — Turkestan; provisional name for the paler bird of Turkestan, in the event that *alexandrovi* proved inseparable from *russicus*.

Eastward and southeastward from southern Poland and Hungary, through southern Russia, Turkey, Iran, and Turkestan, to Tsinghai and Inner Mongolia; eastern Manchuria.

---

[1] Acceptance of Panurinae, rather than the almost universally used name Paradoxornithinae, would seem to be in conflict with the provisions of Article 23(d)(ii) of the Int. Code of Zool. Nomen.—Eds.

[2] MS read by J. Delacour, K. C. Parkes, A. L. Rand, and S. D. Ripley.

GENUS CONOSTOMA HODGSON

Conostoma Hodgson, 1841 or 1842, Journ. Asiat. Soc. Bengal, **10**, p. 856. Type, by original designation and monotypy, *Conostoma oemodium* Hodgson.

cf. Vaurie, 1954, Amer. Mus. Novit., no. 1669, pp. 10-11.

CONOSTOMA OEMODIUM

Conostoma oemodium Hodgson

Conostama (sic) *OEmodius* (sic) Hodgson, 1841 or 1842, Journ. Asiat. Soc. Bengal, **10**, p. 857, pl. — northern Nepal.

Conostoma aemodium bambuseti Stresemann, 1923, Journ. f. Orn., **71**, p. 366 — Wa-shan, near Fulin, Hsikang.

Conostoma oemodium graminicola Deignan, 1950, Zoologica [New York], **35**, p. 127 — Ndamucho, south of Lutien [lat. 27° 12′ N., long. 99° 28′ E.], Yunnan.

Higher Himalayas from Garhwal eastward, through Nepal, Sikkim, Bhutan, southeastern Tibet, northern Assam, and southern Hsikang, to southwestern Szechwan, northwestern Yunnan, and northeastern Burma.

GENUS PARADOXORNIS GOULD

Paradoxornis Gould, 1836, Proc. Zool. Soc. London, p. 17. Type, by original designation and monotypy, *Paradoxornis flavirostris* Gould.

Suthora Hodgson, 1837 (Apr.), India Review, **2**, p. 32. Type, by original designation and monotypy, *Suthora nipalensis* Hodgson. (Subgenus of *Parus* Linnaeus.)

Bathyrynchus McClelland, 1837 (Oct.), Quart. Journ. Calcutta Med. Phys. Soc., [1], no. 4, p. 531. Type, by original designation and monotypy, *Bathyrynchus brevirostris* McClelland.

Bathyrynchus McClelland, 1837 (Nov.), India Review, **2**, p. 513. Type, by original designation and monotypy, *Bathyrynchus brevirostris* McClelland.

Chleuasicus Blyth, 1845, Journ. Asiat. Soc. Bengal, **14**, p. 578. Type, by monotypy, *Chleuasicus ruficeps* Blyth.

Anacrites Gistl, 1848, Naturg. Thierr., p. x. New name for *Paradoxornis* "(! Gould. Isis 1838. 174.)".

Cholornis J. Verreaux, 1870, Nouv. Arch. Mus. Hist. Nat. [Paris], 6, p. 35. Type, by monotypy, *Cholornis paradoxus* J. Verreaux.

Calamornis Gould, 1874, Birds Asia, 3, pt. 26, pl. 73. Type, by monotypy, *Paradoxornis heudei* David. (Subgenus of *Paradoxornis* Gould.)

Scæorhynchus Oates, 1889, Fauna Brit. India, Birds, 1, p. 68. Type, by original designation, *Paradoxornis ruficeps* Blyth. Not *Scaeorhynchus* E. B. Wilson, 1881, Pantopoda.

Psittiparus Hellmayr, 1903, Das Tierreich, Lief. 18, p. 163. New name for *Scaeorhynchus* Oates, preoccupied.

Neosuthora Hellmayr, 1911, in Wytsman, Genera Avium, pt. 18, p. 74. Type, by original designation, *Suthora davidiana* Slater.

Neosuthora Harington, 1914, Journ. Bombay Nat. Hist. Soc., 23, pp. 54, 61. Type, by original designation and monotypy, *Suthora davidiana* Slater.

PARADOXORNIS PARADOXUS

Paradoxornis paradoxus (Verreaux)

Cholornis paradoxa J. Verreaux, 1870, Nouv. Arch. Mus. Hist. Nat. [Paris], 6, p. 35; *idem*, 1871, *ibid.*, 7, p. 34, pl. 1, fig. 1 — "les montagnes du Thibet chinois"; type specimens from Muping [= Paohing], Hsikang, *fide* Verreaux, 1871.

Western Hsikang (Paohing), northwestern Szechwan, southern Kansu, and southern Shensi.

PARADOXORNIS UNICOLOR

Paradoxornis unicolor (Hodgson)

H[*eteromorpha*]. *unicolor* Hodgson, 1843, Journ. Asiat. Soc. Bengal, 12, p. 448 — northern Nepal.

Suthora unicolor canaster Thayer and Bangs, 1912, Mem. Mus. Comp. Zool., 40, p. 171 — Wa-shan, near Fulin, Hsikang.

Paradoxornis unicolor saturatior Rothschild, 1921, Novit. Zool., 28, p. 54 — Shweli-Salween Divide, Yunnan.

Higher regions of Nepal, Sikkim, and Bhutan, through southeastern Tibet and southern Hsikang, to southwestern Szechwan, northwestern Yunnan, and northeastern Burma.

PARADOXORNIS FLAVIROSTRIS

Paradoxornis flavirostris Gould

Paradoxornis flavirostris Gould, 1836, Proc. Zool. Soc. London, p. 17 — probably Nepal.

Bathyrynchus brevirostris McClelland, 1837 (Oct.), Quart. Journ. Calcutta Med. Phys. Soc., [1], p. 531, pl. 1, fig. 2 — Upper Assam.

Bathyrynchus brevirostris McClelland, 1837 (Nov.), India Review, 2, pp. 509 (in list), 513, pl., fig. 1 — Upper Assam.

Himalayas from Nepal eastward to northeastern Assam; hill tracts of Assam south of the Brahmaputra and Chin Hills of western Burma (Mount Victoria).

PARADOXORNIS GUTTATICOLLIS

Paradoxornis guttaticollis David

Paradoxornis guttaticollis David, 1871, Nouv. Arch. Mus. Hist. Nat. [Paris], 7, pp. 8 (in list), 14 — "Setchuan-Moupin"; type specimen from western Szechwan, *fide* David, 1877, in David and Oustalet, Oiseaux Chine, p. 204.

Paradoxornis austeni Gould, 1874, Birds Asia, 3, pt. 26, pl. 73 — near Kuchai, Naga Hills, and Shillong, Khasi Hills, Assam.

Hill tracts of Assam south and east of the Brahmaputra; eastern Burma from Kachin State southward, through Shan States, to Karenni and northernmost Thailand; northwestern Yunnan and western Szechwan; southeastern Yunnan, Tongking, and northern Laos; Kwangtung and Fukien.

PARADOXORNIS CONSPICILLATUS

Paradoxornis conspicillatus conspicillatus (David)

[Suthora] conspicillata David, 1871, Nouv. Arch. Mus. Hist. Nat. [Paris], 7, pp. 9, 14 — easternmost Kokonor.

Easternmost Tsinghai; northwestern Szechwan; southern Kansu; southwestern Shensi.

Paradoxornis conspicillatus rocki (Bangs and Peters)

Suthora conspicillata rocki Bangs and Peters, 1928, Bull. Mus. Comp. Zool., 68, p. 345 — Hsien-tien-tsze, Hupeh. Hupeh.

PARADOXORNIS RICKETTI

Paradoxornis ricketti Rothschild

Paradoxornis webbiana ricketti Rothschild, 1922, Bull. Brit. Orn. Club, 43, p. 11 — Yangtze Valley, northwestern Yunnan.

Northwestern Yunnan. Hybrids between this form and *P. w. brunneus* are reported from the Tali district.[1]

PARADOXORNIS WEBBIANUS

Paradoxornis webbianus mantschuricus (Taczanowski)

Suthora webbiana mantschurica Taczanowski, 1885, Bull. Soc. Zool. France, 10, p. 470 — Alamanovka, Ussuriland.

Eastern Manchuria (Sungari basin) and adjacent area of Ussuriland.

Paradoxornis webbianus fulvicauda (Campbell)

Suthora fulvicauda Campbell, 1892, Ibis, p. 237 — Chemulpo, Korea.

Suthora longicauda Campbell, 1892, Ibis, p. 237 — *ca.* thirty miles east of Seoul, Korea.

Suthora webbiana rosea La Touche, 1923, Bull. Brit. Orn. Club, 43, p. 100 — Shanhaihuan, northeastern Hopeh.

Suthora webbiana pekinensis La Touche, 1923, Bull. Brit. Orn. Club, 43, p. 100 — Peking, Hopeh.

Northeastern China (Hopeh) and southern half of Korea.

Paradoxornis webbianus webbianus (Gould)

Suthora Webbiana Gould, 1852, Birds Asia, pt. 4, pl. 72 — Shanghai, Kiangsu.

Coastal regions of southern Kiangsu and northern Chekiang.

Paradoxornis webbianus suffusus (Swinhoe)

Suthora suffusa Swinhoe, 1871, Proc. Zool. Soc. London, p. 372 — "mountainous sides of the gorges on the upper Yangtsze."

Suthora webbiana fohkienensis La Touche, 1923, Bull.

[1] *Suthora styani* Rippon, 1903, Bull. Brit. Orn. Club, 13, p. 54 — Tali Valley, Western Yunnan.

Brit. Orn. Club, 43, p. 101 — Kuatun, northwestern Fukien.

Shensi (Tsinling Mountains) ; valley of the Yangtze from western Szechwan to Kiangsu and Chekiang (except areas occupied by *webbianus*), and thence southward through coastal provinces to Kwangtung. Hybrids between this form and *P. a. alphonsianus* are reported from western Szechwan (Kwanhsien).

Paradoxornis webbianus bulomachus (Swinhoe)

Suthora bulomachus Swinhoe, 1866, Ibis, p. 300, pl. 9 — Formosa.

Highlands of Formosa.

Paradoxornis webbianus elisabethae (La Touche)

Suthora alphonsiana elisabethæ La Touche, 1922, Bull. Brit. Orn. Club, 42, p. 52 — Loukouchai, southeastern Yunnan.

Paradoxornis webbianus intermedius Delacour and Jabouille, 1930, Oiseau Rev. Franç. Orn., 11, p. 395 — Chapa, Tongking.

Southeastern Yunnan and northwestern Tongking.

Paradoxornis webbianus brunneus (Anderson)

Suthora brunnea Anderson, 1871, Proc. Zool. Soc. London, p. 211 — Momien, western Yunnan.

Northwestern Yunnan and northeastern Burma (Kachin and Northern Shan States).

PARADOXORNIS ALPHONSIANUS

Paradoxornis alphonsianus alphonsianus (Verreaux)

Suthora Alphonsiana J. Verreaux, 1870, Nouv. Arch. Mus. Hist. Nat. [Paris], 6, p. 35; *idem*, 1872, *ibid.*, 8, pl. 3, fig. 2 — "les montagnes du Thibet chinois"; type from Chengtu, Szechwan, *fide* Yen, 1934, Journ. f. Orn., 82, p. 383.

Paradoxornis alphonsiana stresemanni Yen, 1934, Journ. f. Orn., 82, p. 383 — To-pung-shan, Kweichow.

Eastern Hsikang (Kangting), southwestern Szechwan (Kwanhsien to Ipin), and Kweichow.

Paradoxornis alphonsianus yunnanensis (La Touche)
> Suthora webbiana yunnanensis La Touche, 1921, Bull.
> Brit. Orn. Club, 42, p. 31 — Kopaotsun, southeastern
> Yunnan.
>
> Southeastern Yunnan and northwestern Tongking.

PARADOXORNIS ZAPPEYI

Paradoxornis zappeyi (Thayer and Bangs)
> Suthora zappeyi Thayer and Bangs, 1912, Mem. Mus.
> Comp. Zool., 40, p. 171, pl. 4, fig. 2 — Wa-shan, near
> Fulin, Hsikang.
>
> Known only from type locality.

PARADOXORNIS PRZEWALSKII

Paradoxornis przewalskii (Berezowski and Bianchi)
> Suthora Przewalskii Berezowski and Bianchi, 1891, Aves
> expeditionis Potanini per provinciam Gan-su et con-
> finia 1884-1887, p. 67, pl. 2, fig. 1 — Hsiku and Minchow
> districts, southern Kansu.
>
> Southern Kansu.

PARADOXORNIS FULVIFRONS

Paradoxornis fulvifrons fulvifrons (Hodgson)
> T[emnoris]. fulvifrons Hodgson, 1845 (Aug.), Proc. Zool.
> Soc. London, pt. 13, p. 31 — Nepal.
> S[uthora]. fulvifrons "Hodgson" Blyth, 1845 (post
> Sept.), Journ. Asiat. Soc. Bengal, 14, p. 579 — Nepal.
>
> Nepal, Sikkim, and Bhutan.

Paradoxornis fulvifrons chayulensis (Kinnear)
> Suthora fulvifrons chayulensis Kinnear, 1940, Bull. Brit.
> Orn. Club, 60, p. 56 — Lung, Chayul Valley, southeast-
> ern Tibet.
>
> Southeastern Tibet.

Paradoxornis fulvifrons albifacies (Mayr and Birckhead)
> Suthora fulvifrons albifacies Mayr and Birckhead, 1937,
> Amer. Mus. Novit., no. 966, p. 15 — Likiang Range,
> Yunnan.
>
> Northwestern Yunnan and southeastern Hsikang.

Paradoxornis fulvifrons cyanophrys (David)

Sutora (sic) *cyanophrys* David, 1874, Ann. Sci. Nat.,
Zool., sér. 5, 19, art. 9, p. 3 — southwestern Shensi.

Northwestern Szechwan, southern Kansu (?), and southwestern Shensi.

PARADOXORNIS NIPALENSIS

Paradoxornis nipalensis nipalensis (Hodgson)

[*Parus*] *Suthora Nipalensis* Hodgson, 1837, India Review,
2 (1838), p. 32 — central and northern regions of Nepal; restricted to Kathmandu Valley, by Ripley, 1961,
Synops. Birds, India Pakistan, p. 370.

Central Nepal.

Paradoxornis nipalensis humii (Sharpe)

Suthora humii Sharpe, 1883, Cat. Birds Brit. Mus., 7, pp.
486 (in key), 487 — Nepal and Sikkim.

Eastern Nepal, Sikkim, and western Bhutan.

Paradoxornis nipalensis crocotius Kinnear

Suthora poliotis intermedia Kinnear, 1944, in Ludlow,
Ibis, pp. 69, 70 — Yönpu La, near Trashigong, eastern
Bhutan. Not *Paradoxornis webbianus intermedius* Dela-
cour and Jabouille, 1930.

Paradoxornis poliotis crocotius Kinnear, 1954, Ibis, p. 484.
New name for *Suthora poliotis intermedia* Kinnear,
preoccupied.

Southeastern Tibet (7,000 feet) and eastern Bhutan
(8,000-8,500 feet).

Paradoxornis nipalensis poliotis (Blyth)

Suthora poliotis Blyth, 1851, Journ. Asiat. Soc. Bengal,
20, p. 522 — Cherrapunji, Khasi Hills, Assam.

Suthora munipurensis Godwin-Austen and Walden, 1875,
Ibis, p. 250 — near Karakhul, Manipur.

Suthora daflaensis Godwin-Austen, 1876, Ann. Mag. Nat.
Hist., ser. 4, 17, p. 32 — Dafla Hills, Assam.

Eastern Bhutan (4,000 feet) ; hill tracts of Assam, both
north and south of the Brahmaputra (except area occupied
by *patriciae*) ; northeastern Burma; northwestern Yunnan.

Paradoxornis nipalensis patriciae (Koelz)

Suthora poliotis patriciae Koelz, 1954, Contrib. Inst. Regional Explor., no. 1, p. 2 — Blue Mountain, Lushai Hills, Assam.

Southeastern Assam (Lushai Hills).

Paradoxornis nipalensis ripponi (Sharpe)

Suthora ripponi Sharpe, 1905, Bull. Brit. Orn. Club, 15, p. 96 — Mount Victoria, Southern Chin Hills, Upper Burma.

Known only from type locality.

Paradoxornis nipalensis feae (Salvadori)

Suthora Feae Salvadori, 1889, Ann. Mus. Civ. Genova, ser. 2, 7, p. 363 — Karen Hills, Upper Burma.

Hills of southeastern Burma (Karenni State) and northwestern Thailand.

Paradoxornis nipalensis beaulieui Ripley

Paradoxornis verreauxi beaulieui Ripley, 1953, Oiseau Rev. Franç. Orn., 23, p. 90 — Phu Kobo [lat. 19° 16′ N., long. 103° 25′ E.], Laos.

Mountains of northern Laos (Chiang Khwang Province).

Paradoxornis nipalensis craddocki (Bingham)

Suthora craddocki Bingham, 1903, Bull. Brit. Orn. Club, 13, p. 54; *idem*, 1903, Ibis, p. 586, pl. 11, fig. 1 — Loi Pangnao [lat. 21° 20′ N., long. 100° 20′ E.], Upper Burma.

Known only from type locality and northwestern Tongking (Fan Si Pan).

Paradoxornis nipalensis verreauxi (Sharpe)

Suthora gularis J. Verreaux, 1869, Nouv. Arch. Mus. Hist. Nat. [Paris], 5, p. 35; *idem*, 1872, *ibid.*, 8, pl. 3, fig. 1 — "le Thibet oriental"; type specimen from "Setchuan-Moupin," *fide* David, 1871, *ibid.*, 7, p. 9.

Suthora verreauxi Sharpe, 1883, Cat. Birds Brit. Mus., 7, pp. 486 (in key), 488. New name for *Suthora gularis* J. Verreaux, considered preoccupied by *Paradoxornis gularis* G. R. Gray, 1845.

Mountains of eastern Hsikang and southwestern Szechwan.

Paradoxornis nipalensis pallidus (La Touche)

Suthora gularis pallida La Touche, 1922, Bull. Brit. Orn. Club, 43, p. 20 — Kuatun, northwestern Fukien.
Mountains of northwestern Fukien.

Paradoxornis nipalensis morrisonianus (Ogilvie-Grant)

Suthora morrisoniana Ogilvie-Grant, 1906, Bull. Brit. Orn. Club, 16, p. 119 — Mount Morrison, Formosa.
Highlands of Formosa.

PARADOXORNIS DAVIDIANUS

Paradoxornis davidianus davidianus (Slater)

Suthora davidiana Slater, 1897, Ibis, p. 172, pl. 4, fig. 1 — Kuatun, northwestern Fukien.
Highlands of Fukien.

Paradoxornis davidianus tonkinensis (Delacour)

Suthora davidiana tonkinensis Delacour, 1927, Bull. Brit. Orn. Club, 47, p. 167 — Bac Kan [lat. 22° 08′ N., long. 105° 50′ E.], Tongking.
Known only from type locality.

Paradoxornis davidianus thompsoni (Bingham)

Suthora thompsoni Bingham, 1903, Bull. Brit. Orn. Club, 13, p. 63; *idem*, 1903, Ibis, p. 586, pl. 11, fig. 2 — " . . . north-east of Kyatpyin village near the Paunglaung stream," Loilong, Southern Shan State, Upper Burma.
Eastern Burma (Southern Shan State), northwestern Laos, and northwestern portion of eastern plateau of Thailand (Loei Province).

PARADOXORNIS ATROSUPERCILIARIS

Paradoxornis atrosuperciliaris oatesi (Sharpe)

Chl[euasicus]. ruficeps Blyth, 1845, Journ. Asiat. Soc. Bengal, 14, p. 578 — Darjeeling, West Bengal. Not *Paradoxornis ruficeps* Blyth, 1842.

[*Suthora*] *oatesi* Sharpe, 1903, Hand-list Gen. Species Birds, 4, p. 70. New name for *Chleuasicus ruficeps* Blyth, considered preoccupied by *Paradoxornis ruficeps* Blyth, 1842.

Darjeeling District of West Bengal and Sikkim.

Paradoxornis atrosuperciliaris atrosuperciliaris (Godwin-Austen)

Chleuasicus atrosuperciliaris Godwin-Austen, 1877 (May 12), Journ. Asiat. Soc. Bengal, 46, p. 44 — Sadiya, Assam.[1]

Chleuasicus ruficeps, Blyth, var. *atrosuperciliaris* Godwin-Austen, 1877 (June), Proc. Asiat. Soc. Bengal, p. 147 — Sadiya, Assam.

Suthora ruficeps rufina Koelz, 1952, Journ. Zool. Soc. India, 4, p. 37 — Laikul, North Cachar, Assam.

Assam (both north and south of the Brahmaputra), northern Burma, western Yunnan, and northern Laos.

PARADOXORNIS RUFICEPS

Paradoxornis ruficeps ruficeps Blyth

Paradoxornis ruficeps Blyth, 1842, Journ. Asiat. Soc. Bengal, 11, p. 177 — Bhutan.

Himalayas from Nepal, through Bhutan, eastward to eastern Assam (north of the Brahmaputra).

Paradoxornis ruficeps bakeri (Hartert)

Scaeorhynchus ruficeps bakeri Hartert, 1900, Novit. Zool., 7, p. 548 — Hungrum, North Cachar, Assam.

Psittiparus ruficeps psithyrus Koelz, 1954, Contrib. Inst. Regional Explor., no. 1, p. 2 — Sangau, Lushai Hills, Assam.

Psittiparus ruficeps rufitinctus Koelz, 1954, Contrib. Inst. Regional Explor., no. 1, p. 2 — Cherrapunji, Khasi Hills, Assam.

Assam (south of the Brahmaputra) ; northern and eastern Burma (excepting the Shan States ?), southward to the Karenni State.

Paradoxornis ruficeps magnirostris (Delacour)

Psittiparus ruficeps magnirostris Delacour, 1927, Bull. Brit. Orn. Club, 47, p. 166 — Tam Dao [lat. 21° 27′ N., long. 105° 40′ E.], Tongking.

Highlands of central Tongking.

[1] The complete diagnosis reads: "intermediate in size between *Paradoxornis ruficeps* and *Ch. ruficeps.*"

PARADOXORNIS GULARIS

Paradoxornis gularis gularis Gray

Paradoxornis gularis "Horsf." G. R. Gray, 1845, Gen. Birds, 2, p. 389, pl. 94 — no locality; type from Bhutan, *fide* Kinnear, 1937, in Ludlow and Kinnear, Ibis, pp. 27-28.

Himalayas from Sikkim, through Bhutan, eastward into Assam, north of the Brahmaputra.

Paradoxornis gularis transfluvialis (Hartert)

Scæorhynchus gularis transfluvialis Hartert, 1900, Novit. Zool., 7, p. 548 — Guilang, North Cachar, Assam.

Psittiparus gularis schoeniparus Koelz, 1954, Contrib. Inst. Regional Explor., no. 1, p. 2 — Karong, Manipur.

Assam (south of the Brahmaputra); northern and eastern Burma (southward to the Karenni State) ; northwestern Thailand.

Paradoxornis gularis laotianus (Delacour)

Psittiparus gularis laotianus Delacour, 1926, Bull. Brit. Orn. Club, 47, p. 19 — Chiang Khwang [lat. 19° 19' N., long. 103° 22' E.], Laos.

Easternmost Burma (Kengtung) ; eastern portion of northern plateau of Thailand; northern Laos; northwestern Tongking.

Paradoxornis gularis fokiensis (David)

Heteromorpha fokiensis David, 1874, Ann. Sci. Nat., Zool., sér. 5, 19, art. 9, p. 4 — Fukien.

Hill tracts of Fukien and Anhwei.

Paradoxornis gularis hainanus (Rothschild)

Psittiparus gularis hainanus Rothschild, 1903, Bull. Brit. Orn. Club, 14, p. 7 — Mount Wuchi, Hainan.

Hainan.

Paradoxornis gularis rasus (Stresemann)

Psittiparus gularis rasus Stresemann, 1940, in Stresemann and Heinrich, Mitt. Zool. Mus. Berlin, 24, pp. 153 (in list), 180 — Mount Victoria, Southern Chin Hills, Upper Burma.

Western Burma (Chin Hills).

Paradoxornis gularis margaritae (Delacour)

Psittiparus margaritæ Delacour, 1927, Bull. Brit. Orn. Club, 47, p. 167 — Jirinh [lat. 11° 31′ N., long. 108° 01′ E.], Annam.

Highlands of southern Annam.

PARADOXORNIS HEUDEI

Paradoxornis heudei David

Paradoxornis Heudei David, 1872, Compt. Rend. Acad. Sci. Paris, 74, p. 1449 — Kiangsu.

Kiangsu (restricted to reed beds bordering the Yangtze from a few miles above Nanking to a short distance below Chinkiang).

SUBFAMILY PICATHARTINAE[1]

HERBERT G. DEIGNAN

GENUS PICATHARTES LESSON

Picathartes Lesson, 1828, Man. Orn., 1, p. 374. Type, by original designation and monotypy, *Corvus gymnocephalus* Temminck.

PICATHARTES GYMNOCEPHALUS

Picathartes gymnocephalus (Temminck)

Corvus gymnocephalus Temminck, 1825, Pl. col., livr. 55, pl. 327 — locality unknown, but English possessions on coast of Guinea suggested by Temminck.

Forests of western Africa from Sierra Leone eastward to Togoland.

PICATHARTES OREAS

Picathartes oreas Reichenow

Picathartes oreas Reichenow, 1899, Orn. Monatsb., 7, p. 40 — British Cameroons; type alleged to come from Victoria [lat. 4° 00′ N., long. 9° 12′ E.].

Forests of southern British Cameroons and southwestern French Cameroons.

[1] MS read by J. Delacour, B. P. Hall, K. C. Parkes, A. L. Rand, and S. D. Ripley.

Subfamily **POLIOPTILINAE**[1, 2, 3]

Raymond A. Paynter, Jr.

cf. Ridgway, 1904, Bull. U. S. Nat. Mus., 50, p. 3, pp. 710-736 (*Polioptila*; North and Middle America).

——————, 1911, *op. cit.*, pt. 5, pp. 84-90 (*Ramphocaenus, Microbates*; North and Middle America).

Cory and Hellmayr, 1924, Field Mus. Nat. Hist. Publ., Zool. Ser., 13, pt. 3, pp. 205-213 (*Ramphocaenus, Microbates*; North and South America).

Hellmayr, 1934, *op. cit.*, pt. 7, pp. 485-510 (*Polioptila*; North and South America).

Pinto, 1938, Cat. Aves Brasil, pt. 1, Rev. Mus. Paulista, 22, pp. 491-493 (*Ramphocaenus, Microbates*).

——————, 1944, *op. cit.* (Publ. Dept. Zool., São Paulo), pt. 2, pp. 382-386 (*Polioptila*).

Phelps and Phelps, Jr., 1950, Bol. Soc. Venezolana Cienc. Nat., 12, no. 75, pp. 252-255 (Venezuela).

de Schauensee, 1951, Caldasia (Inst. Cienc. Nat. U. Nac. Colombia), 5, no. 25, pp. 924-928 (Colombia).

Rand and Traylor, Jr., 1953, Auk, 70, pp. 334-337 (Systematic position of *Ramphocaenus* and *Microbates*).

Miller, Friedmann, Griscom, and Moore, 1957, Pacific Coast Avifauna (Cooper Ornith. Soc.), no. 33, pp. 201-205; 207-208 (Mexico).

Genus **MICROBATES** Sclater and Salvin[4]

Microbates Sclater and Salvin, 1873, Nomen. Av. Neotrop., p. 155. Type, by original designation and monotypy, *Microbates torquatus* Sclater and Salvin = *Rhamphocaenus collaris* Pelzeln.

[1] MS read by E. R. Blake, J. Bond, K. C. Parkes, H. Sick, and A. Wetmore.

[2] The three genera in this subfamily are presumably not closely related to one another, with the possible exception of *Microbates* and *Ramphocaenus*. Their proper taxonomic placement awaits further study.

[3] For *Psilorhamphus*, now placed in the Rhinocryptidae, see Addenda, p. 456.

[4] Possibly congeneric with *Ramphocaenus* Vieillot, 1819.

MICROBATES COLLARIS

Microbates collaris paraguensis Phelps

Microbates collaris paraguensis Phelps, 1946, Bol. Soc. Venezolana Cienc. Nat., 10, p. 153 — Salto María Espuma, 300 m., Caño Espuma, Alto Río Paragua, Bolívar, Venezuela.

Southeastern Bolívar, Venezuela.

Microbates collaris collaris (Pelzeln)

Rhamphocaenus collaris Pelzeln, 1868, Orn. Brasil., 2, pp. 84, 157 — Barra do Rio Negro, Marabitanas, and Rio Içanna, upper Rio Negro; restricted to Barra do Rio Negro by Berlepsch, 1908, Novit. Zool., 15, p. 156.

Southwestern Venezuela, southeastern Colombia, extreme northwestern Brazil, and French Guiana.

Microbates collaris perlatus Todd

Microbates collaris perlatus Todd, 1927, Proc. Biol. Soc. Washington, 40, p. 161 — Tonantins, Rio Solimões, Brazil.

Left bank of the upper Rio Solimões, Brazil.

MICROBATES CINEREIVENTRIS

Microbates cinereiventris semitorquatus (Lawrence)

Ramphocaenus semitorquatus Lawrence, 1862, Ann. Lyc. Nat. Hist. New York, 7, p. 469 — line of Panama Railroad.

Caribbean lowlands from southern Nicaragua through Costa Rica to western Panama.

Microbates cinereiventris magdalenae Chapman

Microbates cinereiventris magdalenae Chapman, 1915, Bull. Amer. Mus. Nat. Hist., 34, p. 642 — Malena, 1,000 ft., near Puerto Berrío, Magdalena Valley, Antioquia, Colombia.

Caribbean coast of extreme eastern Panama and lowlands of northern Colombia east to the Eastern Andes.

Microbates cinereiventris cinereiventris (Sclater)

Ramphocaenus cinereiventris Sclater, 1855, Proc. Zool. Soc. London, 23, p. 76, pl. 87 — "Pasto," Colombia;

Buenaventura substituted by Cory and Hellmayr, 1924, *op. cit.*, p. 212.

Atrato Valley and entire Pacific coast of Colombia to Guayas, southwestern Ecuador.

Microbates cinereiventris peruvianus Chapman

Microbates cinereiventris peruvianus Chapman, 1923, Amer. Mus. Novit., no. 86, p. 5 — La Pampa, Río Tavara, tributary of Río Inambari, northern Puno, Peru.

Tropical zone of southeastern Nariño, Colombia; eastern Ecuador; Junín and Puno, Peru.

Genus RAMPHOCAENUS Vieillot

Ramphocaenus Vieillot, 1819, Nouv. Dict. Hist. Nat., nouv. éd., 29, p. 5. Type, by monotypy, *Ramphocaenus melanurus*, Vieillot.

cf. Zimmer, 1937, Amer. Mus. Novit., no. 917, pp. 11-16 (South American forms).
Eisenmann, 1953, Auk, 70, pp. 368-369 (nest).
Sick, 1954, Bonn. Zool. Beitr., 5, pp. 179-190 (biology).
Skutch, 1960, Pacific Coast Avifauna (Cooper Ornith. Soc.), no. 34, pp. 54-61 (life history).

RAMPHOCAENUS MELANURUS

Ramphocaenus melanurus rufiventris (Bonaparte)

Scolopacinus rufiventris Bonaparte, 1838, Proc. Zool. Soc. London, 5 (1837), p. 119 — Guatemala.

Tropical southeastern Mexico, except Yucatán Peninsula, from Veracruz and Oaxaca southward through Central America, except northern Guatemala, to southwestern Córdoba, northern Colombia, and eastern Ecuador; apparently absent in eastern Colombia.

Ramphocaenus melanurus ardeleo Van Tyne and Trautman

Ramphocaenus rufiventris ardeleo Van Tyne and Trautman, 1941, Occ. Papers Mus. Zool. Univ. Michigan, no. 439, p. 9 — Chichén Itzá, Yucatán, Mexico.

Yucatán Peninsula of Mexico, in Campeche, Yucatán, and Quintana Roo, and in Petén, Guatemala.

Ramphocaenus melanurus sanctaemarthae Sclater ⋾

Ramphocaenus sanctae[-]*marthae* Sclater, 1861, Proc.
Zool. Soc. London, p. 380 — Santa Marta, Colombia.
Caribbean coast from central Córdoba, northern Colombia,
east to western Zulia, northwestern Venezuela.

Ramphocaenus melanurus griseodorsalis Chapman

Rhamphocaenus rufiventris griseodorsalis Chapman, 1912,
Bull. Amer. Mus. Nat. Hist., 31, p. 145 — Miraflores,
6,800 ft., east of Palmira, west slope of Central Andes,
Valle, Colombia.
Western central Colombia from Antioquia south to Valle.

Ramphocaenus melanurus pallidus Todd

Ramphocaenus melanurus pallidus Todd, 1913, Proc. Biol.
Soc. Washington, 26, p. 172 — El Hacha, Yaracuy,
Venezuela.
Zulia Valley, east of the Andes in northeastern Colombia,
through northern Venezuela from eastern Falcón to
Miranda.

Ramphocaenus melanurus trinitatis Lesson

Ramphocaenus trinitatis Lesson, 1839, Rev. Zool. [Paris],
2, p. 42 — Trinidad (possibly erroneous; *vide* Cory and
Hellmayr, 1924, *op. cit.*, p. 206).
Trinidad, through tropical northeastern, central, and
western Venezuela to Meta, eastern Colombia.

Ramphocaenus melanurus duidae Zimmer

Ramphocaenus melanurus duidae Zimmer, 1937, Amer.
Mus. Novit., no. 917, p. 15 — Esmeralda, 325 ft., [near]
Mt. Duida, Venezuela.
Amazonas and most of Bolívar, southern Venezuela; eastern slopes of the Andes in northeastern Ecuador; presumably in southwestern Colombia, but the species as yet un-
recorded there.

Ramphocaenus melanurus badius Zimmer

Ramphocaenus melanurus badius Zimmer, 1937, Amer.
Mus. Novit., no. 917, p. 11 — mouth of Río Cinipá,
Loreto, Peru.
Northeastern Peru, generally north of Río Marañon, and
southeastern Ecuador.

Ramphocaenus melanurus obscurus Zimmer

Ramphocaenus melanurus obscurus Zimmer, 1931, Amer.
Mus. Novit., no. 509, p. 2 — Santa Rosa, upper Ucayali
(left bank), Loreto, Peru.

Known only from Santa Rosa on the western bank of the
upper Río Ucayali, Loreto, and from Puerto Yessup, on the
upper Río Pichis, Junín, Peru.

Ramphocaenus melanurus amazonum Hellmayr

Ramphocaenus melanurus amazonum Hellmayr, 1907,
Novit. Zool., 14, p. 66 — Teffé, Rio Solimóes, Brazil.

Right bank of the upper Ucayali, Loreto, Peru and northwestern Brazil, south of the Amazon, east to the Rio Tapajós.

Ramphocaenus melanurus sticturus Hellmayr

Ramphocaenus sticturus Hellmayr, 1902, Verh. Zool.-Bot.
Ges. Wien, 52, p. 97 — Villa Bella de Mato Grosso, upper Rio Guaporé, Mato Grosso, Brazil.

Northwestern Mato Grosso, Brazil.

Ramphocaenus melanurus albiventris Sclater

Ramphocaenus albiventris Sclater, 1883, Ibis, p. 95 —
Surinam.

Easternmost Bolívar, Venezuela through British Guiana,
Surinam, and French Guiana and northeastern Brazil, probably south to the left bank of the Amazon.

Ramphocaenus melanurus austerus Zimmer

Ramphocaenus melanurus austerus Zimmer, 1937, Amer.
Mus. Novit., no. 917, p. 12 — Pedral, Baiáo, right bank
of Rio Tocantins, Pará, Brazil.

Eastern Pará, south of Tocantins, and northern Maran-
háo, Brazil.

Ramphocaenus melanurus melanurus Vieillot

Ramphocaenus melanurus Vieillot, 1819, Nouv. Dict. Hist.
Nat., nouv. éd., 29, p. 6 — Brazil [= vicinity of Rio de
Janeiro].

Coastal region of northern and eastern Brazil from Pernambuco to São Paulo.

GENUS **POLIOPTILA** SCLATER

Polioptila Sclater, 1855, Proc. Zool. Soc. London, 23, p. 11.
Type, by subsequent designation (Baird, 1864, Rev.
Amer. Birds, 1, p. 67), *Motacilla caerulea* Linnaeus.

cf. Zimmer, 1942, Amer. Mus. Novit., no. 1168, 7 pp.
(*P. plumbea; P. guianensis*).
Brodkorb, 1944, Journ. Washington Acad. Sci., 34, pp.
312-316 (*P. albiloris*).
Bent, 1949, Bull. U. S. Nat. Mus., 196, pp. 344-383 (life
histories of *P. caerulea* and *P. melanura*).
Skutch, 1960, Pacific Coast Avifauna (Cooper Ornith.
Soc.), no. 34, pp. 43-53 (life history of *P. plumbea*).

POLIOPTILA CAERULEA

Polioptila caerulea caerulea (Linnaeus)

Motacilla caerulea Linnaeus, 1766, Syst. Nat., ed. 12, 1,
p. 337 — Pennsylvania [= Philadelphia, Pennsylvania].
Culicivora mexicana Bonaparte, 1851, Consp. Av., 1
(1850), p. 316 — Mexico [= Ciudad Oaxaca; *vide* Stresemann, 1954, Condor, 56, p. 90].

Breeds from eastern Nebraska, central Minnesota, south-
ern Wisconsin, southern Michigan, southern Ontario, central
New York, and southern Massachusetts south to Gulf Coast
from southern Texas east to central Florida, and in the
Bahamas; casual visitor in coastal New England north to
Maine. Winters from central (Michoacán) and northeastern
(Nuevo León) Mexico, the Gulf states, and Virginia to Gua-
temala, the Yucatán Peninsula (probably), Cuba, and the
Bahamas.

Polioptila caerulea amoenissima Grinnell

Polioptila caerulea amoenissima Grinnell, 1926, Proc. California Acad. Sci., ser. 4, 15, p. 494 — Pleasant Valley,
600 ft., Mariposa County, California.

Breeds from northern California, central Nevada, southern Utah, and Colorado south to northernmost Baja California, northern Sonora, Chihuahua, Coahuila, and Nuevo
León. Winters from southern California, southern Nevada,
central Arizona, southern New Mexico, western Texas, Coahuila, Nuevo, León, and Tamaulipas south to central Baja

California, Colima, Michoacán, Morelos, and Puebla; ? Yucatán.

Polioptila caerulea obscura Ridgway

Polioptila caerulea obscura Ridgway, 1883, Proc. U. S. Nat. Mus., 5 (1882), p. 535 — San José del Cabo, Baja California, Mexico.

Baja California from about lat. 28° N. south to the Cape district.

?Polioptila caerulea gracilis van Rossem and Hachisuka

Polioptila caerulea gracilis van Rossem and Hachisuka, 1937, Proc. Biol. Soc. Washington, 50, p. 109 — Rancho Santa Barbara, *ca.* 5,000 ft., 20 mi. NE. Guirocoba, SE. Sonora, Mexico.

Breeds in foothills of southeastern Sonora; winters in lowlands to west and on Isla San Esteban.

Polioptila caerulea deppei van Rossem

Polioptila caerulea deppei van Rossem, 1934, Bull. Mus. Comp. Zool., 77, p. 402 — Río Lagartos, Yucatán, Mexico.

Southeastern Mexico from San Luis Potosí and Guanajuato through Hidalgo, Tlaxcala, Puebla, Veracruz, Tabasco, and northern Chiapas to Yucatán Peninsula.

Polioptila caerulea cozumelae Griscom

Polioptila caerulea cozumelae Griscom, 1926, Amer. Mus. Novit., no. 236, p. 10 — Isla Cozumel, Yucatán [= Quintana Roo], Mexico.

Isla Cozumel, Quintana Roo, Mexico.

Polioptila caerulea nelsoni Ridgway

Polioptila nelsoni Ridgway, 1903, Proc. Biol. Soc. Washington, 16, p. 109 — Ciudad Oaxaca, Oaxaca, Mexico.

Southwestern and southern Mexico in Guerrero, Oaxaca, and central and southern Chiapas.

POLIOPTILA MELANURA

Polioptila melanura californica Brewster

Polioptila californica Brewster, 1881, Bull. Nuttall Orn. Club, 6, p. 103 — Riverside, San Bernardino Co. [= Riverside Co.], California.

Southwestern California from lower Santa Clara Valley south to northwestern Baja California, west of Sierra San Pedro Mártir, to about lat. 30° N.

Polioptila melanura lucida van Rossem

Polioptila melanura lucida van Rossem, 1931, Condor, 33, p. 36 — 10 miles north of Guaymas, Sonora, Mexico.

Deserts from southeastern California, southernmost Nevada, and central Arizona southward in northeastern Baja California to about lat. 31° N., and in Sonora, Chihuahua, and northwestern Durango.

Polioptila melanura melanura Lawrence

Polioptila melanura Lawrence, 1857, Ann. Lyc. Nat. Hist. New York, 6, p. 168 — Texas [= Rio Grande Valley].

Central western Nevada through Río Grande Valley, Texas southward in Mexico to Durango (except northwest), Jalisco, Guanajuato, San Luis Potosí, and Tamaulipas.

Polioptila melanura pontilis van Rossem

Polioptila melanura pontilis van Rossem, 1931, Proc. Biol. Soc. Washington, 44, p. 99 — San Francisquito Bay, Baja California, Mexico. New name for *P. m. nelsoni* van Rossem, 1931, Condor, 33, p. 36; preoccupied.

Central Baja California, from about lat. 28° 26′ N. to about lat. 27° N.

Polioptila melanura margaritae Ridgway

Polioptila margaritae Ridgway, 1904, Bull. U. S. Nat. Mus., 50, pt. 3, p. 733 — [Santa] Margarita Island, Baja California, Mexico.

Baja California from about lat. 27° N. southward and on Islas Santa Margarita and Espíritu Santo.

Polioptila melanura curtata van Rossem

Polioptila melanura curtata van Rossem, 1932, Trans. San Diego Soc. Nat. Hist., 7, p. 140 — Petrel Bay, east side of Isla Tiburón, Sonora, Mexico.

Restricted to Isla Tiburón, Sonora.

POLIOPTILA LEMBEYEI

Polioptila lembeyei (Gundlach)

Culicivora lembeyei Gundlach, 1858, Ann. Lyc. Nat. Hist. New York, 6, p. 273 — eastern Cuba.

Confined to eastern Cuba, in provinces of Oriente and Camagüey.

POLIOPTILA ALBILORIS

Polioptila albiloris restricta Brewster

Polioptila nigriceps restricta Brewster, 1889, Auk, 6, p. 97 — Alamos, Sonora, Mexico.

Lowlands of Sonora and adjacent Chihuahua.

Polioptila albiloris nigriceps Baird

Polioptila nigriceps Baird, 1864, Rev. Amer. Birds, 1, p. 69 — Mazatlán, Sinaloa, Mexico.

Western Durango and Pacific coast from Sinaloa through Nayarit and Jalisco to Colima.

Polioptila albiloris vanrossemi Brodkorb

Polioptila albiloris vanrossemi Brodkorb, 1944, Journ. Washington Acad. Sci., 34, p. 312 — Quiotepec, District of Cuicatlán, Oaxaca, Mexico.

Pacific lowlands and interior of Michoacán, Puebla, Guerrero, Oaxaca and Chiapas.

Polioptila albiloris albiventris Lawrence

Polioptila albiventris Lawrence, 1885, Ann. New York Acad. Sci., 3, p. 273 — Temax, Yucatán, Mexico.

Northernmost Yucatán; doubtful record from Isla Cozumel, Quintana Roo.

Polioptila albiloris albiloris Sclater and Salvin

Polioptila albiloris Sclater and Salvin, 1860, Proc. Zool. Soc. London, p. 298 — Motagua Valley, Zacapa, Guatemala.

Polioptila bairdi Ridgway, 1903, Proc. Biol. Soc. Washington, 16, p. 110 — San Juan del Sur, Nicaragua.

Interior of Guatemala and Honduras, El Salvador, and Pacific lowlands of Nicaragua and northwestern Costa Rica.

POLIOPTILA PLUMBEA[1]

Polioptila plumbea superciliaris Lawrence

Polioptila superciliaris Lawrence, 1861, Ann. Lyc. Nat.

---

[1] The ranges and relationships of the South American populations of *Polioptila* are poorly known. The arrangement of this list is tentative.

Hist. New York, 7, p. 304 — New Grenada, Isthmus of
Panama [= Atlantic slope, near Panama Railroad; *fide*
Lawrence, *loc. cit.*, p. 322].
Polioptila superciliaris magna Ridgway, 1903, Proc. Biol.
Soc. Washington, 16, p. 110 — Cartago, Costa Rica.
Quintana Roo and southern Campeche, Mexico through
Central America to Panama.

?Polioptila plumbea cinericia Wetmore

Polioptila plumbea cinericia Wetmore, 1957, Smiths. Misc.
Coll., 134 (9), p. 80 —Isla Coiba, Panama.
Isla Coiba, off Pacific coast of Veraguas, Panama.

Polioptila plumbea bilineata (Bonaparte)

Culicivora bilineata Bonaparte, 1851, Consp. Av., 1
(1850), p. 316 — Cartagena, Colombia.
From the west base of Sierra Nevada de Santa Marta
along the Caribbean and Pacific coasts of northern and western Colombia south through western Ecuador to the coast
and western slopes of the western Andes of western Peru,
south to department of Libertad.

Polioptila plumbea plumbiceps Lawrence

Polioptila plumbiceps Lawrence, 1865, Proc. Acad. Nat.
Sci. Philadelphia, 17, p. 37 — Venezuela; restricted to
Tucacas, Falcón, by Zimmer, 1942, Amer. Mus. Novit.,
no. 1168, p. 5.
Northern Colombia, from eastern base of Sierra Nevada
de Santa Marta and the eastern side of the Eastern Andes
in Norte de Santander, through northern Venezuela to Sucre, and south to the Llanos de Guárico in Portuguesa and
Barinas.

Polioptila plumbea anteocularis Hellmayr

Polioptila nigriceps anteocularis Hellmayr, 1900, Novit.
Zool., 7, p. 537 — "Bogotá," Colombia.
Upper Magdalena Valley, Colombia.

Polioptila plumbea daguae Chapman

Polioptila livida daguae Chapman, 1915, Bull. Amer. Mus.
Nat. Hist., 34, p. 648 — Los Cisneros, Río Dagua, Valle,
Colombia.

Region of the Río Dagua and the upper Río Patía, Valle, Colombia.

Polioptila plumbea innotata Hellmayr

Polioptila buffoni innotata Hellmayr, 1901, Novit. Zool., 8, p. 359 — Forte de São Joaquim, upper Rio Branco, Brazil.

From extreme eastern Colombia, along the Orinoco through southern Venezuela, and south to British Guiana and the Rio Branco region of extreme northern Brazil.

Polioptila plumbea plumbea (Gmelin)

Todus plumbeus Gmelin, 1788, Syst. Nat., 1, pt. 1, p. 444 — Surinam.

Surinam, French Guiana, and northern Brazil west to the Rio Tapajóz and south to northern Maranhão.

Polioptila plumbea maior[1] Hellmayr

P(olioptila) nigriceps maior Hellmayr, 1900, Novit. Zool., 7, p. 538 — Sueccha [= Succha], 3,000 m., Huamachuco, Libertad, Peru.

Northern Peru, in catchment basin of the upper Río Marañon and its tributaries, in departments of Piura, Lambayeque, Cajamarca, and Libertad.

Polioptila plumbea parvirostris Sharpe

Polioptila parvirostris Sharpe, 1885, Cat. Birds Brit. Mus., 10, pp. 441, 448 — Chamicuros, Río Amazonas, Peru.

Eastern Peru on the upper Amazon and southern tributaries (Huallaga; Ucayali) of the Marañon; range poorly known.

Polioptila plumbea atricapilla (Swainson)

Culicivora atricapilla Swainson, 1832, Zool. Ill., 2nd ser., 2, pl. 57 — no locality; probably Bahia, Brazil (*vide* Hellmayr, 1934, *ibid.*, p. 495).

Northeastern Brazil from the interior of Maranhão through Piauí, Ceará, Pernambuco, and Bahia.

---

[1] *P. p.* "*major*" is an unjustified emendation.

POLIOPTILA LACTEA[1]

Polioptila lactea Sharpe

Polioptila lactea Sharpe, 1885, Cat. Birds Brit. Mus., 10, p. 453 — South America; Rio de Janeiro suggested by Hellmayr, 1934, Field Mus. Nat. Hist. Publ., Zool. Ser., 13, pt. 7, p. 494.

Southeastern Brazil, in Rio de Janeiro, São Paulo, and Paraná, eastern Paraguay, and northeastern Argentina, in Misiones.

POLIOPTILA GUIANENSIS[2]

Polioptila guianensis facilis Zimmer

Polioptila guianensis facilis Zimmer, 1942, Amer. Mus. Novit., no. 1168, p. 6 — Solano, Río Casiquiare, Amazonas, Venezuela.

South-central Amazonas, Venezuela and adjacent northeastern Brazil, in region of upper Rio Negro.

Polioptila guianensis guianensis Todd

Polioptila guianensis Todd, 1920, Proc. Biol. Soc. Washington, 33, p. 72—Tamanoir, Mana River, French Guiana.

British Guiana, Surinam, and French Guiana.

Polioptila guianensis paraensis Todd

Polioptila paraensis Todd, 1937, Ann. Carnegie Mus., 25, p. 255 — Benevides, Pará, Brazil.

Pará, Brazil; known only from the type locality, near Belém, and from Caxiricatuba, on the Rio Tapajóz.

POLIOPTILA SCHISTACEIGULA

Polioptila schistaceigula Hartert

Polioptila schistaceigula Hartert, 1898, Bull. Brit. Orn. Club, 7, p. 30 — Cachabi [= Cachaví], 500 ft., Esmeraldas, Ecuador.

Eastern Panama; Colombia on the Pacific coast and the interior to the lower Cauca and middle Magdalena Valleys; western Ecuador.

[1] Possibly conspecific with *P. plumbea*.
[2] Possibly conspecific with *P. schistaceigula*.

POLIOPTILA DUMICOLA

Polioptila dumicola berlepschi Hellmayr

Polioptila berlepschi Hellmayr, 1901, Novit. Zool., 8, p. 356 — Cuyabá, Mato Grosso, Brazil.

Central Brazil, in states of Goiaz, western Minas Gerais, São Paulo, and Mato Grosso (except southwest), and eastern Bolivia.

Polioptila dumicola dumicola (Vieillot)

Sylvia dumicola Vieillot, 1817, Nouv. Dict. Hist. Nat., nouv. éd., 11, p. 170 — Paraguay.

Rio Grande do Sul and southwestern Mato Grosso, Brazil, west to central and southern Bolivia, and southward through Paraguay and Uruguay to states of La Rioja, Córdoba, and Buenos Aires, Argentina.

Polioptila dumicola saturata Todd

Polioptila dumicola saturata Todd, 1946, Proc. Biol. Soc. Washington, 59, p. 155 — Samaipata, Bolivia.

Highlands of Bolivia.

ADDENDA

FAMILY **RHINOCRYPTIDAE** (PART)

RAYMOND A. PAYNTER, JR.

The affinities of *Psilorhamphus* have long been obscure. Cory and Hellmayr (1924, Field Mus. Nat. Hist. Publ., Zool. Ser., 13, pt. 3, p. 204) placed the genus with the Formicariidae. Peters (1951, Check-list, 7, p. 213) intended to place it with the Sylviidae. Sick (1954, Bonn. Zool. Beitr., 5, pp. 179-190) provisionally considered it to be one of the Formicariidae, but later Plótnick (1958, Physis, 21, pp. 130-136) presented convincing anatomical evidence that *Psilorhamphus* is actually a representative of the Rhinocryptidae. Wetmore (*in litt.*) suggests placing the genus near *Scytalopus*.

GENUS **PSILORHAMPHUS** SCLATER

Psilorhamphus Sclater, 1855, Proc. Zool. Soc. London, 23, p. 90. New name for *Leptorhynchus* Ménétriés, 1835, preoccupied by *Leptorhynchus* Clift, 1829. Type, by monotypy, *Leptorhynchus guttatus* Ménétriés.

cf. Cory and Hellmayr, 1924, Field Mus. Nat. Hist. Publ., Zool. Ser., 13, pt. 3, pp. 204-205.
Pinto, 1938, Cat. Aves Brasil, pt. 1, Rev. Mus. Paulista, 22, pp. 490-491.
Partridge, 1954, Rev. Inst. Nac. Invest. Cienc. Nat., 3, pp. 146-147 (Argentina).

PSILORHAMPHUS GUTTATUS

Psilorhamphus guttatus (Ménétriés)

Leptorhynchus guttatus Ménétriés, 1835, Mém. Acad. Imp. Sci. St. Pétersbourg, ser. 6, 3, pt. 2 (sci. nat.), p. 516, pl. 10, fig. 1 — "Cuyaba," [? Minas Gerais,] Brazil.

Southeastern Brazil in Espírito Santo, Rio de Janeiro, São Paulo, Paraná, and ? Minas Gerais, and in Misiones, northeastern Argentina.

INDEX

INDEX

CPSIA information can be obtained
at www.ICGtesting.com
Printed in the USA
LVHW010226231220
674915LV00023B/392